Rod Machado's Private Pilot Workbook

Written by
Rod Machado

Published by The Aviation Speakers Bureau

Workbook updated with each printing.

Please visit (and *bookmark*) our web site for any additional book updates: *www.rodmachado.com.*

Published by:
The Aviation Speakers Bureau, P.O. Box 6030, San Clemente, CA 92674-6030.

All rights reserved. The contents of this manual are protected by copyright throughout the world under the Berne Union and the Universal Copyright Convention.

No part of this publication may be reproduced in any manner whatsoever—electronic, photographic, photocopying, facsimile—or stored in a retrieval system without the prior written permission of the author; Rod Machado, care of The Aviation Speakers Bureau.

Printed and bound in the United States of America.

Nothing in this text supersedes any operational documents or procedures issued by the Federal Aviation Administration (FAA), the aircraft and avionics manufacturers, the Pilot Operating Handbook (POH), flying schools or the operators of the aircraft.

The author has made every effort in the preparation of this book to ensure the accuracy of the information. However, the information is sold without warranty either expressed or implied. Neither the author nor the publisher will be liable for any damages caused or alleged to be caused directly, indirectly, incidentally or consequentially by the information in this book.

The opinions in this book are solely those of the author and not the publisher.

Don't even think about using any performance chart in this book for performance computations in your airplane. Use a performance chart appropriate for your airplane. Also, don't even think about using this book for navigation. In other words, there are aeronautical chart excerpts in this book but none of them should in any way be used in lieu of actual charts for any type of information. The charts, graphs and tables in this book are for training purposes only.

Cover layout by **Diane Titterington**
Front cover artwork by **Erik Hildebrandt** (See page S-11)
All illustrations in this book designed and drawn by **Rod Machado**
Photographs (unless marked otherwise or in the public domain) by **Rod Machado**

Copyright 2001 by Rod Machado

CONTENTS

Copyright Information..ii
Contents..iii
About the Author...iv
Introduction...v
Foreword..vi

1 **Chapter One** - Pages A1-2
Airplane Components: Getting to Know Your Airplane

2 **Chapter Two** - Pages B1-14
Aerodynamics: The Wing is the Thing

3 **Chapter Three** - Pages C1-14
Engines: Knowledge of Engines is Power

4 **Chapter Four** - Pages D1-6
Electrical Systems: Knowing What's Watt

5 **Chapter Five** - Pages E1-10
Flight Instruments: Clocks, Tops & Toys

6 **Chapter Six** - Pages F1-20
Federal Aviation Regulations: How FAR Can We Go?

7 **Chapter Seven** - Pages G1-10
Airport Operations: No Doctor Needed

8 **Chapter Eight** - Pages H1-6
Radio Operations: Aviation Spoken Here

9 **Chapter Nine** - Pages I1-14
Airspace: The Wild Blue, Green & Red Yonder

10 **Chapter Ten** - Pages J1-6
Aviation Maps: The Art of the Chart

11 **Chapter Eleven** - Pages K1-12
Radio Navigation: The Frequency Flyer Program

12 **Chapter Twelve** - Pages L1-22
Understanding Weather: Looking for Friendly Skies

13 **Chapter Thirteen** - Pages M1-22
Weather Charts & Briefings: PIREPS, Progs & METARS

14 **Chapter Fourteen** - Pages N1-18
Flight Planning: Getting There From Here

15 **Chapter Fifteen** - Pages O1-14
Airplane Performance Charts: Know Before You Go

16 **Chapter Sixteen** - Pages P1-8
Weight and Balance: Let's Wait and Balance

17 **Chapter Seventeen** - Pages Q1-10
Pilot Potpourri: Neat Aeronautical Information

Part 141 Approved Syllabus..........................R1-R14
Stage One Exam...R15-R18
Stage Two Exam...R19-R24
Stage Three Exam...R25-R38
Appendix (Airport/Facility Directory Legend)......S1-S4
The Senior Editor..S5
Aviation Speakers Bureau/Ongoing Editor................S6
Rod Machado's Products.....................................S7-S10
Cover Photographer & Pilot......................................S11

About THE AUTHOR

Rod Machado traded his motorcycle for flying lessons at the age of 16. Captivated by the adventure of flying in a Taylorcraft L-2 at Amelia Reid Aviation in San Jose, Rod has remained hooked ever since. He may be one of the few ATP-rated pilots who still gets excited by a Cessna 150 fly-by.

Rod is a professional speaker, educator and humorist who travels to all 50 states and Europe sharing his upbeat and lively presentations. His unusual talent for simplifying the difficult and adding humor to help the lessons stick has made him a popular lecturer and author.

For six years, Rod wrote and coanchored *ABC's Wide World of Flying*. He is an instructor on the Cessna *Cleared For Takeoff* Private Pilot CD-ROM Course, and he is AOPA's National CFI Spokesman. You can fly with Rod as your flight instructor on Microsoft's *Flight Simulator*.

Rod is a National Aviation Safety Counselor appointed by the FAA in Washington D. C. He began flying in 1970 and has over 8,000 hours of flight time earned the hard way, one CFI hour at a time. Since 1976, Rod has taught hundreds of flight instructor revalidation clinics and safety seminars and was named the 1991 Western Region Flight Instructor of the Year. Rod has written for a number of publications and is a columnist for *AOPA Pilot Magazine* and *Flight Training Magazine*.

Rod's eclectic interests are reflected by his equally varied academic credentials. In addition to having an Airline Transport Pilot license and all fixed wing flight instructor ratings, he holds a degree in aviation science and degrees in psychology.

Rod believes you must take time to exercise or you'll have to take time to be sick. He gets his exercise from practicing and teaching martial arts. Rod holds black belts in the Korean disciplines of tae kwon do and hapkido and a ranking in Gracie jujitsu. He also runs 20 miles a week, and claims it's uphill both ways.

Acknowledgments

The author wishes to acknowledge the help and or support of the following individuals, companies or groups:

Brian Weiss of WORD'SWORTH, Santa Monica, CA; Dr. Barry Wallis of Delphi System; Celia Vanderpool; Jeff Broomall Diane Titterington of The Aviation Speakers Bureau, San Clemente, CA; The New Piper Aircraft Corporation. Charts and graphs provided by Piper are to be used for information purposes only (*A Pilot Operating Handbook is the only true source of information*); The Cessna Aircraft Corporation. Cessna authorized the use of their materials with the understanding that they are to be used for training purposes only, not the actual operation of an aircraft; Danny Mortensen, President of Airline Ground Schools; Captain Ralph Butcher; Tim Peterson and the many others who kindly lent their assistance during the creation of this project.

INTRODUCTION

Thank you for purchasing my *Private Pilot Workbook*. This book complements my *Private Pilot Handbook*. Using them together will give you a chance to gauge how well you understand the material presented. Here's how this *Workbook* is constructed and how to use it properly.

First, you'll notice that the questions in the workbook use the same sequence and black-on-white topic headings as the *Private Pilot Handbook*. When you've completely read a particular topic in the *Handbook*, you should proceed to the *Workbook* and answer the questions listed under that topic. This gives you immediate feedback on how well you understood the information you just read. I suggest that you write your answers on a separate piece of paper instead of marking the *Workbook*. When you've finished reading the *Private Pilot Handbook*, then you might want to answer all the *Workbook* questions once again in preparation for the FAA Knowledge exam. This would be a good time to mark your answers to each question directly in the *Workbook*.

To the right of the number for each *Workbook* question is a sequence of characters that looks something like this: **[E17/3/2].** This code tells you where to find the explanation of the answer to this particular question in my *Handbook*. The first two characters represent the page number within a particular chapter, the second number represents the column and the last number represents the paragraph (counting down from the top of the column) where the explanation is found. Therefore, if you want the explanation for this question, it will be found on page E17, column 3, second paragraph from the top. The answers to the individual questions are found at the end of each chapter in the *Workbook*.

If you're interested in tips for taking the FAA Knowledge exam, please visit the *Book & Slide Updates* section of my web site at (www.rodmachado.com). Additionally, when reading my *Handbook* or *Workbook*, you might want to use a highlighter to aid you in reviewing later.

If this manual is being used as part of a Part 141 ground school, then it should be used in conjunction with *Rod Machado's FAA approved, Part 141 ground training syllabus* located at the back of this book. You'll find the approved syllabus for a nine-week, 51.5 hour ground school starting on page R1. If you are enrolled in a Part 141 ground school using my *Handbook* then look at the syllabus to determine the course schedule and homework assignments. This approved ground school requires that three stage exams be given during the course of instruction. The exams are found beginning on page R15.

I hope you have a wonderful time learning about aviation.

Have fun!

Laugh & Learn,

Rod Machado

Rod Machado

FOREWORD

With only some six hundred flight hours, I feel like the least qualified person possible to introduce one of Rod Machado's publications. On the other hand, the twelve years I "flew" with *Microsoft Flight Simulator* before actually taking lessons in 1996 gave me a perspective that those with thousands of hours may not have. It certainly made me appreciate the need for clear instructions, good visualization, and practice that provides usable feedback.

Dr. Barry Wallis

When I began to take lessons I chose *Rod's Machado's Private Pilot Handbook* because it offered the clearest explanations, using the best graphics, that I found among the many materials available. The *Handbook* enabled me to understand the lessons, visualize what was occurring during flight, and react to changes with confidence. But my experience in the training field left me wishing for more practice and for testing to see if I had learned the material.

I was delighted to learn that Rod was working to answer that need, and I was honored to be asked to help evaluate *Rod Machado's Private Pilot Workbook*. The *Workbook* provides self-testing on every topic covered by the *Handbook*, many practice problems, references to the *Handbook*, and more.

I wish I'd had the *Workbook* when I was preparing for my Knowledge and Practical tests. Now, having seen it, I would say that any flight student who does not use it is making a conscious decision to lengthen their flight training.

Dr. Barry Wallis is the president of Delphi Systems, a training company in Downers Grove, Illinois. His background includes video production and custom training solutions to end-user training and support problems.

DV 20 KATANA
Katana DA20

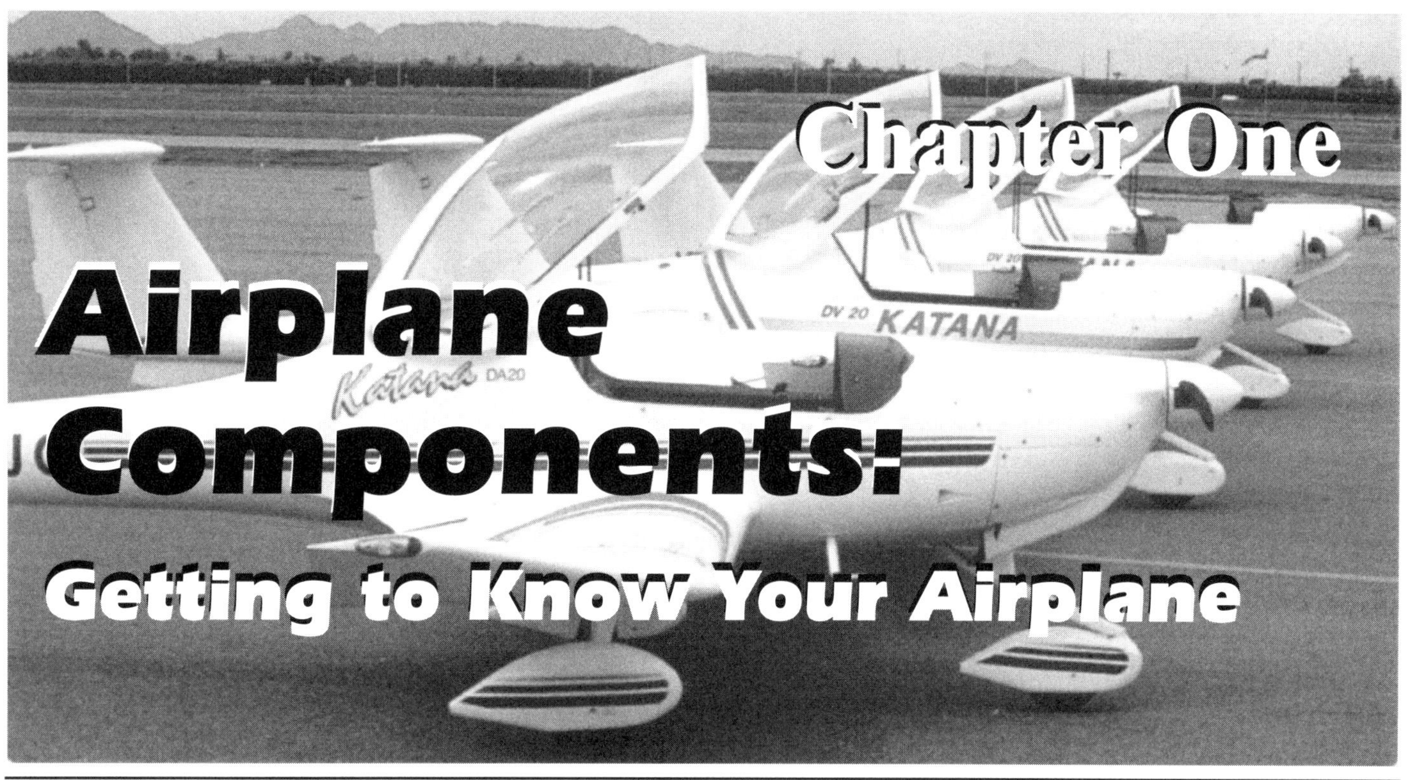

Chapter One

Airplane Components:

Getting to Know Your Airplane

Label the individual parts of the airplane:

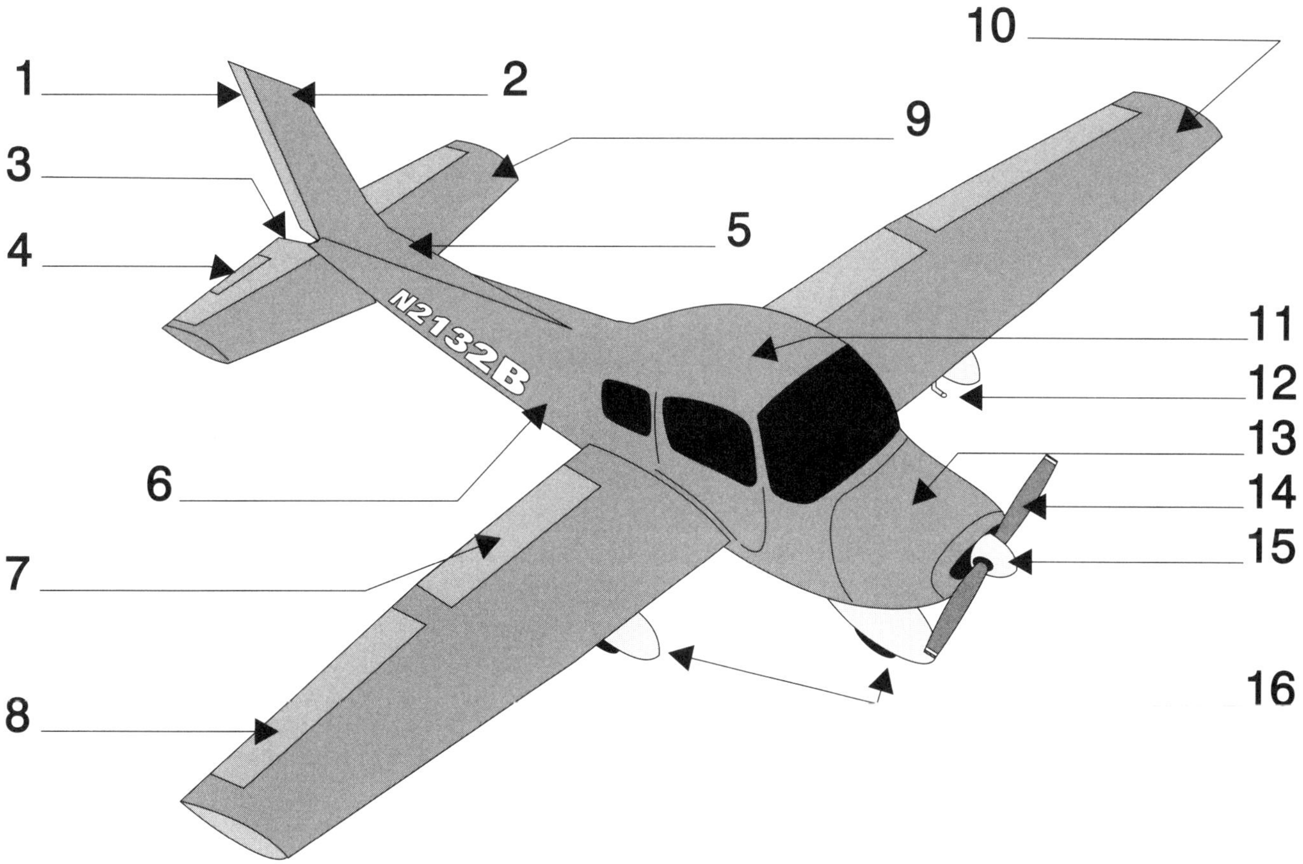
1
2
3
4
5
6
7
8
9
10
11
12
13
14
15
16
N2132B

Chapter One Answers

1. Rudder
2. Vertical stabilizer
3. Elevator
4. Trim tab
5. Empennage
6. Fuselage
7. Flap
8. Aileron
9. Horizontal stabilizer
10. Wing
11. Cockpit
12. Pitot tube
13. Engine cowling
14. Propeller
15. Spinner
16. Landing gear

Note: To ensure that you have the most current answers to these questions, please check the *Book & Slide Updates* section at Rod Machado's web site: www.rodmachado.com

Chapter Two

Aerodynamics: The Wing Is the Thing

May the Four Forces Be With You

1. [B1/3/2]
The four forces acting on an airplane in flight are
A. lift, weight, thrust, and drag.
B. lift, weight, gravity, and thrust.
C. lift, gravity, power, and friction.

2. [B1/Figure 1] Fill in the four forces:

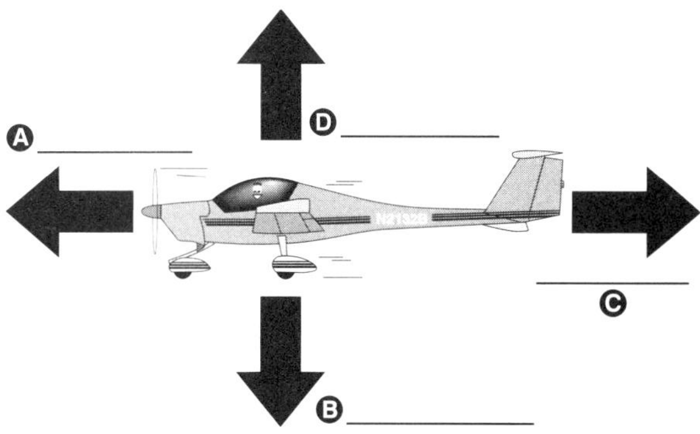

3. [B2/2/2]
When are the four forces that act on an airplane in equilibrium?
A. During unaccelerated flight.
B. When the aircraft is accelerating.
C. When the aircraft is in a stall.

4. [B2/2/2 & B2/3/2]
What is the relationship of lift, drag, thrust, and weight when the airplane is in straight-and-level flight?
A. Lift equals weight and thrust equals drag.
B. Lift, drag, and weight equal thrust.
C. Lift and weight equal thrust and drag.

Climbs

5. [B3/2/1]
Airplanes climb because of _____
A. excess lift.
B. excess thrust.
C. reduced weight.

6. [B4/Figure 5]
Lift acts at a ______degree angle to the relative wind.
A. 180
B. 360
C. 90

7. [B4/Figure 5] Fill in the blanks for the forces in a climb:

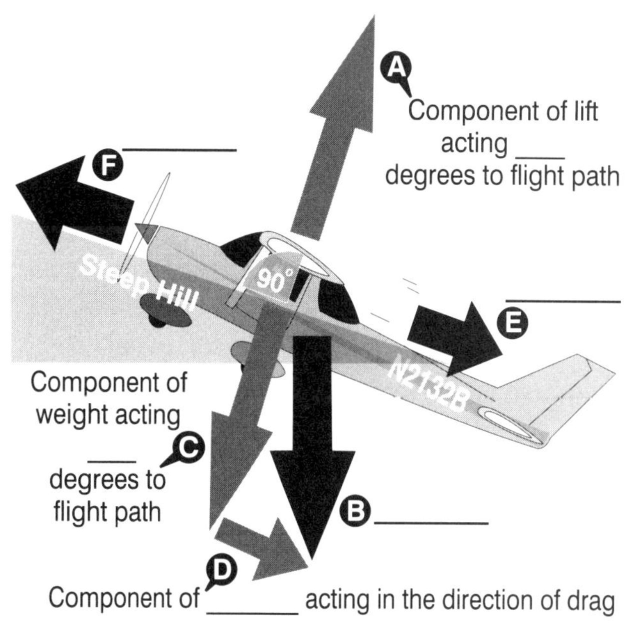

8. [B4/3/3]
The minimum forward speed of the airplane is called the _____ speed.
A. certified
B. stall
C. best rate of climb

9. [B5/3/1]
You can determine the proper climb attitude for your airplane by referring to the
A. attitude indicator.
B. airspeed indicator.
C. vertical speed indicator.

Descents

10. [B6/Figure 9] Fill in the blanks for the forces in a descent:

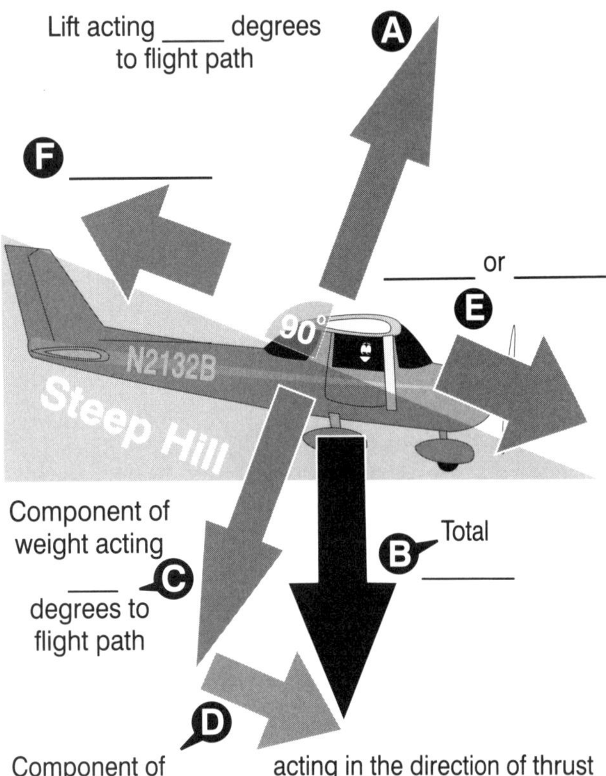

The Wing and Its Things

Defining the Wing

11. [B7/Figure 10] Fill in the parts of the wing:

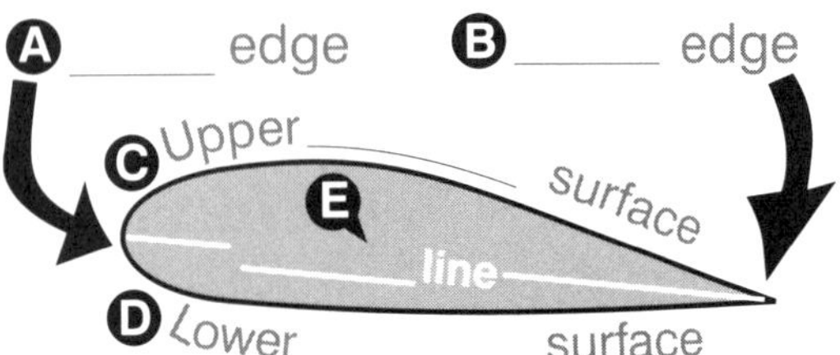

12. [B7/1/4]
The chord line is an imaginary line connecting the
A. trailing edge of the wing with the leading edge.
B. leading edge of the wing with the trailing edge.
C. wing root with the wing tip.

13. [B7/2/1]
The chord line is used to represent
A. the general shape of the wing.
B. the sound the wing makes when it moves in air.
C. the average width of the wing.

How the Wing Works

14. [B7/3/1]
The definition of chord enables us to understand
A. how the wing moves through the air.
B. how to preflight the airplane.
C. the angle the wind meets a wing that may vary in size and shape.

Relative Wind

15. [B8/1/2]
Relative wind results from the motion of the
A. airplane's thrust.
B. airplane through the air.
C. wind blowing on the airplane.

16. [B8/1/2]
Relative wind is called *relative* because it
A. results from the motion of the airplane.
B. is perpendicular to the airplane's flight path.
C. is independent of airplane motion.

17. [B8/1/2]
Relative wind is _____ and _____ to the airplane's motion.
A. tangential to, equal
B. opposite to, perpendicular
C. opposite to, equal

18. [B8/3/2]
Relative wind is _____ which way the airplane's nose is pointed.
A. dependent on
B. indifferent to
C. independent of

Attacking the Air

19. [B9/2/1]
The term "angle of attack" is defined as the angle
A. between the wing chord line and the relative wind.
B. between the airplane's climb angle and the horizon.
C. formed by the longitudinal axis of the airplane and the chord line of the wing.

20. [B9/1/2]
An important principle to understand when dealing with angle of attack is that the nose (therefore the wing) can be pointed on an incline that's _____ the actual climb path.
A. different from
B. always the same as
C. always parallel to

21. [B9/Figure 14] Label wing and wind components:

22. [B9/2/1]
Angle of attack is defined as the angle between the chord line of an airfoil and the
A. direction of the relative wind.
B. pitch angle of an airfoil.
C. rotor plane of rotation.

How Lift Develops

23. [B9/3/2]
Wings are expressly built to plow through air molecules separating them either above or below while offering little resistance in the _____ direction.
A. vertical
B. horizontal
C. perpendicular

Impact Versus Pressure Lift

24. [B10/1/3]
Wind deflected downward by the airfoil creates a/an _____ movement of the wing.
A. downward
B. sideways
C. upward

Bending the Wind with the Wing

25. [B11/1/2]
Bernoulli figured out that the faster the air flows over a surface, the _____ pressure it exerts on that surface.
A. less
B. more
C. higher the

26. [B11/2/1] Fill in the blank:
Air flowing faster over a curved surface causes a slight _____________ in pressure on that surface.

27. [B11/2/1]
High velocity airflow over the wing causes a slight decrease in pressure on the wing's upper surface. In other words, the pressure on _____ of the wing is now _____ than the pressure on bottom of the wing.
A. the side, greater
B. top, less
C. top, greater

28. [B12/1/1]
Since high pressure always moves toward low pressure, the wing (which just happens to be in the way) is pushed _____ in the process.
A. horizontally
B. downward
C. upward

29. [B12/1/2]
Because of the wing's shape, even at a very small angle of attack, a cambered wing still adds a slight curve and _____ to the wind.
A. acceleration
B. deceleration
C. crossflow parameter

Angle of Attack and the Generation of Lift

30. [B12/2/1]
At a relatively slow speed (such as during takeoff), the wing's engineered curve isn't capable of curving or deflecting enough air _____ to produce the necessary lift for flight.
A. upward
B. sideways
C. downward

31. [B12/1/2]
Raising the nose slightly increases the angle of attack which forces the air to undergo an additional _____ greater than that which the _____ of the airfoil can produce.
A. curve, engineered shape
B. deceleration, creator
C. acceleration, pilot

32. [B12/2/2]
An increased angle of attack permits the airplane to produce the necessary lift for flight at a _____ airspeed.
A. faster
B. constant
C. slower

33. [B12/2/2]
As the angle of attack increases, an airplane can fly at a _____ speed and still develop the necessary lift for flight.
A. slower
B. constant
C. faster

STALLS

Stall, Angle of Attack & How the Nose Knows

34. [B14/1/3]
As the angle of attack exceeds approximately 18 degrees the air molecules flowing over the wing don't negotiate the turn very well. When this happens, they spin off or burble into the free air, no longer providing a uniform, high-velocity, laminar airflow over the wing. The wing _____
A. develops lift.
B. experiences an increase in drag.
C. stalls.

35. [B14/1/3]
When the critical angle of attack is exceeded, the airplane will
A. stall.
B. ascend.
C. descend.

36. [B14/1/6]
All wings
A. have a critical angle of attack.
B. produce equal amounts of lift at all angles of attack.
C. have a bird that they belong to.

37. [B15/2/3]
If the wing stalls, you need to do one very important thing:
A. apply back pressure to reduce the angle of attack.
B. increase the angle of attack.
C. reduce the angle of attack to less than its critical value.

38. [B15/2/3]
You can unstall a wing by _____ the angle of attack.
A. increasing
B. reducing
C. ignoring

Stall at Any Attitude or Airspeed

39. [B15/3/1]
You should realize that an airplane can be _____ at any attitude or any airspeed.
A. taxied
B. stalled
C. maneuvered

40. [B15/3/1]
Whether an airplane exceeds its *critical angle of attack* is independent of
A. attitude or airspeed.
B. relative wind.
C. the angle between the chord line and relative wind.

41. [B15/3/3]
If an airplane stalls, the first step in recovering is to decrease the angle of attack by moving the elevator control _____ or releasing _____ on the elevator control.
A. forward, side pressure
B. backward, back pressure
C. forward, back pressure

42. [B16/1/2]
Once the airplane is no longer stalled it should be put back in the desired attitude while making sure you don't _____ again.
A. stall
B. fly
C. accelerate

Five Stall Warning Signs

43. [B16/2/All, B16/3/All, B17/1/2 & 3]
Which of the following may indicate the onset of a stall?
A. Improved control response, low nose attitude, noticeable buffeting.
B. Diminished control response, stall horn silent, airspeed in the green.
C. Stall horn sounding, diminished control response, noticeable buffet.

Stalling Speed, Gee Whiz and G-Force

44. [B17/2/1]
When the weight of an airplane is increased, the airplane stalls at _____ indicated speed.
A. a higher
B. a lower
C. the same

45. [B17/2/2]
An increase in weight (apparent or real) means the wings must develop more _____ to remain airborne.
A. lift
B. angle beyond its critical value
C. drag

46. [B17/3/2]
The critical angle of attack at which the wing stalls _____
A. is always changing, based on the airplane's weight.
B. never changes, regardless of airplane weight.
C. changes, based on the airplane's speed.

47. [B17/3/2]
The angle of attack at which an airplane wing stalls will
A. increase if the CG is moved forward.
B. change with an increase in gross weight.
C. remain the same regardless of gross weight.

48. [B17/3/2] Fill in the blank:
Increasing an airplane's weight will not affect the _____________ of attack at which the airplane stalls.

49. [B18/2/3]
Which basic flight maneuver increases the load factor on an airplane as compared to straight-and-level flight?
A. A climb.
B. A turn.
C. A stall.

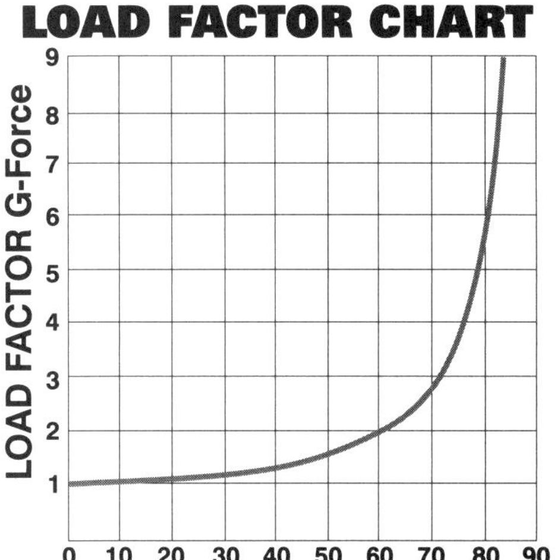

50. [B18/3/2]
Referring to the load factor chart above, if an airplane weighs 2,300 pounds, what approximate weight would the airplane structure be required to support during a 60 degree banked turn while maintaining altitude?
A. 2,300 pounds.
B. 3,400 pounds.
C. 4,600 pounds.

51. [B18/3/2]
Referring to the load factor chart above, if an airplane weighs 3,300 pounds, what approximate weight would the airplane structure be required to support during a 30 degree banked turn while maintaining altitude?
A. 1,200 pounds.
B. 3,100 pounds.
C. 3,960 pounds.

52. [B18/3/2]
Referring to the load factor chart above, if an airplane weighs 4,500 pounds, what approximate weight would the airplane structure be required to support during a 45 degree banked turn while maintaining altitude?
A. 4,500 pounds.
B. 6,750 pounds.
C. 7,200 pounds.

53. [B18/3/3]
If the airplane "feels" twice as heavy as it actually is, then the lift must _____ if the airplane is to maintain altitude.
A. remain the same
B. decrease
C. double

54. [B19/1/2,3]
An increased load factor will cause the airplane to
A. stall at a higher airspeed.
B. have a tendency to spin.
C. be more difficult to control.

55. [B19/2/3 & Figure 32]
Based on the stall speed and bank angle chart below, at a 60 degree bank in level flight, the stall speed increases by
A. 67%
B. 40%
C. 2%

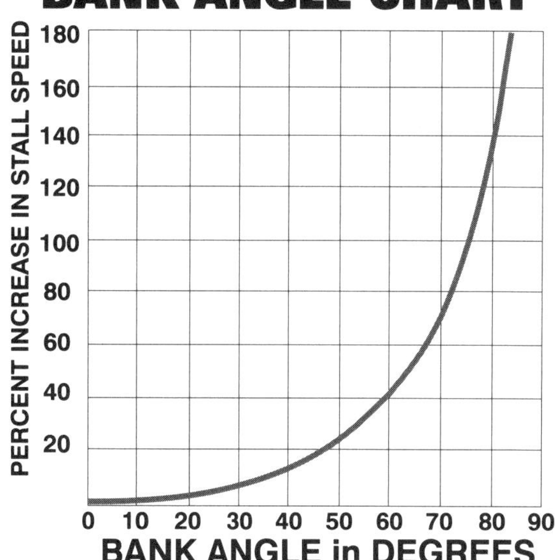

56. [B20/1/2]
When the bank increases, the nose wants to _____
A. raise up automatically, which puts the airplane near a stall if the pilot doesn't do something.
B. pitch forward, which automatically sends the airplane into a diving, unrecoverable spiral.
C. pitch down, which results in the pilot pulling on the elevator to maintain altitude, causing an increased angle of attack.

57. [B20/1/3]
Why it important for a pilot to be especially sensitive to the amount of G-force he or she is experiencing while maneuvering the airplane?
A. G-force always causes the airplane to move in a different trajectory than planned by the pilot.
B. Increasing G-force always means an increase in stall speed.
C. G-force can cause embarrassment by pulling a student's dentures out of their mouth.

58. [B20/1/3]
What are the most important parts of your anatomy for avoiding stalls?
A. Your brain, for planning to avoid steep turns near the ground and your derrière for sensing G-force which helps alert you to an increase in stall speed.
B. Your derrière for thinking and your brain for feeling G-force.
C. Your hands, since they are the things that pulled back on the elevator and got you in trouble in the first place.

DRAG

What a Drag

59. [B20/2/3]
Drag is the airplane's natural response to an object's movement through the
A. slipstream.
B. wing's downwash.
C. air.

Horizontal and Vertical Movement of Air

60. [B20/3/3]
Wings are designed to deflect air _____ while offering very little _____ resistance.
A. horizontally, vertical
B. vertically, horizontal
C. sideways, diagonal

61. [B20/3/4]
The two basic forms of drag are:
A. parasite and induced drag.
B. planform and interference drag.
C. good and bad drag.

62. [B21/1/2]
Parasite drag is the result of
A. friction.
B. the development of lift.
C. small bugs living on the wing.

63. [B21/1/2]
As airspeed doubles, parasite drag _____.
A. doubles
B. triples
C. quadruples

64. [B21/1/3]
Induced drag is resistance to motion induced by the wing turning some of its _____ into _____.
A. drag, lift
B. thrust, upwash
C. lift, drag

Total Drag and Your Go Far Speed

65. [B21/Figure 36] Label the three drag curves below:

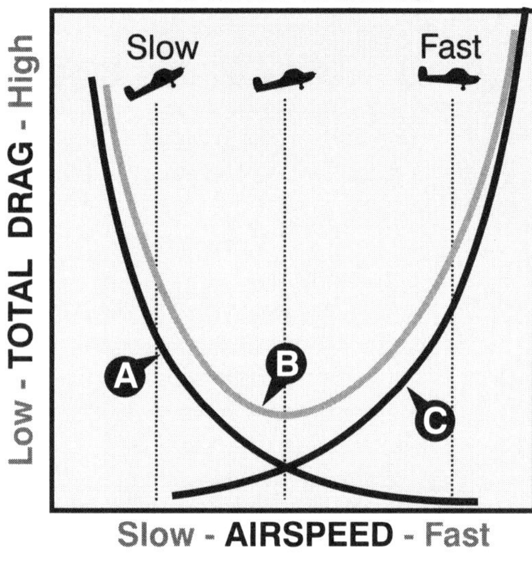

A ______________ Drag
B ______________ Drag
C ______________ Drag

66. [B21/2/3]
As the airplane speeds up, induced drag _____ while parasite drag _____.
A. increases, decreases
B. decreases, increases
C. remains the same, increases

67. [B21/2/4]
When the induced and parasite drag curves are added together, they produce the _____ curve.
A. interference drag
B. straight line
C. total drag

68. [B21/2/4]
The lowest spot in the total drag curve is your magic number, a specific airspeed known as the
A. best power speed.
B. best L/D speed.
C. best cruise speed.

69. [B21/2/4 & B21/3/1,2]
The sum of the parasite and induced drag curves reveals a point on the total drag curve (curve B above) where drag is at a minimum. The speed associated with this point is
A. the airplane's maximum power-off glide range.
B. the minimum speed to use for turbulence penetration.
C. the speed that results in maximum fuel consumption in forward flight.

70. [B21/3/2]
The most important rule to remember in the event of a power failure after becoming airborne is to
A. immediately establish the proper gliding attitude and airspeed.
B. quickly check the fuel supply for possible fuel exhaustion.
C. determine the wind direction to plan for the forced landing.

Stretching the Glide, Saving the Hide

71. [B22/3/2]
In a power-off glide, the best L/D speed allows the airplane to glide a _____ forward distance with a _____ amount of altitude loss.
A. minimum, maximum
B. maximum, maximum
C. maximum, minimum

Ground Effect

72. [B23/1/4]
Ground effect allows an airplane flying close to the runway to become or remain airborne at a slightly _____ speed.
A. lower-than-normal
B. higher-than-normal
C. higher and lower

73. [B23/3/1 & B23/3/3]
High pressure on the bottom of the wing causes air molecules to move sideways (toward the wingtip) in the direction of _____ pressure on top of the wing. This action is responsible for the creation of _____.
A. higher, wingtip vortices
B. lower, known life in the universe
C. lower, wingtip vortices

74. [B23/3/3 & Figure 40]
Wingtip vortices rotate which way about the wingtip?
A. Outward, upward and inward.
B. Downward, outward and inward.
C. Upward, inward and outward

75. [B23/3/3]
Wingtip vortex action increases with an increase in
A. airspeed.
B. angle of attack.
C. thrust.

76. [B24/1/1]
The wingtip vortex not only spirals around the wingtip, it also adds a/an _____ to the air behind and along the wing's span.
A. upward flow
B. downward flow
C. acutely agonizing flow

77. [B/24/1/1]
At higher angles of attack the downward bending of the relative wind in the vicinity of the wing changes the direction of the _____ wind. This newly bent relative wind is often called the _____ relative wind.
A. relative, perpendicular
B. induced, local
C. relative, local

78. [B24/1/1]
Recalling that effective lift always acts _____ to the relative wind, when the angle of attack increases, the total lift force tilts _____ slightly.
A. parallel, forward
B. perpendicular, rearward
C. perpendicular, forward

79. [B24/1/2]
Floating caused by the phenomenon of ground effect will be most realized during an approach to land when at
A. less than the length of the wingspan above the surface.
B. twice the length of the wingspan above the surface.
C. a higher-than-normal angle of attack.

80. [B24/1/3]
Ground effect results from a/an _____ in induced drag.
A. increase
B. decrease
C. tilting change

Where to Use Caution

81. [B24/1/5]
Ground effect is most likely to result in which situation?
A. Settling to the surface abruptly during landing.
B. Becoming airborne before reaching recommended takeoff speed.
C. Inability to get airborne even though airspeed is sufficient for normal takeoff needs.

82. [B25/1/2]
When operating in ground effect
A. wingtip vortices increase, creating wake turbulence problems for arriving and departing aircraft.
B. induced drag decreases; therefore, any excess speed at the point of flare may cause considerable floating.
C. a full stall landing will require less up elevator deflection than a full stall when done free of ground effect.

83. [B25/1/3]
If you're approaching at a speed above the normal approach speed, make it a point to _____ before entering ground effect to prevent an excessive landing roll.
A. speed up
B. get as close to the runway as possible
C. slow down

Pitch Changes In and Out of Ground Effect

84. [B25/See *Different Designs* sidebar]
What causes an airplane (except a T-tail) to pitch nose down when power is reduced and controls are not adjusted?
A. The CG shifts forward when thrust and drag are reduced.
B. The downwash on the elevators from the propeller slipstream is reduced and elevator effectiveness is reduced.
C. When thrust is reduced to less than weight, lift is also reduced and the wings can no longer support the weight.

85. [B25/2/1]
As the airplane becomes airborne and flies out of ground effect, the wing's downwash _____.
A. decreases
B. remains the same
C. increases

86. [B25/2/1]
It's possible, when attempting to climb out of ground effect, to become airborne without sufficient climb speed, then attempt to climb and have the nose _____ slightly.
A. pitch down
B. pitch up
C. pitch sideways

87. [B25/2/2]
During landing, as the airplane enters ground effect and the downwash diminishes, the nose tends to pitch _____.
A. upward
B. sideways
C. downward

88. [B25/2/2]
Low wing airplanes experience _____ ground effect than their high wing cousins.
A. much less
B. less
C. more

Flap Over Flaps

89. [B25/2/3]
Extending or retracting flaps changes the wing's _____ and _____ characteristics.
A. masculine, feminine
B. lift, drag
C. weight, thrust

90. [B25/2/4]
Lowering flaps lowers the trailing edge of the wing, _____ the angle the chord line makes with the relative wind. This increases the wing's lift.
A. eliminating
B. increasing
C. decreasing

91. [B25/2/4]
When the flaps are lowered, the lowered trailing edge _____ the curvature on part of the wing, resulting in increased air velocity over the wing's upper surface.
A. eliminates
B. decreases
C. increases

92. [B26/1/1]
Because of the larger angle of attack and greater curvature, flaps provide you with _____ lift for a given airspeed.
A. more
B. less
C. similar

Flap Varieties

93. [B26/1/3 & 4] Name the four basic varieties of flaps:

TYPES OF FLAPS

A ____________ C ____________

B ____________ D ____________

Why Use Flaps?

94. [B26/1/5]
What's the reason for putting flaps on airplanes?
A. To create the lift necessary to maintain flight at slower airspeeds.
B. To allow the airplane to fly at cruise speeds with less power.
C. To prevent excessive overspeed conditions in turbulence.

95. [B26/2/2]
If the wind is gusty, you might use _____ flap extension than in non-gusty conditions.
A. the same
B. more
C. less

96. [B26/2/3 & Figure 45]
The beginning of the white arc (B) as shown on the airspeed indicator at the top of the next page is known as the
A. power-off, full-flap stalling speed.
B. power-on, full-flap stalling speed.
C. power-off, no-flap stalling speed.

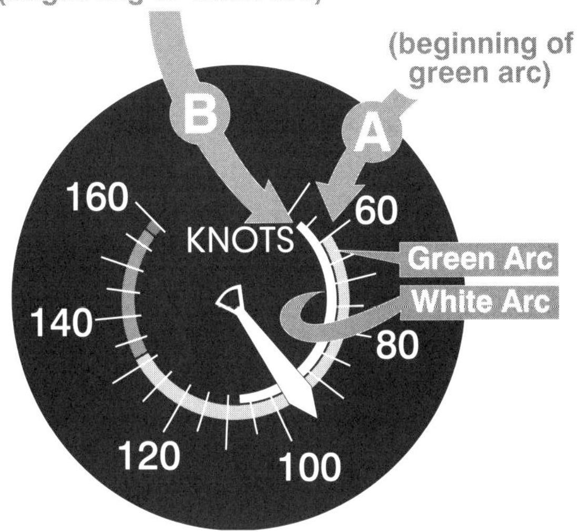

97. [B26/2/3 & Figure 45]
In the airspeed indicator shown above, the airplane will fly when _____ knots of wind blow over the wings (full flaps extended) if the wings are below their critical angle of attack.
A. 46
B. 60
C. 53

98. [B26/2/4 & Figure 45]
In the airspeed indicator shown above, the high speed end of the white arc is the maximum speed you may fly with flaps
A. fully extended.
B. fully retracted.
C. partially extended.

99. [B27/1/2 & 3]
What is one purpose of wing flaps?
A. To enable the pilot to make steeper approaches to a landing without increasing the airspeed.
B. To relieve the pilot of maintaining continuous pressure on the controls.
C. To decrease wing area to vary the lift.

100. [B27/1/3]
One of the main functions of flaps during approach and landing is to
A. decrease the angle of descent without increasing the airspeed.
B. permit a touchdown at a higher indicated airspeed.
C. increase the angle of descent without increasing the airspeed.

101. [B27/3/1]
During a go-around, retract the flaps
A. immediately to their fully-retracted position.
B. in increments.
C. at your convenience when the airplane is stable.

How Airplanes Turn

102. [B28/1/6]
What force makes an airplane turn?
A. The horizontal component of lift.
B. The vertical component of lift.
C. Centrifugal force.

103. [B28/Figure 46B] Write in the names of the force vectors an airplane experiences in a turn:

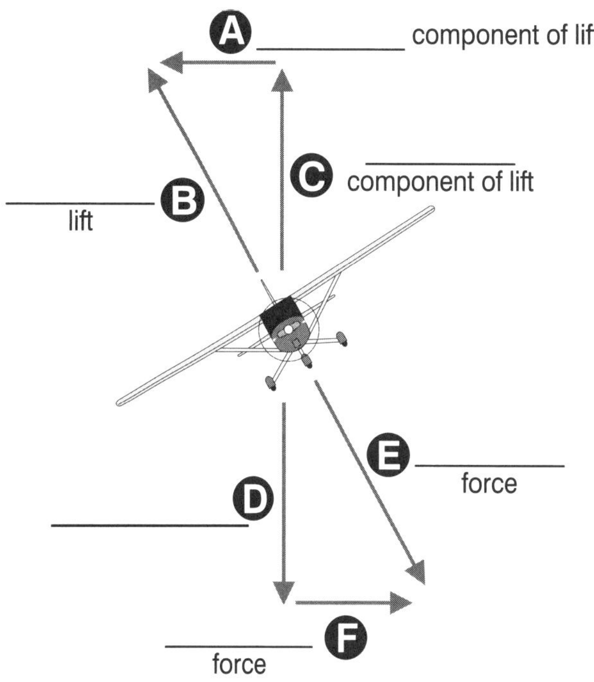

104. [B28/1/7]
Tilting the total lift force while in a turn means _____ lift is available to act vertically against the airplane's weight.
A. more
B. less
C. horizontal

105. [B28/Figure 47] Label the axes below:

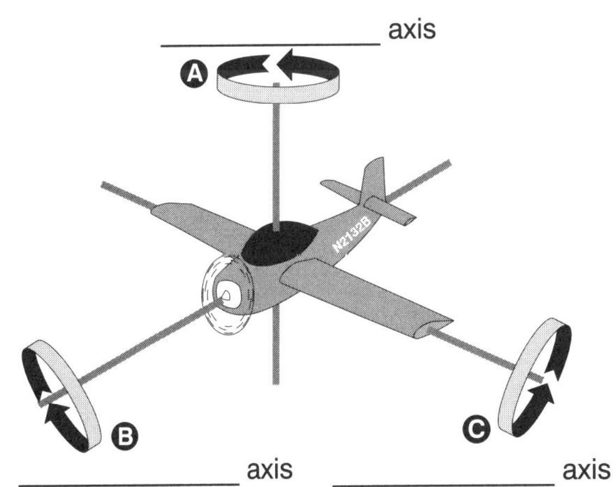

Flight Controls

106. [B29/1/2]
Which axis runs through the centerline of the airplane from nose to tail?
A. Lateral axis.
B. Vertical axis.
C. Longitudinal axis.

107. [B29/1/2]
Airplanes roll about what axis?
A. The lateral axis.
B. The vertical axis.
C. The longitudinal axis.

108. [B29/1/3]
Which axis runs from wing tip to wing tip, sideways through the airplane?
A. The lateral axis.
B. The vertical axis.
C. The longitudinal axis.

109. [B29/1/3]
Airplanes _____ about their lateral axis.
A. roll
B. pitch
C. yaw

Ailerons

110. [B29/1/5]
The purpose of ailerons is to _____ the airplane in the direction you desire to turn.
A. yaw
B. bank
C. pitch

111. [B29/1/5]
Ailerons function to allow the right and left wings to develop _____ lift.
A. the same amount of
B. different amounts of
C. negative degrees of

112. [B29/Figure 49]
When the control wheel (or stick) is turned to the right or left, the ailerons simultaneously move in _____.
A. the same direction
B. different directions
C. variable directions (depends on other factors)

Adverse Yaw

113. [B30/2/2]
Adverse yaw is
A. a desirable characteristic used in the design of wings.
B. an undesirable byproduct of turning.
C. an artifact of wing design that plays no role in normal flight.

Rudders

114. [B30/3/1]
What is the purpose of the rudder on an airplane?
A. To control yaw.
B. To control overbanking tendency.
C. To control roll.

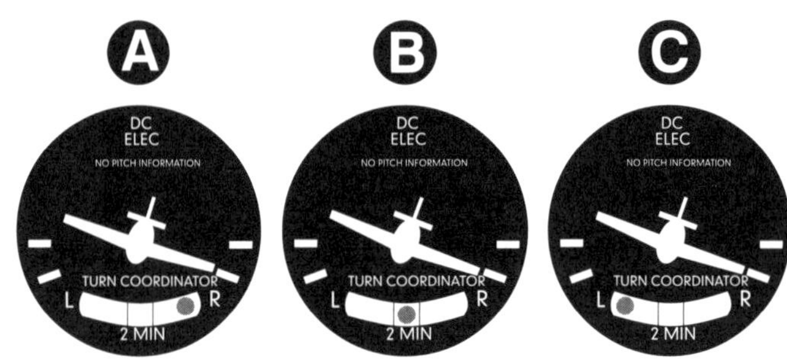

115. [B31/Figure 55]
Which of the illustrations above depicts the excessive use of right rudder during the entry of a right turn?
A. Instrument B.
B. Instrument A.
C. Instrument C.

116. [B32/1/4]
Not using the rudder during a turn will result in _____ turn.
A. a complex
B. a coordinated
C. an uncoordinated

117. [B32/1/5]
As the airplane slows down and enters a stall, which of the following three control surfaces is the last to lose control authority?
A. Aileron.
B. Elevator.
C. Rudder.

Elevator

118. [B33/1/1&2]
Applying forward pressure on the control wheel (or stick) deflects the elevator _____ causing the tail to _____.
A. downward, rise
B. upward, lower
C. in the opposite direction, yaw

Trim Tabs

119. [B33/1/3]
The purpose of the trim tab is to _____ control pressure required of the pilot.
A. eliminate
B. increase
C. change the direction of the

120. [B33/2/1]
Which direction does the trim tab move relative to the primary control surface it affects?
A. The same direction.
B. Rearward.
C. Opposite.

Left Turning Tendencies (not political)

121. [B34/1/3]
In what flight condition is torque effect the greatest in a single-engine airplane?
A. Low airspeed, high power.
B. Low airspeed, low power.
C. High airspeed, high power.

122. [B35/1/3]
The left turning tendency of an airplane caused by P-factor is the result of the
A. clockwise rotation of the engine and the propeller turning the airplane counter-clockwise.
B. propeller blade descending on the right producing more thrust than the ascending blade on the left.
C. gyroscopic forces applied to the rotating propeller blades acting 90 degrees in advance of the point to which the force was applied.

123. [B35/1/3]
P-factor is more likely to cause the airplane to yaw to the left
A. at low angles of attack.
B. at high angles of attack.
C. at high airspeeds.

Postflight Briefing 2-1: How A Spin Occurs

124. [B20 & B37/See *How a Spin Occurs*]
In what flight condition must an aircraft be placed in order to spin?
A. Partially stalled with one wing low.
B. In a steep diving spiral.
C. Stalled.

125. [B20 & B37/See *How a Spin Occurs*]
During a spin to the left, which wing(s) is/are stalled?
A. Both wings are stalled.
B. Neither wing is stalled.
C. Only the left wing is stalled.

126. [B37/2/2]
A typical situation that often results in a spin occurs when
A. the pilot over-aggressively leans the airplane's engine.
B. a pilot overshoots the turn to the final approach and applies rudder to align the nose with the runway while holding the bank angle constant with aileron.
C. the pilot applies full flaps at too high an airspeed.

Postflight Briefing 2-2: Your Airplane in Drag

127. [B38/1/3]
Parasite drag is caused by
A. the effect of decreasing temperatures, but increasing relative humidity on the airplane's surfaces.
B. the pilot's failure to perform an adequate preflight inspection.
C. friction of protruding airplane parts with the air.

128. [B38/1/4]
Generally speaking, parasite drag
A. is a result of the design of an airplane and there is little the pilot can do to reduce it.
B. is a result of careless maintenance.
C. has little real-world effect on general aviation pilots.

129. [B38/3/2]
As an airplane's airspeed doubles, parasite drag
A. remains constant.
B. is significantly reduced.
C. quadruples.

130. [B39/3/2]
Induced drag is
A. the rearward pull of the total lifting force.
B. the drag caused by necessary items such as struts, antennas, etc.
C. always a result of poor piloting technique.

131. [B40/Figures 67 & 68]
At small angles of attack
A. induced drag is at its greatest.
B. induced drag is at its lowest.
C. induced drag is always greater than parasite drag.

Postflight Briefing 2-3: Maximum Endurance & Range

132. [B41/1/1]
If you are trying to reach a destination on limited fuel
A. fly at maximum cruise speed in order to arrive before you run out of fuel.
B. fly at 65% power, leaning the fuel mixture in strict accordance with the POH or owner's manual.
C. fly at the airplane's maximum range speed.

133. [B41/1/3]
A good reason to use maximum endurance speed is
A. to minimize fuel consumption, e.g., while waiting for the weather to clear.
B. to minimize the time to get to your destination.
C. to go the greatest distance possible on your available fuel.

134. [B41/2/2,3]
The maximum endurance speed is always _____ the maximum range speed.
A. greater than
B. within 1% or 2% of
C. less than

135. [B42/1/2 & Figure 72]
In the "region of reversed command," slower level flight requires
A. less power.
B. more power.
C. It depends on the airplane and engine.

Postflight Briefing 2-4: Weight, Glide & the Ride

136. [B43/1/1]
To maintain an airplane's best L/D (lift over drag ratio), a decrease in weight requires
A. an increase in airspeed.
B. a decrease in airspeed.
C. a continuous airspeed.

137. [B43/2/3]
You can determine an airplane's best glide speed
A. by experimentation in the region of reversed command.
B. by reference to the airspeed indicator.
C. by referring to the POH or the owner's manual.

Postflight Briefing 2-5: A Different Look at Maneuvering Speed

138. [B44/1/1]
To prevent structural damage to the airplane during turbulence,
A. maintain flight at or below the airplane's design maneuvering speed.
B. carry no more than 50% of maximum capacity of fuel.
C. use at least 10 degrees of flaps.

139. [B44/1/2]
Which V-speed represents maneuvering speed?
A. V_a
B. V_{lo}
C. V_{ne}

140. [B44/1/4]
With respect to the certification of aircraft, which are categories of aircraft?
A. Normal, utility, acrobatic.
B. Airplane, rotorcraft, glider.
C. Landplane, seaplane.

141. [B44 & B45/1/3]
The amount of excess load that can be imposed on the wing of an airplane depends upon the
A. position of the CG.
B. speed of the airplane.
C. engine power.

142. [B45/1/4]
When the airplane's weight decreases, the maneuvering speed _____.
A. remains the same
B. increases
C. decreases

143. [B44, Postflight Briefing #2-5]
What is an important airspeed limitation that is not color coded on airspeed indicators?
A. Never-exceed speed.
B. Maximum structural cruising speed.
C. Maneuvering speed.

144. [B44, Postflight Briefing #2-5]
Upon encountering severe turbulence, which flight condition should the pilot attempt to maintain?
A. Constant altitude and airspeed.
B. Constant angle of attack.
C. Level flight attitude.

Postflight Briefing 2-6: Frost

145. [B46/See *Frost* insert]
How will frost on the wings of an airplane affect takeoff performance?
A. Frost will disrupt the smooth flow of air over the wing, adversely affecting its lifting capability.
B. Frost will change the camber of the wing, increasing its lifting capability.
C. Frost will cause the airplane to become airborne with a higher angle of attack, decreasing the stall speed.

146. [B46/See *Frost* insert]
Why is frost considered hazardous to flight?
A. Frost changes the basic aerodynamic shape of the airfoils, thereby decreasing lift.
B. Frost slows the airflow over the airfoils, thereby increasing control effectiveness.
C. Frost spoils the smooth flow of air over the wings, thereby decreasing lifting capability.

147. [B46/See *Frost* insert]
How does frost affect the lifting surfaces of an airplane on takeoff?
A. Frost may prevent the airplane from becoming airborne at normal takeoff speed.
B. Frost will change the camber of the wing, increasing lift during takeoff.
C. Frost may cause the airplane to become airborne with a lower angle of attack at a lower indicated airspeed.

148. [B46/See *Frost* insert]
Frost on the airfoil _____ the airplane's stalling speed.
A. increases
B. decreases
C. has no effect on

Chapter Two Answers

1. A
2. A/Thrust
   B/Weight
   C/Drag
   D/Lift
3. A
4. A
5. B
6. C
7. A/90
   B/weight
   C/90
   D/weight,
   E/drag
   F/thrust
8. B
9. B
10. A/90
    B/weight
    C/90
    D/weight,
    E/thrust, weight
    F/Drag
11. A/leading
    B/trailing
    C/cambered,
    D/cambered
    E/Chord
12. B
13. A
14. C
15. B
16. A
17. C
18. C
19. A
20. A
21. A/attack
    B/motion
    C/relative
22. A
23. B
24. C
25. A
26. decrease
27. B
28. C
29. A
30. C
31. A
32. C
33. A
34. C
35. A
36. A
37. C
38. B
39. B
40. A
41. C
42. A
43. C
44. A
45. A
46. B
47. C
48. angle
49. B
50. C
51. C
52. B
53. C
54. A
55. B
56. C
57. B
58. A
59. C
60. B
61. A
62. A
63. C
64. C
65. A/induced
    B/total
    C/parasite
66. B
67. C
68. B
69. A
70. A
71. C
72. A
73. C
74. A
75. B
76. B
77. C
78. B
79. A
80. B
81. B
82. B
83. C
84. B
85. C
86. B
87. C
88. C
89. B
90. B
91. C
92. A
93. A/fowler
    B/plain
    C/slotted
    D/split
94. A
95. C
96. A
97. C
98. A
99. A
100. C
101. B
102. A
103. A/horizontal
     B/total
     C/vertical,
     D/weight
     E/resultant
     F/centrifugal
104. B
105. A/vertical
     B/longitudinal,
     C/lateral
106. C
107. C
108. A
109. B
110. B
111. B
112. B
113. B
114. A
115. C
116. C
117. C
118. A
119. A
120. C
121. A
122. B
123. B
124. C
125. A
126. B
127. C
128. A
129. C
130. A
131. B
132. C
133. A
134. C
135. B
136. B
137. C
138. A
139. A
140. A
141. B
142. C
143. C
144. C
145. A
146. C
147. A
148. A

Note: To ensure that you have the most current answers to these questions, please check the *Book & Slide Updates* section at Rod Machado's web site: www.rodmachado.com

Chapter Three

ENGINES: Knowledge of Engines Is Power

The Airplane Engine

1. [C2/1/2] Fill in the blank:
Most general aviation airplane engines are of the ____________ opposed variety.

2. [C2/1/3] Fill in the blanks:
Horizontally opposed cylinder arrangements pack a lot of engine into a __________ amount of space. Less space used by the engine means less overall __________.

Four Cycle Engine

3. [C2/2/2] Fill in the blanks:
Name the four cycles of an airplane engine:
__________, __________, __________, __________ .

4. [C2/2/3] Fill in the blank:
While the piston is in its downward journey, the ________ valve opens, and a mixture of fuel and air rushes in.

5. [C2/2/4] Fill in the blank:
The ____________ cycle occurs when the intake valve closes and the piston rises.

6. [C2/2/6] Fill in the blank:
Just before the cylinder hits the top of its return journey, the ____________ ____________ fire.

7. [C2/2/7] Fill in the blank:
The burning mixture pushes the piston downward. This is the ____________ stroke.

8. [C3/Figure 3] Label the four strokes:

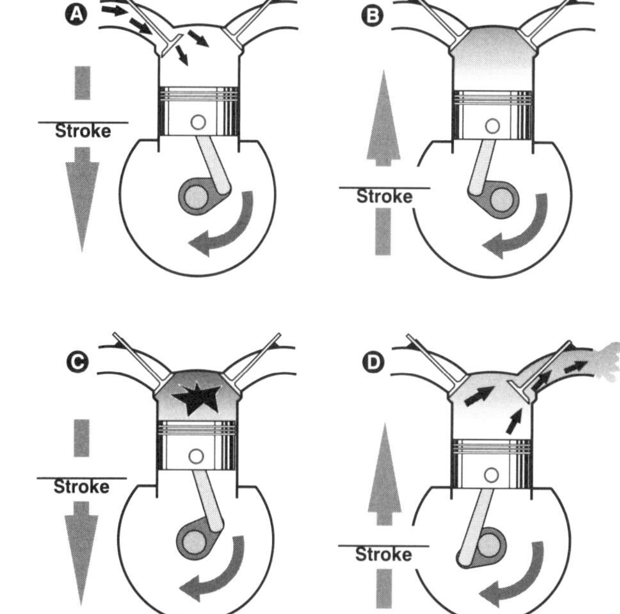

The Ignition System

Dual Ignition Systems

9. [C4/1/2]
One purpose of the dual ignition system on an aircraft engine is to provide for
A. improved engine performance.
B. uniform heat distribution.
C. balanced cylinder head pressure.

Meet Mister Magneto

10. [C4/1/4] Fill in the blanks:

Magnetos contain spinnable magnets, housed in metal cases. When the internal magnets are spun, they generate electricity for the ___________ ___________.

11. [C4/2/2] Fill in the blank:

What's particularly interesting about magnetos is that they are self-contained spark generators and require no outside source of electrical energy to work other than the _____________ motion of the airplane engine.

Impulse Coupling

12. [C5/1/2]

What is the purpose of the magneto's impulse coupling?

A. To charge the battery.
B. To provide extra spin energy for the magneto's internal magnets.
C. To move the more massive gear system of large electrical starters.

Selecting Magnetos

13. [C5/1/4]

Normally, airplanes are operated with the magneto switch in the _____ position.

A. right

B. left

C. both

14. [C5/2/1]

Is it permissible to operate the engine on one magneto?

A. No, never. Not under any circumstances.
B. Only if an instructor is on board and he or she appears calm.
C. Yes. If one magneto goes bad, it's permissible to switch to a single magneto.

15. [C5/3/1] Fill in the blank:

In many cases we're concerned not only with the RPM drop on each magneto, but with the _____________ in RPM between each mag drop.

The P-Lead

16. [C5/3/3]

Selecting the right or left magneto _____ the other mag by grounding it to the airframe.

A. activates
B. deactivates
C. thermonuclearizes

17. [C5/3/4]

The mag is grounded to the airframe via a wire called the _____.

A. P-lead
B. grounding wire
C. wing strut

18. [C6/2/2]

Mechanics (and flight instructors) sometimes recommend doing a P-lead security check just before shutting down. With the engine idling, quickly turn the mag switch from _____ to *off,* then immediately back to _____ again.

The Exhaust System

19. [C6/3/2]

Exhaust gases sometimes have an afterlife. They can be put to use spinning a _____ or indirectly heating the _____ or cabin.

A. propeller, carburetor
B. turbocharger, carburetor
C. nosewheel, runway

The Induction System

20. [C7/1/7]

The _____, _____ and _____ manifold (pipes connected to each cylinder) make up the induction system on carburetor equipped airplanes

A. cowling, carburetor, exhaust
B. air filter, carburetor, exhaust
C. air filter, carburetor, intake

The Carburetor

21. [C8/1/6]

Airplane carburetors located underneath the engine are called _____ type carburetors because air and fuel must be drawn upward toward each cylinder.

A. downdraft

B. injection

C. updraft

22. [C8/2/2]

The operating principle of float-type carburetors is based on the

A. automatic metering of air at the venturi as the aircraft gains altitude.
B. difference in air pressure at the venturi throat and the air inlet.
C. increase in air velocity in the throat of a venturi causing an increase in air pressure.

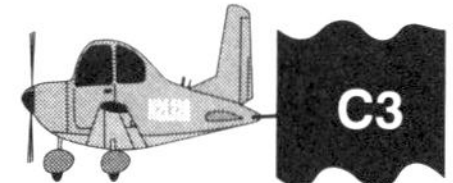

23. [C8/Figure 10] Label the carburetor's components:

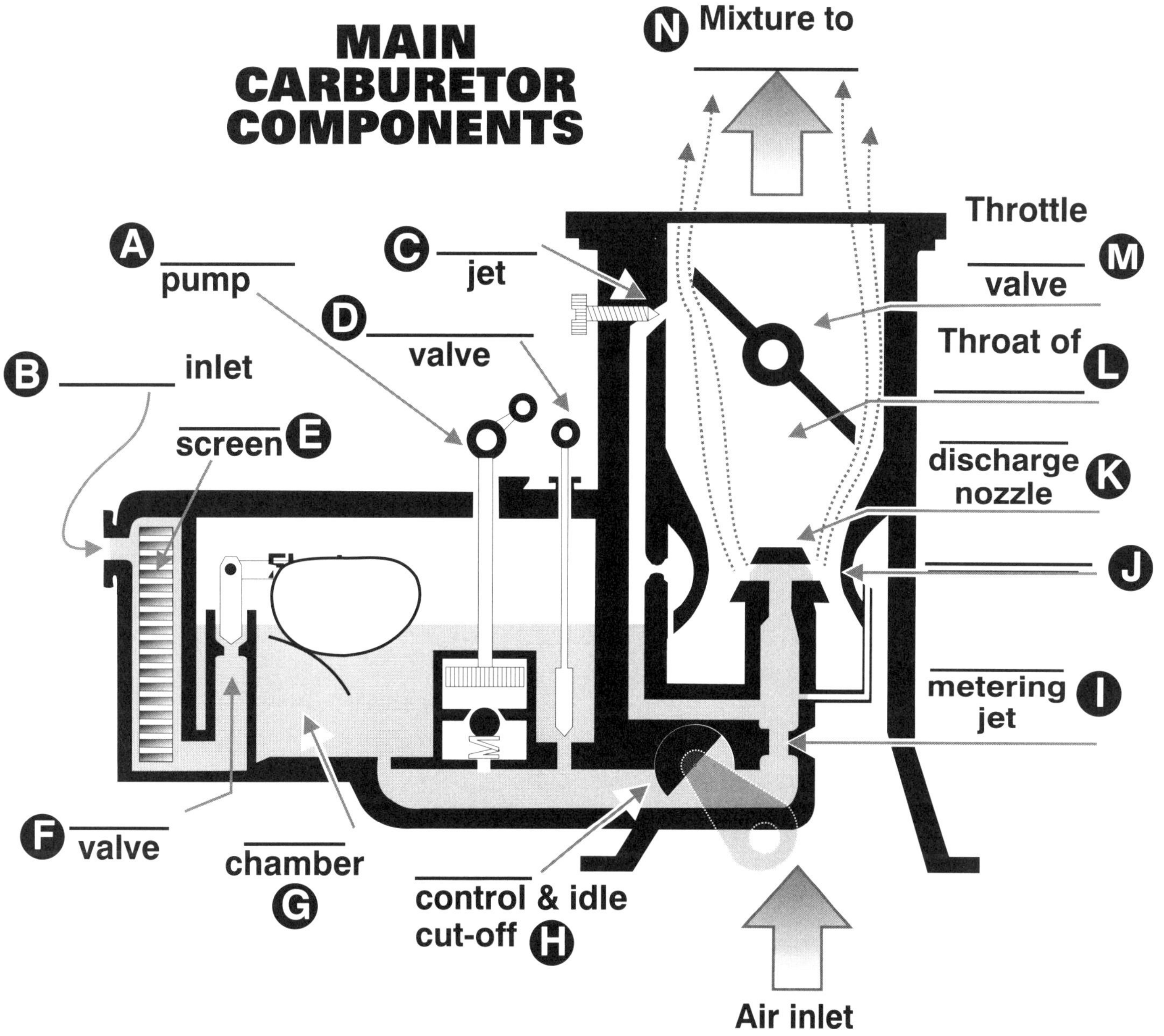

MAIN CARBURETOR COMPONENTS

N Mixture to ________

A ________ pump

C ________ jet

D ________ valve

B ________ inlet

________ screen E

Throttle ________ valve M

Throat of ________ L

________ discharge nozzle K

________ J

________ metering jet I

F ________ valve

________ chamber G

________ control & idle cut-off H

Air inlet

The Idling System

24. [C9/2/3] Fill in the blank:
The ____________ jet is that portion of the carburetor that allows the engine to run when the throttle is pulled full aft.

The Accelerator Pump

25. [C10/1/3]
Why doesn't the engine quit or falter when the throttle is opened abruptly?
A. The idling jet keeps a constant flow of fuel into the engine.
B. The accelerator pump supplies a shot of fuel into the throat of the carburetor along with the inrushing air.
C. Gravity supplies the pressure to keep fuel entering the throat of the carburetor when the throttle is opened.

26. [C10/1/4]
Assuming the airplane has an updraft carburetor with an accelerator pump, what happens when the engine isn't running and the throttle is pumped?
A. Fuel will rush up into the engine.
B. Nothing, unless you make the "Vroom, Vroom" sound.
C. Fuel may fall to the bottom of the carburetor and soak the air filter, creating an opportunity for an engine fire.

Atomization of Fuel

27. [C11/1/1] Fill in the blanks:
Fuel-air ratios of approximately ______________ part(s) fuel to ______________ parts of air are the most efficient for combustion.

Your Carburetor, the Ice Maker

28. [C11/1/4]

Temperature drops of as much as _____ within the carburetor's throat are not uncommon.
A. 10°F
B. 550°F
C. 70°F

29. [C11/2/2]

Your carburetor is a fine ice maker. Because of the considerable drop in temperature caused by the atomization and evaporation of fuel, any _____ present can and will freeze.
A. moisture
B. gasoline
C. ice

30. [C11/2/4]

Be prepared for carburetor ice to form at almost any outside air temperature, though it's most likely to occur between outside temperatures of _____.
A. 20°C to 70°C
B. -120°C to 70°F
C. 20°F to 70°F

Ice: Just Your Type

31. [C11/3/2] Fill in the blanks:

Impact ice occurs when ______ ______ is present and the outside air temperature (OAT) is at or below freezing.

32. [C12/1/1]

When is it possible to have an air filter freeze over while nowhere near a cloud?
A. If the air is moist and temperatures are low, water can accumulate on the air filter's membrane and freeze.
B. Only when the airplane is operated in freezing rain.
C. If the air is between -10°C and -20°C, regardless of the moisture present.

33. [C12/1/2]

Another occasion where impact ice is likely is during _____ (to be discussed further in Chapter 12).
A. thick fog
B. ice-o-cumulus clouds
C. freezing rain

34. [C12/1/2]

If your air filter ever becomes clogged by impact ice, you have a remedy at hand. It's called the _____ control.
A. air traffic
B. throttle
C. carburetor heat

35. [C12/1/3]

Fuel ice forms _____ of the main discharge nozzle.
A. in the slipstream
B. upstream
C. downstream

36. [C12/2/2]

Which condition is most favorable for the development of carburetor icing?
A. Any temperature below freezing and a relative humidity of less than 50 percent.
B. Temperature between 32°F and 50°F and low humidity.
C. Temperature between 20°F and 70°F and high humidity.

37. [C12/2/2]

Fuel ice can occur at outside air temperatures as high as _____ and at humidities as low as _____.
A. 85°F, 50%
B. 10°F, 30%
C. 50°F, 85%

38. [C12/2/3]

Throttle ice forms on the _____ side of the throttle valve. It is more likely to occur when the throttle is in a _____ position.
A. rear, partially closed
B. front, open
C. rear, fully open

39. [C12/Figure 17] Label the two types of ice:

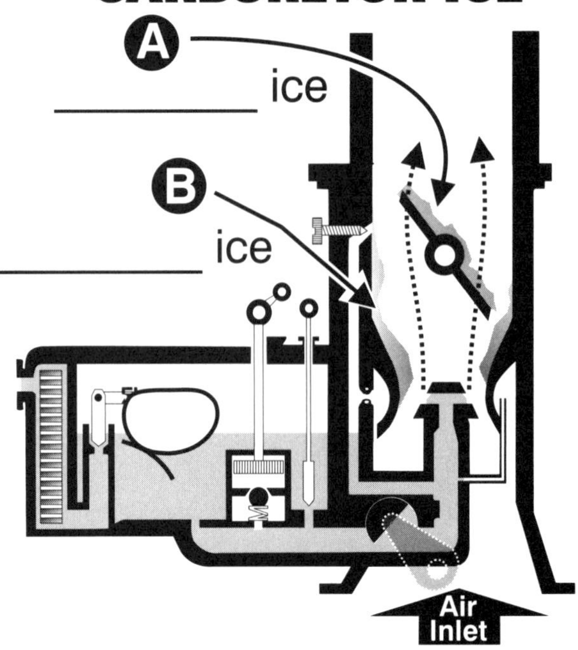

The Ice Eater: The Carburetor Heater

40. [C12/3/3]
Pulling the carburetor heat lever allows heated air to enter the carburetor, raising the air temperature within its throat as much as _____.
A. 30°F
B. 90°C
C. 90°F

Carb Ice Symptoms

41. [C13/1/2]
If an aircraft is equipped with a fixed pitch propeller and a float-type carburetor, the first indication of carburetor ice would most likely be
A. a drop in oil temperature and cylinder head temperature.
B. engine roughness.
C. loss of RPM.

42. [C13/2/3]
The presence of carburetor ice in an aircraft equipped with a fixed pitch propeller can be verified by applying carburetor heat and noting
A. an increase in RPM and then a gradual decrease in RPM.
B. a further decrease in RPM and then a constant RPM indication.
C. a further decrease in RPM and then a gradual increase in RPM.

43. [C13/Figure 20]
Circle the reference below (A, B, C or D) that best represents the sequence of identifying carb ice, applying carb heat, having the ice dissipate and then removing carb heat.

44. [C14/2/2]
Generally speaking, the use of carburetor heat tends to
A. decrease engine performance.
B. increase engine performance.
C. have no effect on engine performance.

45. [C14/2/2]
Applying carburetor heat will
A. result in more air going through the carburetor.
B. enrich the fuel/air mixture.
C. not affect the fuel/air mixture.

46. [C14/2/2]
What change occurs in the fuel/air mixture when carburetor heat is applied?
A. A decrease in RPM results from the leaner mixture.
B. The fuel/air mixture becomes richer.
C. The fuel/air mixture becomes leaner.

Apply Carb Heat as a Precautionary Measure

47. [C14/3/1]
A carburetor air temperature gauge allows you to identify the _____ temperature range where carburetor ice is most likely to occur.
A. critical
B. noncritical
C. humidity

Carburetor Icing Potential in Different Engines

48. [C15/1/1]
All engines _____ have the same carburetor icing potential.
A. do
B. do not
C. almost always

Fuel: Going With the Flow

The Mixture Control

49. [C15/2/3]
Pulling out (toward you) on the mixture control _____the amount of fuel for a given amount of air entering the engine.
A. doesn't change
B. increases
C. decreases

The Fuel/Air Mixture

50. [C15/3/2]
With an increase in altitude the air becomes thinner and doesn't _____ as much for a given volume.
A. weigh
B. count
C. vary

51. [C15/3/2]
The basic purpose of adjusting the fuel/air mixture at altitude is to
A. decrease the amount of fuel in the mixture in order to compensate for increased air density.
B. decrease the fuel flow in order to compensate for decreased air density.
C. increase the amount of fuel in the mixture to compensate for the decrease in pressure and density of the air.

When to Lean

52. [C16/2/2]
Most engine manufacturers recommend leaning the mixture whenever you're operating at or below _____ of the engine's maximum power output (check your POH to be sure).
A. 25%
B. 90%
C. 75%

53. [C16/3/1]
How might you estimate that you're operating at power levels greater than 75% in a non-turbocharged engine?
A. When the throttle extends no more than 1/4 of its maximum travel from its full-in position.
B. If the airplane is operating with full throttle at less than 5,000 feet MSL.
C. If the RPM is less than 2,500, then you're operating at less than 75% power.

54. [C17/1/1]
Remember, failure to lean appropriately means you'll use up _____ portion of fuel unnecessarily.
A. an extra
B. a lesser
C. more than an equal

How to Lean

55. [C17/3/2]
Airplanes with fixed pitch propellers (propellers having one pitch that can't be changed in flight) and float-type carburetors can be leaned by reference to the _____.
A. ailerons
B. tachometer
C. leaner

56. [C17/3/2]
While leaning with reference to the tachometer, the RPM peaks. This means that you are at the fuel-air ratio that produces maximum _____ for a given air density and throttle setting.
A. torque
B. leaning
C. power

Too Rich and Too Lean

57. [C18/2/2]
A mixture that is too rich causes engine _____.
A. roughness
B. purring
C. heat

58. [C18/2/3]
A fouled spark plug in flight can sometimes be detected by _____.
A. an increase in the pilot's heart beat
B. an increase in EGT
C. a decrease in EGT

59. [C18/2/4]
An excessively rich mixture contributes to _____, _____, _____ and _____.
A. a rough running engine, high fuel consumption, less range, smaller fuel reserves
B. a smooth running engine, low fuel consumption, more range, larger fuel reserves
C. a barely running engine, high alcohol consumption, less radar range, smaller food reserves

60. [C18/2/4]
During the run-up at a high elevation airport, a pilot notes a slight engine roughness that is not affected by the magneto check but grows worse during the carburetor heat check. Under these circumstances, what would be the most logical initial action?
A. Check the results obtained with a leaner setting of the mixture.
B. Taxi back to the flight line for a maintenance check.
C. Reduce manifold pressure to control detonation.

61. [C18/3/2]
The biggest danger with an excessively lean mixture is that it _____.
A. doesn't burn
B. burns hot
C. burns cool

62. [C18/3/2]
What is one procedure to aid in cooling an engine that is overheating?
A. Enrichen the fuel mixture.
B. Increase the RPM.
C. Reduce the airspeed.

63. [C18/3/3]
High cylinder head temperatures also lead to something known as _____.
A. pre-ignition
B. detonation
C. combustion

Leaning & High Altitude Takeoffs for Non-turbocharged Airplanes

64. [C16-C18]
While cruising at 9,500 feet MSL, the fuel/air mixture is properly adjusted. What will occur if a descent to 4,500 feet MSL is made without readjusting the mixture?
A. The fuel/air mixture may become excessively lean.
B. There will be more fuel in the cylinders than is needed for normal combustion, and the excess fuel will absorb heat and cool the engine.
C. The excessively rich mixture will create higher cylinder head temperatures and may cause detonation.

65. [C19/Figure 30/See *EGT Gauge Setting...* sidebar]
For *best power* (most useable power per unit of air), enrich the mixture until the temperature _____ . (Check your POH to ensure the proper procedure for your aircraft.)
A. increases 50°F from peak EGT
B. drops 50°F from peak EGT
C. decreases 125°F from peak EGT

The Fuel System

Components

66. [C20/3/2]
Water is the most frequent contaminant found in fuel. Water, weighing approximately _____ pounds per gallon, _____ than fuel, which weighs approximately _____ pounds per gallon.
A. 8, is heavier, 6
B. 6, is lighter, 8
C. 8, tastes better, 10

67. [C21/1/1]
If it's present, water rests on the _____ of fuel tanks, where it's often the first thing to go to the engine.
A. top
B. bottom
C. outside

68. [C20/Figure 32] Label the fuel system components shown below:

TYPICAL GRAVITY FED (HIGH WING) FUEL SYSTEM

1. Fuel tank _________ drain
2. Fuel line _________
3. Fuel _________ valve
4. Fuel _________ & valve
5. Engine driven _________ pump
6. _________
7. Fuel _________ lines
8. _________
9. _________
10. _________ control
11. Fuel _________
12. Tank _________
13. Fuel _________
14. Rheostat _________
15. Fuel _________
16. Fuel tank _________

69. [C21/2/2]
Filling the fuel tanks after the last flight of the day is considered a good operating procedure because this will
A. force any existing water to the top of the tank away from the fuel lines to the engine.
B. prevent expansion of the fuel by eliminating airspace in the tanks.
C. prevent moisture condensation by eliminating airspace in the tanks.

70. [C21/2/3]
Fuel tanks have sump drains found at the _____ part of the tank. The sumps should be drained after every _____.
A. lowest, refueling
B. highest, refueling
C. lowest, annual inspection

Fuel Colors

71. [C21/3/5]
Different grades of aviation fuel (called AVGAS) are dyed for easy identification. Red colored fuel is of the ____ octane variety (very hard to find nowadays). _____ fuel is of the 100LL or low lead variety and is one of the most common aviation fuels in use today. _____ fuel is 100 octane and is also very hard to find.
A. 80, green, blue
B. 100LL, blue, green
C. 80, blue, green

72. [C22/1/2]
What type fuel can be substituted for an aircraft if the recommended octane is not available?
A. The next higher octane aviation gas.
B. The next lower octane aviation gas.
C. Unleaded automotive gas of the same octane rating.

Fuel Vents

73. [C22/2/1] Fill in the blank:
As the fuel pump sucks fuel from the tank, air must replace the departing fuel or a _____________ forms.

Auxiliary Fuel Pumps

74. [C22/3/2]
Boost pumps are often used during _____ to pressurize the fuel system. This helps purge air trapped within the fuel lines. After the engine starts, the electric boost pump is _____ to see if the _____ pump is operational and is pressurizing the system.
A. engine failures, kept on, vacuum
B. engine start, located, mechanical
C. engine start, turned off, mechanical

75. [C23/1/2]
On aircraft equipped with fuel pumps, the practice of running a fuel tank dry before switching tanks is considered unwise because
A. the engine-driven fuel pump or electric fuel boost pump may draw air into the fuel system and cause vapor lock.
B. the engine-driven fuel pump is lubricated by fuel and operating on a dry tank may cause pump failure.
C. any foreign matter in the tank will be pumped into the fuel system.

Prime Time

76. [C23/2/2]
The primer usually squirts fuel in the _____ portions of the induction system, _____ the carburetor completely.
A. upper, bypassing
B. lower, bypassing
C. most forward, involving

Fuel Gauges

77. [C24/1/1]
The regulations require fuel gauges to be accurate in only two conditions: when the tank is _____ and when it's _____.
A. 1/4 full, empty (no useable fuel on board)
B. half-full, empty (no useable fuel on board)
C. full, empty (no useable fuel on board)

How Much Is Enough?

78. [C24/3/1]
Remember Rod Machado's fuel axiom No. 13: If you smell fuel on takeoff, go back and land. You probably left the cap off your _____.
A. tractor
B. toothpaste
C. tank

The Oil System

79. [C25/See *Sudden Loss of Oil* sidebar]
If an airplane's oil cap is left off or comes loose during flight, oil may be _____ the engine.
A. expelled from
B. consumed by
C. overheated by

80. [C25/1/2]
For internal cooling, reciprocating aircraft engines are especially dependent on
A. a properly functioning thermostat.
B. air flowing over the exhaust manifold.
C. the circulation of lubricating oil.

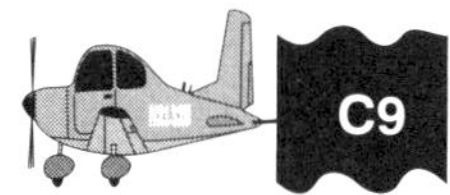

Malfunctions in the Oil System

81. [C25/2/3]
What should be the first action after starting an aircraft engine?
A. Adjust for proper RPM and check for desired indications on the engine gauges.
B. Place the magneto or ignition switch momentarily in the OFF position to check for proper grounding.
C. Test each brake and the parking brake.

82. [C26/1/1]
Engines are generally preheated in extreme cold. Lycoming, for example, recommends the engine be preheated at temperatures below _____ to prevent engine damage during startup. Continental says _____.
A. 10°F, 20°F
B. 30°F, 50°F
C. -20°F, -60°F

83. [C26/1/2]
An abnormally high engine oil temperature indication may be caused by
A. the oil level being too low.
B. operating with a slightly higher viscosity oil.
C. operating with an excessively rich mixture.

84. [C26/1/2]
Excessively high engine temperatures will
A. cause damage to heat-conducting hoses and warping of the cylinder cooling fins.
B. cause loss of power, excessive oil consumption, and possible permanent internal engine damage.
C. not affect an aircraft engine.

85. [C26/1/2]
If the engine oil temperature and cylinder head temperature gauges have exceeded their normal operating range, the pilot may have been operating with
A. the mixture set too rich.
B. higher-than-normal oil pressure.
C. too much power and with the mixture set too lean.

86. [C26/1/2]
What action can a pilot take to aid in cooling an engine that is overheating during a climb?
A. Reduce rate of climb and increase airspeed.
B. Reduce climb speed and increase RPM.
C. Increase climb speed and increase RPM.

87. [C26/1/3]
Inside most airplane engines is something known as an _____. This valve provides an alternate path for the flow of oil if the pressure within the system becomes dangerously _____.
A. oil pressure temperature value, low
B. oil pressure relief valve, high
C. oil pressure relief valve, low

The Engine Cooling System

88. [C26/1/5]
Most modern airplane engines are _____ cooled. Water cooling may be more efficient, but it would mean carrying considerably more _____.
A. air, weight
B. water, weight
C. oil, canteens

89. [C26/ See *It's Best Not to Be That Cool* sidebar]
In order to avoid shock-cooling a high performance engine during a cruise descent, make small power changes. A good rule of thumb is to make no more than a _____ RPM per minute reduction in engine speed on small engines and no more than ____ of manifold pressure per minute on high performance airplanes. Always leave enough power on so as to not cool the cylinder head temperatures below allowable limits.
A. 500, 5 inches
B. 100, 1 inch
C. 1000, .5 inches

90. [C26/2/2]
Keep in mind that engine cooling is least effective at_____ power settings and _____ airspeeds, where a limited amount of air enters the engine cowling.
A. low, high
B. high, low
C. medium, medium

91. [C26/2/5]
Keeping the _____flaps closed helps maintain engine temperatures during the descent.
A. wing
B. cow
C. cowl

92. [C27/1/1]
During a descent, your job is to maintain stable cylinder head temperatures (CHT) and oil temperatures. On some airplanes, _____ extension or even _____ extension at high speeds can be used in lieu of large power reductions to start a descent (check your POH).
A. gear, cowl flap
B. arm, head
C. gear, partial flap

The Propeller

93. [C27/1/2]
In an airplane with a fixed pitch prop, one lever, the _____ controls both power and propeller blade RPM (revolutions per minute).
A. primer
B. propeller lever
C. throttle

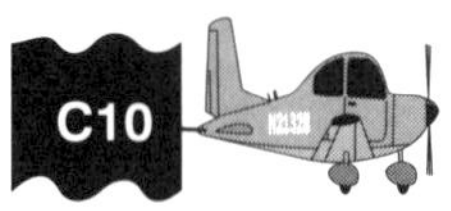

94. [C27/1/5]
As you move up into higher performance airplanes you encounter constant speed (controllable pitch) propellers. Airplanes with these propellers usually have both a _____ and a/an _____ control, so you manage engine power and propeller RPM separately.
A. throttle, propeller
B. throttle, ignition
C. propeller, cowl flap

95. [C27/1/6]
What is an advantage of a constant speed propeller?
A. It permits the pilot to select and maintain a desired cruising speed.
B. It permits the pilot to select the blade angle for the most efficient performance.
C. It provides smoother operation with stable RPM and eliminates vibrations.

96. [C27/1/6]
On airplanes with constant speed propellers, movement of the _____ determines the amount of fuel and air reaching the cylinders.
A. throttle
B. propeller control
C. ignition key

97. [C27/1/6]
Simply stated, the _____ determines how much power the engine can develop.
A. propeller
B. throttle
C. flap lever

98. [C27/1/6]
Movement of the propeller control changes the propeller's _____.
A. pitch
B. tone
C. ignition timing

99. [C27/1/6]
On an airplane with a constant speed propeller, the _____ determines engine power while the _____ determines how efficiently that power is used.
A. propeller control, turbocharger
B. throttle, propeller's pitch
C. throttle, elevator

100. [C27/1/7]
Forward movement of the propeller control causes both halves of the propeller to rotate about their axes and attack the wind at a _____ angle.
A. larger
B. smaller
C. negative lift

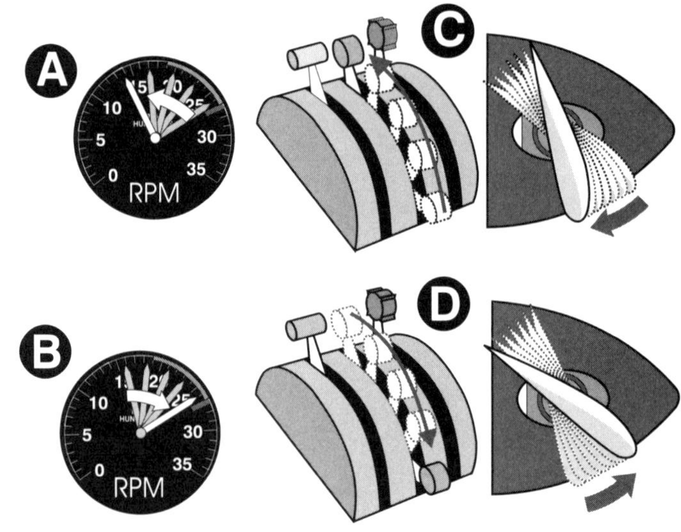

101. [C27/Figure 46]
Referring to the figure above, determine which tachometer goes with which throttle-propeller combination.
A. Tachometer A with C, B with D.
B. Tachometer A with D, B with C.
C. Neither tachometer goes with C or D.

102. [C28/1/1]
Pulling the propeller control rearward causes the propeller to attack the wind at a _____ angle of attack.
A. larger
B. equidistant
C. smaller

103. [C28/2/1]
On an airplane with a constant speed propeller, the _____ tells you how fast the propeller spins (its RPM), and the manifold pressure gauge gives you an approximate measure of engine _____.
A. tachometer, power
B. tachometer, speed
C. tachometer, temperature

104. C28/3/3]
Opening the throttle slightly causes an increase in _____ pressure. More air and fuel are drawn inside the engine, and power _____. Eventually, as the throttle is fully opened, the pressure downstream of the throttle valve approaches that of the _____.
A. manifold, increases, atmosphere
B. atmospheric, decreases, intake manifold
C. manifold, decreases, atmosphere

105. [C29/1/1] Fill in the blank:
Under normal conditions (assuming a non-turbocharged engine), the engine's manifold pressure can't rise above _____________ pressure.

106. [C29/3/2]
As a non-turbocharged airplane climbs, you'll notice the manifold pressure gauge shows _____ even though the throttle is fully opened.
A. an increase
B. a decrease
C. no change

107. [C30/2/3]
Why have a propeller that can change its pitch in flight in the first place?
A. It increases the airplane's resale value.
B. Controllable pitch propellers allow more efficient cooling of the engine.
C. Airplanes with controllable pitch propellers can select the optimum angle of attack for climb and cruise.

108. [C31/1/2]
Engine power is dependent on RPM. For an engine to develop its maximum power, it must be operated at its _____ allowable RPM. At any lower RPM the engine develops only a fraction of its total horsepower. That's why on takeoff we want the propeller set to its _____ pitch (highest RPM) position.
A. highest, lowest
B. lowest, highest
C. maximum, highest

109. [C31/3/2]
During cruise flight, is it necessary to always fly with maximum power and maximum RPM?
A. Yes.
B. No.

Why Constant Speed Propellers?

110. [C32/1/2]
Controllable pitch propellers on general aviation airplanes are of the constant speed variety. Once the RPM is established, changes in manifold pressure (by moving the throttle) _____ affect engine speed.
A. will
B. won't
C. always

How to Make Power Changes

111. [C33/1/3]
When you want to increase both the manifold pressure and RPM, change the _____ first, then increase the _____.
A. RPM, manifold pressure
B. manifold pressure, RPM
C. RPM, mixture

112. [C33/1/3]
A precaution for the operation of an engine equipped with a constant speed propeller is to
A. avoid high RPM settings with high manifold pressure.
B. avoid high manifold pressure settings with low RPM.
C. always use a rich mixture with high RPM settings.

113. [C33/1/4]
Follow the same philosophy when decreasing manifold pressure and RPM. Pull the _____ back first, followed by the _____.
A. propeller control, throttle
B. throttle, propeller control
C. door handle, ripcord

Propeller Tips and Ideas

114. [C33/1/5]
Be aware that the propeller governor starts working only when the engine is operating above a specific _____ and not below.
A. flap setting
B. manifold pressure
C. RPM

Up, Up and Away

Postflight Briefing 3-1: Detonation & Preignition

115. [C34/1/3]
Detonation occurs in a reciprocating aircraft engine when
A. the spark plugs are fouled or shorted out or the wiring is defective.
B. hot spots in the combustion chamber ignite the fuel/air mixture in advance of normal ignition.
C. the unburned charge in the cylinders explodes instead of burning normally.

116. [C34/3/4]
Which would most likely cause the cylinder head temperature and engine oil temperature gauges to exceed their normal operating ranges?
A. Using fuel that has a lower-than-specified fuel rating.
B. Using fuel that has a higher-than-specified fuel rating.
C. Operating with higher-than-normal oil pressure.

117. [C34/3/7]
If a pilot suspects that an airplane engine with a fixed-pitch propeller is detonating during climbout after takeoff, the initial corrective action to take would be to
A. lean the mixture.
B. lower the nose slightly to increase airspeed.
C. apply carburetor heat.

118. [C34/3/4]
If the grade of fuel used in an aircraft engine is lower than specified for the engine, it will most likely cause
A. a mixture of fuel and air that is not uniform in all cylinders.
B. lower cylinder head temperatures.
C. detonation.

119. [C35/1/1]
The uncontrolled firing of the fuel/air charge in advance of normal spark ignition is known as
A. combustion.
B. preignition.
C. detonation.

120. [C35/1/1]
Preignition causes peak pressures within the cylinder to occur before the beginning of the _____ cycle.
A. power
B. intake
C. compression

Postflight Briefing 3-2: Fuel Injection Systems

121. [C35/1/3]
Fuel injection is a process in which fuel is directly _____ and _____ to each cylinder without the use of a carburetor.
A. metered, distributed
B. atomized, sent
C. mixed with air, sent

122. [C35/1/4]
The FCU (fuel control unit) regulates both the volume of _____ entering the engine and the quantity of _____ delivered to the FMU (fuel manifold unit).
A. fuel, spark
B. air, fuel
C. fuel vapor, air

123. [C35/2/1]
You can't get carburetor ice in a fuel injection system for one very important reason: there is _____.
A. no fuel in this system
B. no water in this system
C. no carburetor associated with this system

124. [C35/2/1]
With regard to carburetor ice, float-type carburetor systems in comparison to fuel injection systems are generally considered to be
A. more susceptible to icing.
B. equally susceptible to icing.
C. susceptible to icing only when visible moisture is present.

125. [C36/1/3]
A major advantage of fuel injection is
A. improved control of fuel-air ratio.
B. less uniform delivery of the fuel-air mixture to each cylinder.
C. a greater chance of vaporization ice (fuel ice).

126. [C36/2/1]
A major disadvantage of fuel injection is
A. increased engine efficiency.
B. contamination of dirt and water can more easily affect fuel injected engines due to the small orifices of injector nozzles.
C. instant acceleration of engine without tendency to hesitate.

Postflight Briefing 3-3: Advanced Airplane Systems - Turbocharging & Pressurization

127. [C36/3/2]
Airplanes with turbochargers are able to produce _____ power at high altitudes.
A. sea level
B. less
C. political

128. [C36/3/3]
Turbocharging compresses air into the intake manifold by utilizing the normally unused _____ exiting the engine.
A. fuel vapor
B. ram air
C. exhaust gases

129. [C37/1/1]
The _____valve closes (either automatically or manually, depending on your airplane) allowing more _____ to flow over the turbine. This spins the _____, allowing compressed air to be pumped into the induction system.
A. exhaust, heat, engine
B. wastegate, exhaust, compressor,
C. intake, exhaust, propeller

130. [C37/2/1]
Pressurizing an airplane allows you to fly at very high altitudes while remaining in a cabin environment with near _____ pressures. This is accomplished by directing _____ air from the _____ (or from an auxiliary compressor) into the cabin.
A. sea-level, compressed, turbocharger
B. cabin, turbocharged, exhaust
C. hot air, exhaust fumes, engine

Chapter Three Answers

1. horizontally
2. small, drag
3. intake, compression, power, exhaust
4. intake
5. compression
6. spark plugs
7. power
8. A/intake
   B/compression
   C/power,
   D/exhaust
9. A
10. spark plug
11. turning
12. B
13. C
14. C
15. difference
16. B
17. A
18. both, both
19. B
20. C
21. C
22. B
23. A/accelerator, B/fuel
    C/idling, D/economizer
    E/fuel, F/needle, G/float
    H/mixture, I/main, J/venturi
    K/main, L/carburetor
    M/butterfly, N/cylinder
24. idling
25. B
26. C
27. one, 13
28. C
29. A
30. C
31. visible moisture
32. A
33. C
34. C
35. C
36. C
37. A
38. A
39. A/throttle, B/fuel
40. C
41. C
42. C
43. C
44. A
45. B
46. B
47. A
48. B
49. C
50. A
51. B
52. C
53. B
54. A
55. B
56. C
57. A
58. B
59. A
60. A
61. B
62. A
63. B
64. A
65. C
66. A
67. B
68. 1/sump, 2/strainer
    3/selector, 4/strainer
    5/fuel, 6/carburetor
    7/primer, 8/primer
    9/throttle, 10/mixture
    11/gauges, 12/interconnect
    13/rheostat, 14/float
    15/cap, 16/vent
69. C
70. A
71. C
72. A
73. vacuum
74. C
75. A
76. A
77. C
78. C
79. A
80. C
81. A
82. A
83. A
84. B
85. C
86. A
87. B
88. A
89. B
90. B
91. C
92. C
93. C
94. A
95. B
96. A
97. B
98. A
99. B
100. B
101. B
102. A
103. A
104. A
105. atmospheric
106. B
107. C
108. A
109. B
110. B
111. A
112. B
113. B
114. C
115. C
116. A
117. B
118. C
119. B
120. A
121. A
122. B
123. C
124. A
125. A
126. B
127. A
128. C
129. B
130. A

Note: To ensure that you have the most current answers to these questions, please check the *Book & Slide Updates* section at Rod Machado's web site: www.rodmachado.com

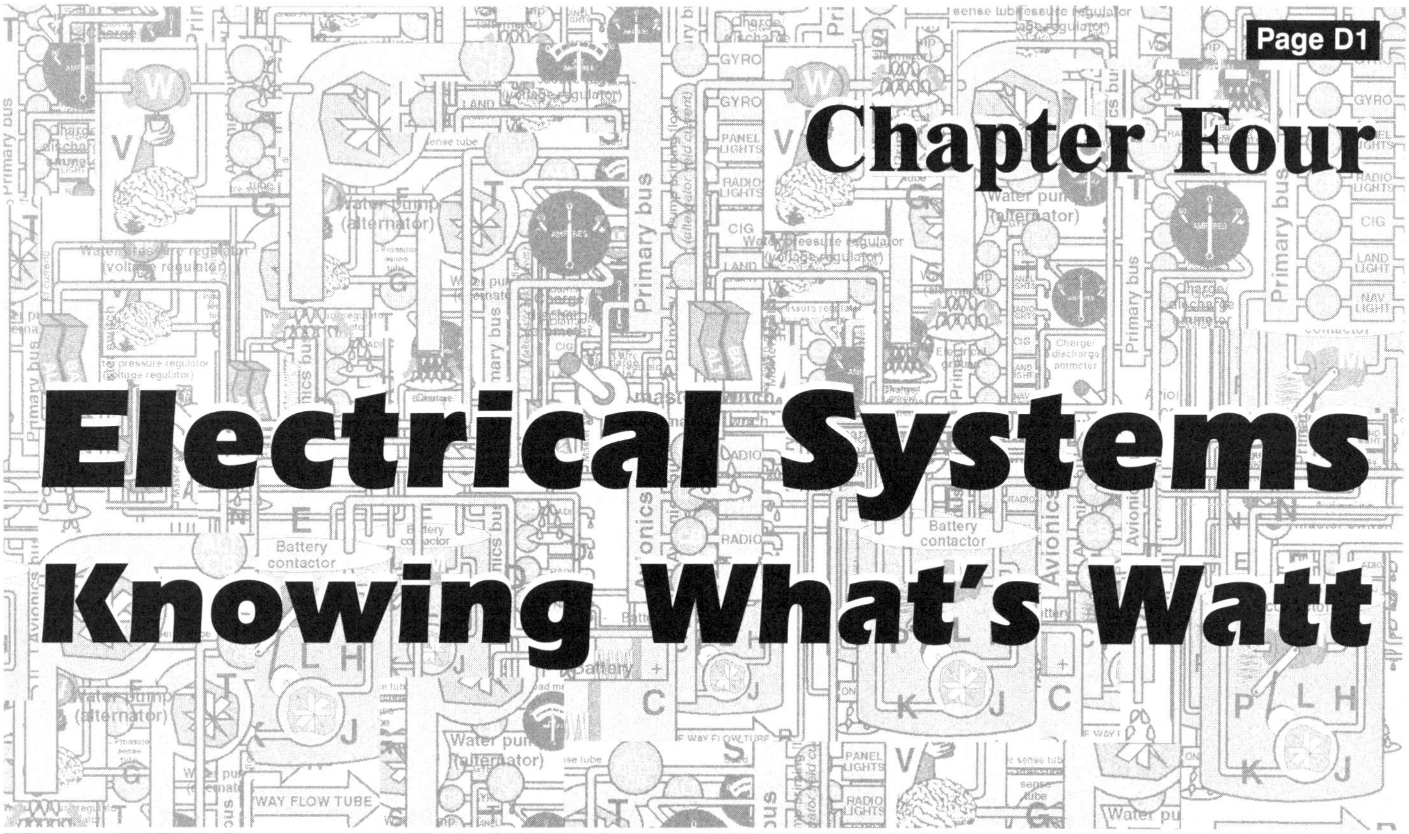

Chapter Four

Electrical Systems Knowing What's Watt

Electricity and Water

1. [D2/1/1]
Electrons flow from the _____ to the _____ side of a battery.
A. negative, positive
B. positive, negative
C. positive, neutral

2. [D2/1/2]
Electricity involves many terms that can be directly related to water or plumbing. Voltage is comparable to, and can be thought of as, the _____ that pushes water through pipes. Current is the amount of _____ actually flowing through the pipes at any given time.
A. temperature, water
B. pressure, pressure
C. pressure, water

3. [D2/1/2]
The greater the _____ (voltage) in a line, the greater the amount of _____ (electric current) that flows.
A. water pressure, water
B. back flow, air
C. restriction, molecules

4. [D2/2/2]
Using the water analogy, the alternator can be thought of as the _____ which provides water pressure to the _____ and the _____.
A. battery, electrical ground, propeller
B. pump, primary bus, avionics bus
C. pump, radio, fuel tank

The Water Pump

5. [D2/2/5]
A spinning alternator causes electric _____ to flow into the primary bus and through the electrical equipment. It then returns to its _____ source in a manner similar to the tray in our water analogy. This tray is called the _____.
A. current, original, electrical ground
B. force, battery, oil pan
C. pressure, alternator, battery

6. [D3/Figure 5] Label the individual components of the water analogy circuitry:

WATER ANALOGY OF ELECTRICITY (SCHEMATIC)

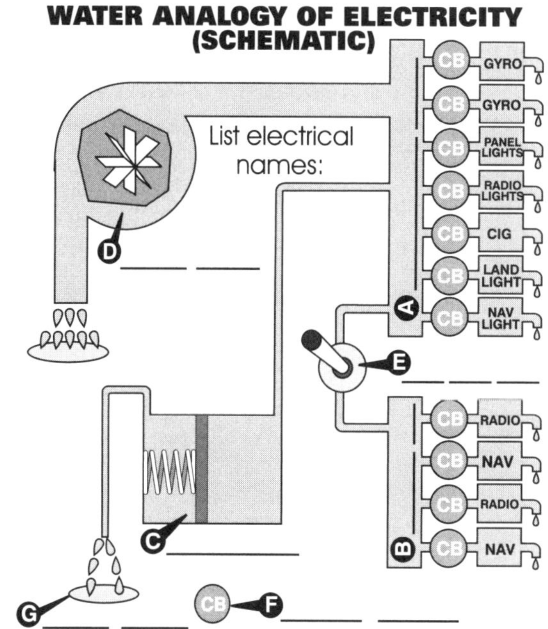

The Electrical Ground

7. [D3/3/2]
The electrical ground is not the ground upon which you stand. It's usually the airplane's _____.
A. battery
B. metal frame
C. ground wire

Load Meter

8. [D4/1/2]
A load meter is located between the _____ and the _____ of our water (electrical) system.
A. pump (alternator), starter
B. pump (alternator), battery
C. pump (alternator), primary bus

9. [D4/1/2]
The load meter shows the _____ placed on the pump (alternator) by the airplane's water (electrical) system.
A. pressure (voltage)
B. water load (current)
C. resistance

10. [D4/1/3]
A zero reading on the load meter (a full left deflection) means that the water pump (alternator) _____ providing pressure to the primary bus.
A. is
B. isn't
C. Both of the above.

11. [D4/Figure 7]
Based on the water analogy circuit below, the load meter would be located at _____.
A. position Z
B. position X
C. position Y

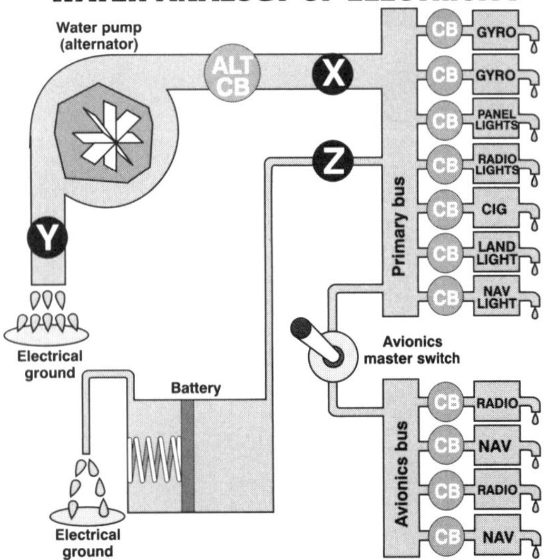

12. [D4/1/3]
A zero reading on the load meter means the alternator _____ providing electricity to the primary bus or there is ____ electrical current drawn by the airplane's electrical equipment. This can happen when the _____ fails or all the airplane's electrical equipment is turned off.
A. isn't, no, alternator
B. is, little, battery
C. isn't, no, ground wire

13. [D4/3/1]
Amps are a measure of _____ flow.
A. voltage
B. current
C. water pressure

14. [D4/3/2]
Circuit breakers protect electrical equipment and wires from receiving more _____ than they can safely handle.
A. current
B. voltage
C. resistance

The Battery

15. [D5/2/2]
In the actual electrical system, the battery _____ stores the electrical charge provided by the alternator.
A. spring
B. mechanically
C. chemically

16. [D5/2/2]
It's the _____ which allows you to operate the airplane's electrical equipment when the engine isn't running or when the alternator fails in flight.
A. alternator
B. propeller
C. battery

Battery Potential

17. [D5/3/2]
A 35 amp hour battery should (in theory) provide _____ amps of continuous current to the electrical system for one hour.
A. 35
B. 17.5
C. 3.5

18. [D5/3/4]
While airplane batteries are rated at 12 or 24 volts, airplane electrical systems (their alternators) are rated for ____ or ____ volts.
A. 12, 24
B. 14, 28
C. 7, 14

The Charge-Discharge Ammeter

19. [D6/1/1]
Between the positive terminal of the battery and the primary bus is an ammeter, called a _____ ammeter.
A. charge-discharge
B. battery
C. load

20. [D6/1/1]
As the name implies, the charge-discharge ammeter tells you if electrical current is flowing into or out of the _____.
A. load meter
B. alternator
C. battery

21. [D6/1/2]
A positive deflection of the charge-discharge ammeter usually implies that the battery is being _____ .
A. charged
B. discharged
C. neutralized

22. [D6/1/2]
A negative needle deflection on the charge-discharge ammeter usually implies that the _____ is supplying the primary bus with electrical current.
A. ammeter
B. regulator
C. battery

23. [D6/Figure 12]
Based on the water analogy circuit below, the charge-discharge ammeter would be located at _____.
A. position Y
B. position X
C. position Z

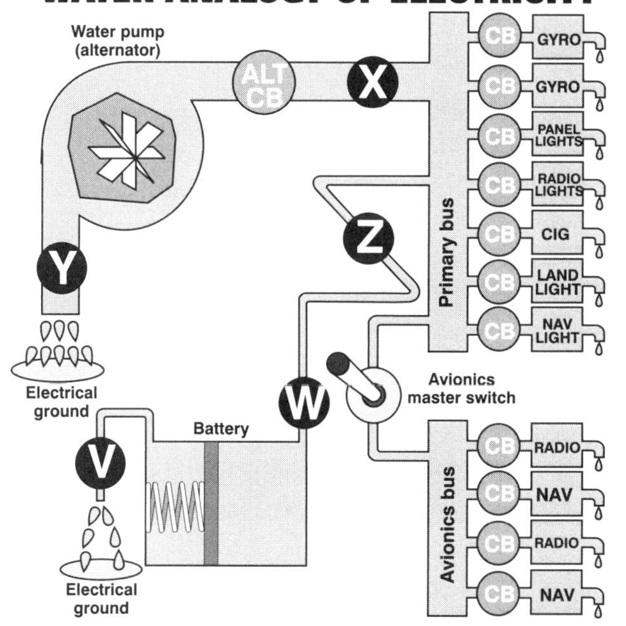

24. [D6/1/3]
Normally, the needle of the charge-discharge ammeter should be resting near the _____ or _____ mark. This implies that the battery is neither being charged nor discharged (a good sign).
A. zero, center
B. full left, full right
C. 0, 60

25. [D6/2/2]
After startup, the battery is sure to be slightly drained. You can expect to see a _____ needle deflection of five, maybe six or seven needle widths on the ammeter right after engine start.
A. positive (+)
B. neutral (+/-)
C. negative (-)

26. [D7/1/2]
Most airplane operation manuals suggest that after approximately _____ of cruising flight, the ammeter needle should return to within a two-needle-width deflection from center on the positive (+) or charging side.
A. 1 minute
B. 10 minutes
C. 30 minutes

27. [D6/Figure 14 & D7/Figure15]
Based on the schematic below, which ammeter (Y or Z) goes with which circuit (W or X)?
A. Y goes with circuit W, Z goes with circuit X.
B. Y goes with circuit X, Z goes with circuit W.
C. Y goes with circuit W, Z goes with circuit W.

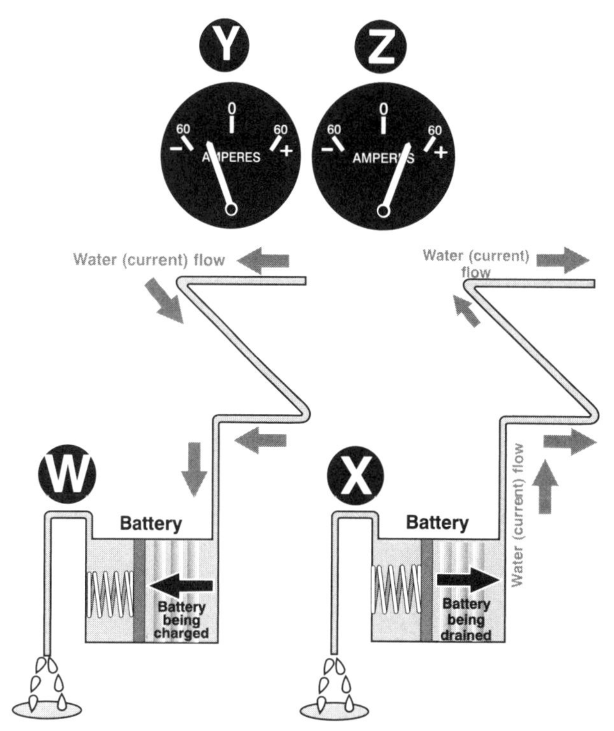

28. [D7/1/2]
Excess voltage resulting from a battery overcharge can _____ battery fluid (electrolyte), damaging the battery and possibly causing a battery ____.
A. replenish, freeze
B. boil off, fire
C. boil off, charge

29. [D7/1/3]
A needle deflection on the _____ side means current is flowing out of the battery onto the primary bus
A. 0 or 60
B. positive (+)
C. negative (-)

Load Meters

30. [D7/3/3]
Load meters provide important indications about the health of the airplane's electrical system. Unlike charge-discharge ammeters, they are calibrated to reflect the actual _____ load placed on the alternator.
A. ampere
B. voltage
C. resistance

31. [D8/1/1]
Load meters with a zero or full-left deflection indicate the alternator _____ providing current to the primary bus.
A. is
B. isn't
C. should be

32. [D8/1/1]
During flight with electrical equipment in use, a full left deflection of the load meter needle is similar to a charge-discharge ammeter reading pointing to the _____ side of its scale.
A. negative (-)
B. neutral
C. positive (+)

33. [D8/1/2]
A load meter needle deflection to the right of the zero index represents the _____ drain on the alternator.
A. electrical voltage
B. electrical current
C. electrical resistance

34. [D8/1/2]
If you add all the electrical current used by the active electrical equipment, this sum should be equal to the amount of the load meter _____ deflection.
A. needle's
B. voltage
C. negative

35. [D8/Figure 17]

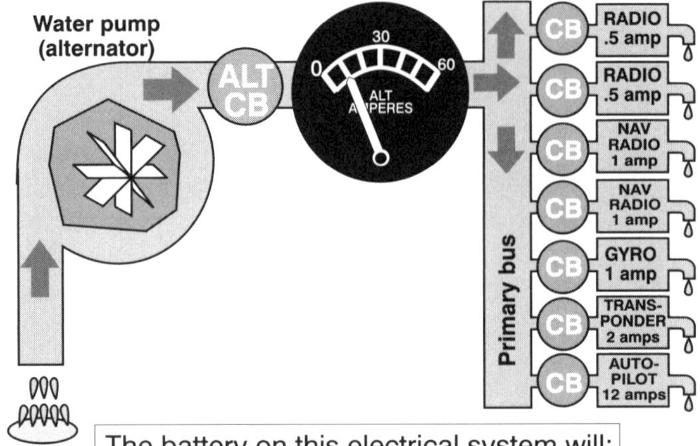

The battery on this electrical system will:
A. overcharge.
B. undercharge.
C. remain unaffected.

36. [D8/Figure 18]

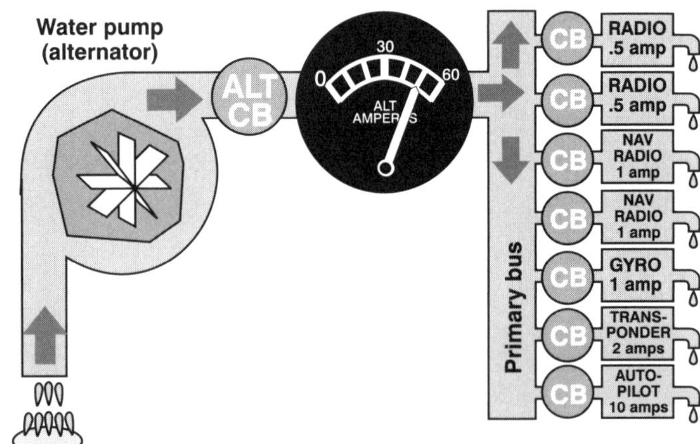

The battery on this electrical system will:
A. overcharge.
B. undercharge.
C. remain unaffected.

Electrical Drain

37. [D8/1/4]
Radios typically consume _____ amp(s) of current while receiving and about _____ amps(s) while transmitting.
A. one-half, 5
B. 5, one-half
C. 1, 2

38. [D8/2/2]
With two radios, two nav radios, one electric gyro, a transponder and an autopilot in use, a 16 amp deflection should be shown on the load meter. A needle deflection less than 16 amps implies that the _____ isn't providing enough current to run the equipment.
A. resistor
B. battery
C. alternator

39. [D8/3/1]
Load meter needle deflections less than the summed amperage of active and properly working electrical equipment imply that the _____ will eventually be _____.
A. alternator, drained
B. battery, drained
C. electrical ground, shorted

40. [D8/3/2]
Suppose the load meter's needle deflection is greater than the needs of the electrical equipment. This can lead to
A. engine failure.
B. pilot failure.
C. evaporating battery fluid or battery and electrical fires.

The Voltage Regulator

41. [D9/3/1]
Voltage regulators or alternator control units (they're essentially the same thing) regulate the _____ output.
A. engine's
B. alternator's
C. battery's

42. [D9/3/1]
Voltage regulators help alternators maintain a _____ voltage output under varying RPM conditions.
A. constant
B. low
C. high

43. [D10/1/1]
Alternators need a little bit of electricity running through them before they'll start producing electricity. This small amount of electrical prime is called the _____.
A. battery field current
B. primary bus current
C. alternator field current

Problems With Brains

44. [D11/1/2]
If your airplane has a low-voltage light, it can illuminate
A. during low engine idle.
B. when the alternator has been taken offline.
C. Both of the above.

45. [D11/1/4]
If your airplane has a high-voltage warning light, it can activate:
A. when the alternator puts out excessive voltage.
B. if the voltage regulator malfunctions.
C. Both of the above.

46. [D11/1/5]
If you need the equipment that was disabled by a popped circuit breaker, a good recommendation is to never reset the circuit breaker _____.
A. more than two times
B. more than one time
C. at all

47. [D11/2/1]
Manually taking the alternator offline becomes an important consideration if _____ condition occurs and the circuitry doesn't _____ isolate the alternator.
A. a battery full, fully
B. an overvoltage, automatically
C. an isolated voltage, reset and

Making Connections

48. [D12/1/3]
If the battery is dead, the _____ isn't going to work.
A. magneto
B. engine
C. alternator

49. [D12/1/5]
If the voltage regulator goes bad and the alternator produces too much voltage, you can take the alternator offline by pulling the _____ circuit breaker.
A. main alternator
B. battery terminals off the
C. avionics

50. [D12/1/5]
Turning off the alternator half of the master switch on many airplanes will shut off the _____ field current flow, thus deactivating the _____.
A. alternator, alternator
B. engine's, battery
C. alternator, engine

51. [D13/1/2]
If faced with an errant voltage regulator in flight which forces you to deactivate the alternator, unload the system by turning off _____ equipment.
A. all essential electrical
B. all mechanical
C. all nonessential electrical

Postflight Briefing 4-1: How the Battery Contactor Works

52. [D14/Figure 25 & D15/Figure 27]
Compare the water analogy circuitry with the airplane's electrical circuit below.

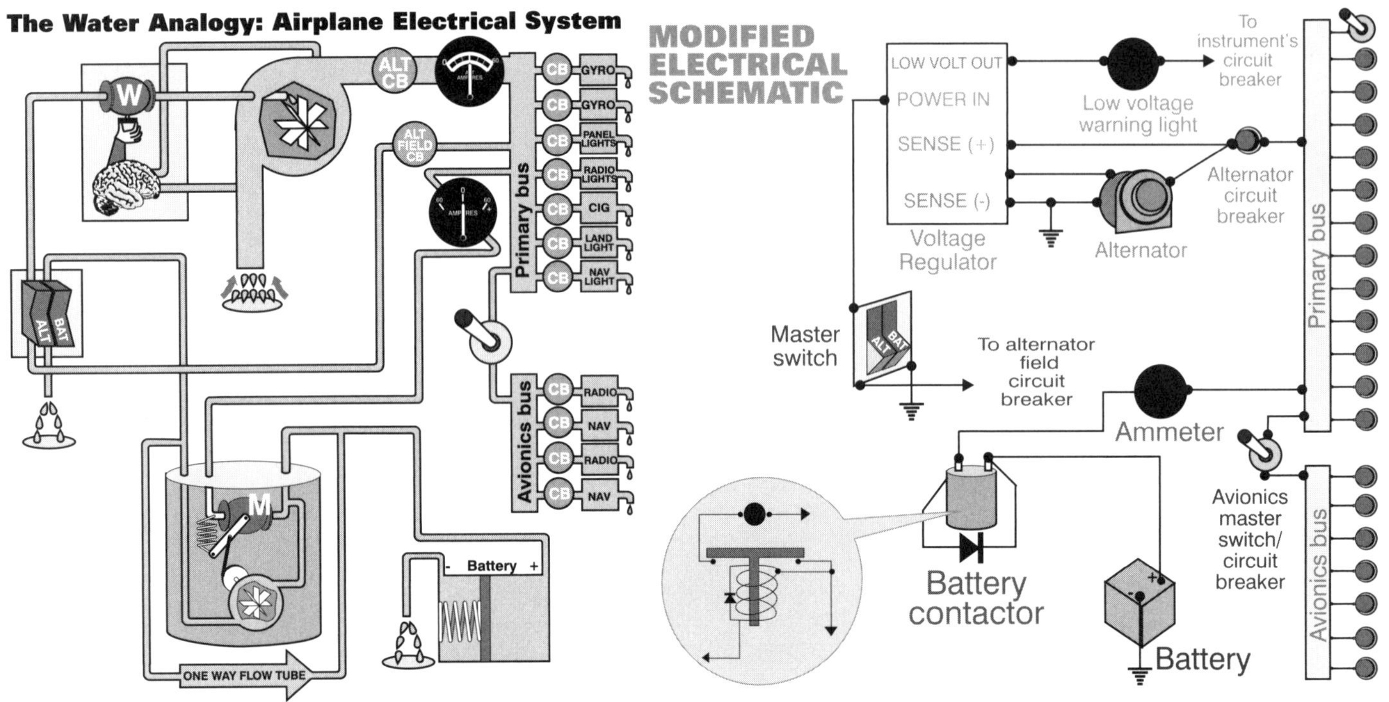

Chapter Four Answers

1. A
2. C
3. A
4. B
5. A
6. A/primary bus, B/avionics bus, C/battery, D/alternator, E/avionics master switch, F/circuit breaker, G/electrical ground
7. B
8. C
9. B
10. B
11. B
12. A
13. B
14. A
15. C
16. C
17. A
18. B
19. A
20. C
21. A
22. C
23. C
24. A
25. A
26. C
27. B
28. B
29. C
30. A
31. B
32. A
33. B
34. A
35. B
36. A
37. A
38. C
39. B
40. C
41. B
42. A
43. C
44. C
45. C
46. B
47. B
48. C
49. A
50. A
51. C

Note: To ensure that you have the most current answers to these questions, please check the *Book & Slide Updates* section at Rod Machado's web site: www.rodmachado.com

Chapter Five

Flight Instruments: Clocks, Tops & Toys

1. **[E1/Figure 1]** Label the six main flight instruments:

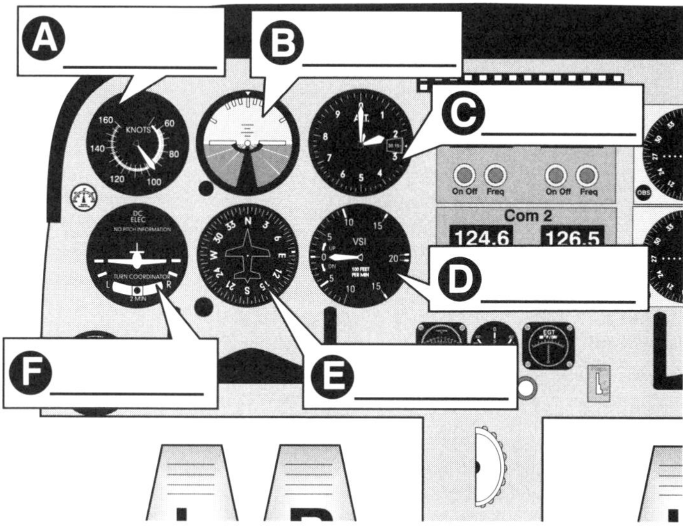

Non-Gyro Instruments

Airspeed Indicator

2. **[E2/Figure 3]** Label the parts of the airspeed indicator:

INSIDE A BASIC AIRSPEED INDICATOR

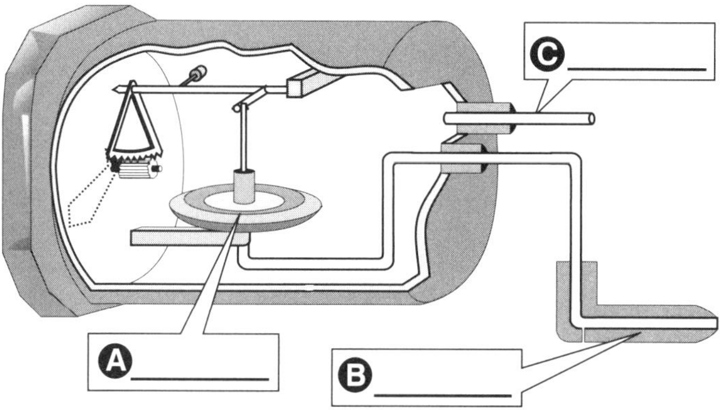

3. **[E3/3/2]**
Atmospheric pressure ____ with a gain in altitude.
A. decreases
B. increases
C. equalizes

4. **[E2/3/2 & E3/3/3]**
If the pitot tube and outside static vents become clogged, which instruments would be affected?
A. The altimeter, airspeed indicator, and turn-and-slip indicator.
B. The altimeter, airspeed indicator, and vertical speed indicator.
C. The altimeter, attitude indicator, and turn-and-slip indicator.

Static Pressure

5. **[E3/3/3]**
In addition to the airspeed indicator, two additional instruments also require static air pressure to function: the ____ and the ____.
A. vertical speed indicator, altimeter
B. directional gyro, altimeter
C. attitude indicator, vertical speed indicator

6. **[E3/3/3]**
Which instrument(s) will read incorrectly or become inoperative if the static vents become clogged?
A. Airspeed only.
B. Altimeter only.
C. Airspeed, altimeter, and vertical speed.

Pitot Tubes

7. **[E4/1/2]**
Which instrument will become inoperative if the pitot tube becomes clogged?
A. Altimeter.
B. Vertical speed indicator.
C. Airspeed indicator.

The Airspeed Indicator's Face

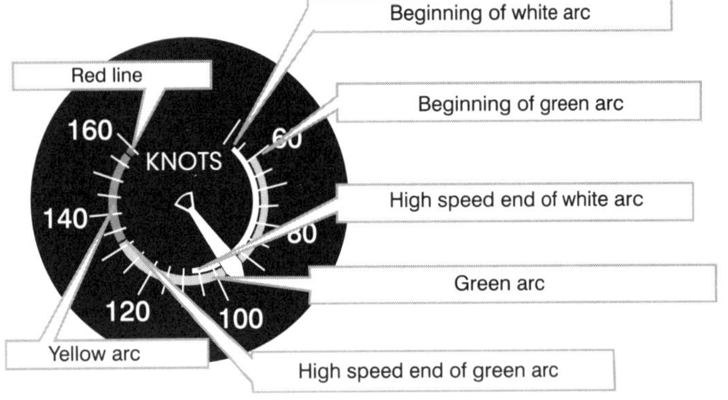

8. [E4/3/1]
The white arc of the airspeed indicator represents the airplane's flap operating range. The beginning or low-speed end of the white arc is the _____ stall speed in the landing configuration (i.e., with flaps fully extended and gear down). This is called ____.
A. power-on, V_{s1}
B. power-off, V_{so}
C. power-off, V_{s1}

9. [E4/3/1&2]
V_{so} is defined as the
A. stalling speed or minimum steady flight speed in the landing configuration.
B. stalling speed or minimum steady flight speed in a specified configuration.
C. stalling speed or minimum takeoff safety speed.

10. [E5/1/2]
The high-speed end of the airspeed indicator's white arc is called the _____ speed, or _____.
A. maximum flap retraction, V_{fe}
B. minimum flap retraction, V_{fe}
C. maximum flap extended, V_{fe}

11. [E5/1/3]
V_{s1} is defined as the
A. stalling speed or minimum steady flight speed in the landing configuration.
B. stalling speed or minimum steady flight speed in a specified configuration.
C. stalling speed or minimum takeoff safety speed.

12. [E4/Figure 8] Refer to the figure above. What is the full flap operating range for the airplane?
A. 53 to 107 knots.
B. 60 to 132 knots.
C. 132 to 157 knots.

13. [E4/Figure 8] Refer to the figure at the top/left of this page. What is the maximum flaps-extended speed?
A. 53 knots.
B. 107 knots.
C. 132 knots.

14. [E5/1/2] Refer to the figure at the top/left of this page. Which color identifies the normal flap operating range?
A. The lower limit of the white arc to the upper limit of the green arc.
B. The green arc.
C. The white arc.

15. [E5/1/2]
Which V speed represents maximum flap extended speed?
A. V_{fe}
B. V_{lof}
C. V_{fc}

16. [E5/1/3] Refer to the figure at the top/left of this page. V_{s1} is the beginning of the green arc represents the power-off stalling speed or _____ flight speed in a _____ configuration.
A. maximum steady, specified
B. minimum steady, specified
C. minimum steady, variable

17. [E5/1/3] Refer to the figure at the top/left of this page. Which color identifies the power-off stalling speed with wing flaps and landing gear in the landing configuration?
A. Upper limit of the green arc.
B. Upper limit of the white arc.
C. Lower limit of the white arc.

18. [E5/1/3] Refer to the figure at the top/left of this page. Which color identifies the power-off stalling speed in a specified configuration?
A. Upper limit of the green arc.
B. Upper limit of the white arc.
C. Lower limit of the green arc.

19. [E5/1/4]
V_{no} is defined as the
A. normal operating range.
B. never-exceed speed.
C. maximum structural cruising speed.

20. [E5/1/2]
Which V-speed represents the maximum structural cruising speed?
A. V_{fe}
B. V_{no}
C. V_{cs}

21. [E5/1/2] Refer to the figure at the top/left of this page. What is the maximum structural cruising speed?
A. 107 knots.
B. 132 knots.
C. 157 knots.

22. [E5/1/4]
V_{no} is defined as the
A. normal operating range.
B. never-exceed speed.
C. maximum structural cruising speed.

23. [E5/2/1] Refer to the figure at the top/left of the previous page. What is the caution range of the airplane?
A. 53 to 107 knots.
B. 60 to 132 knots.
C. 132 to 157 knots.

24. [E5/2/1] Refer to the figure at the top/left of the previous page. The maximum speed at which the airplane can be operated in smooth air is
A. 107 knots.
B. 132 knots.
C. 157 knots.

25. [E5/2/3]
What does the red line on an airspeed indicator represent?
A. Maneuvering speed.
B. Turbulent or rough-air speed.
C. Never-exceed speed.

26. [E5/2/3] Refer to the figure at the top/left of the previous page. Which color identifies the never-exceed speed?
A. Lower limit of the yellow arc.
B. Upper limit of the white arc.
C. The red line.

27. [E5/3/3]
What is an important airspeed limitation that is not color coded on airspeed indicators?
A. Never-exceed speed.
B. Maximum structural cruising speed.
C. Maneuvering speed.

28. [E5/3/3]
Which V-speed represents the maneuvering speed?
A. V_a
B. V_{lo}
C. V_{ne}

29. [E6/1/1]
Which V-speed represents the maximum landing gear extended speed?
A. V_{le}
B. V_{lo}
C. V_{fe}

30. [E6/1/1]
Which V-speed represents the maximum landing gear operating speed?
A. V_{le}
B. V_{lo}
C. V_{fe}

Indicated Airspeeds

31. [E7/1/2]
The number showing on the airspeed indicator is the _____.
A. indicated airspeed
B. calibrated airspeed
C. true airspeed

Calibrated Airspeeds

32. [E7/1/4 See *Air Error*]
Calibrated airspeed is _____ airspeed corrected for _____ or_____ errors.
A. true, installation, position
B. indicated, installation, position
C. indicated, temperature, coriolis

True Airspeed

33. [E8/1/3]
Airplanes flying at higher altitudes actually move _____ through the air for a given power setting because of the _____ in density.
A. faster, increase
B. slower, decrease
C. faster, decrease

34. [E9/See *TAS & IAS High Altitude Airports*]
As altitude increases the indicated airspeed at which a given airplane stalls in a particular configuration will
A. decrease as the true airspeed decreases.
B. decrease as the true airspeed increases.
C. remain the same regardless of altitude.

35. [E9/See *TAS & IAS High Altitude Airports*]
When making an approach at a high altitude airport, you should:
A. approach at a lower than normal indicated airspeed.
B. approach at a higher than normal indicated speed.
C. approach at a normal indicated speed.

Dense Doings

36. [E9/1/3]
Two atmospheric factors that affect air density, which in turn affects true airspeed, are: _____ and _____.
A. pressure, temperature
B. pressure, air movement
C. temperature, engine power

The Altimeter

37. [E10/1/4]
What is true altitude?
A. The vertical distance of the aircraft above sea level.
B. The vertical distance of the aircraft above the surface.
C. The height above the standard datum plane.

38. [E10/1/5]
What is absolute altitude?
A. The altitude read directly from the altimeter.
B. The vertical distance of the aircraft above the surface.
C. The height above the standard datum plane.

39. [E10/1/6]
An altimeter works by measuring the difference between sea level _____ and pressure at the airplane's present _____.
A. temperature, altitude
B. pressure, airspeed
C. pressure, altitude

40. [E10/Figure 18] Label the altimeter's components:

INSIDE A BASIC ALTIMETER

A ________

B ________

41. [E11/1/2]
For the purpose of altimeter calibration, a column of mercury will change approximately ____ inch(es) in height for every thousand-foot change in altitude.
A. one
B. 10
C. one-half

Pressure Variations and the Altimeter

42. [E12/2/1]
Whatever pressure value you set in the Kollsman window, the altimeter assumes this is the new _____ pressure. Now the altimeter measures the difference between the pressure value set in the Kollsman window and the _____ pressure to obtain your height above sea level.
A. present altitude, dynamic
B. sea level, outside static
C. sea level, pitot tube

43. [E13/1/1 & E17/Figure 29]
Prior to takeoff, the altimeter should be set to which altitude or altimeter setting?
A. The current local altimeter setting, if available, or the departure airport elevation.
B. The corrected density altitude of the departure airport.
C. The corrected pressure altitude for the departure airport.

44. [E13/1/1]
The altimeter setting is the value to which the barometric pressure scale of the altimeter is set so the altimeter indicates
A. calibrated altitude at field elevation.
B. absolute altitude at field elevation.
C. true altitude at field elevation.

45. [L25/3/2]
What causes variations in altimeter settings between weather reporting points?
A. Unequal heating of the earth's surface.
B. Variation of terrain elevation.
C. Coriolis force.

46. [E10-E18, All]
Under what condition is indicated altitude the same as true altitude?
A. If the altimeter has no mechanical error.
B. When at sea level under standard conditions.
C. When at 18,000 feet MSL with the altimeter set at 29.92.

47. [E14/1/3 & E15/Figure 26]
If it is necessary to set the altimeter from 29.15 to 29.85, what change occurs?
A. 70-foot increase in indicated altitude.
B. 70-foot increase in density altitude.
C. 700-foot increase in indicated altitude.

48. [E14/1/3 & E15/Figure 26]
If a pilot changes the altimeter setting from 30.11 to 29.96, what is the approximate change in indication?
A. Altimeter will indicate .15" Hg higher.
B. Altimeter will indicate 150 feet higher.
C. Altimeter will indicate 150 feet lower.

49. [E15/1/2 & Figure 27]
If a flight is made from an area of low pressure into an area of high pressure without the altimeter setting being adjusted, the altimeter will indicate
A. the actual altitude above sea level.
B. higher than the actual altitude above sea level.
C. lower than the actual altitude above sea level.

50. [E15/2/1 & Figure 25]
If a flight is made from an area of high pressure into an area of lower pressure without the altimeter setting being adjusted, the altimeter will indicate
A. lower than the actual altitude above sea level.
B. higher than the actual altitude above sea level.
C. the actual altitude above sea level.

Temperature Variation and the Altimeter

51. [E16/2/4 & Figure 28]
Under what condition will true altitude be lower than indicated altitude?
A. In colder than standard air temperature.
B. In warmer than standard air temperature.
C. When density altitude is higher than indicated altitude.

52. [E16/2/5 & Figure 28]
Which condition would cause the altimeter to indicate a lower altitude than true altitude?
A. Air temperature lower than standard.
B. Atmospheric pressure lower than standard.
C. Air temperature warmer than standard.

53. [E16/Figure 28]
How do variations in temperature affect the altimeter?
A. Pressure levels are raised on warm days and the indicated altitude is lower than true altitude.
B. Higher temperatures expand the pressure levels and the indicated altitude is higher than true altitude.
C. Lower temperatures lower the pressure levels and the indicated altitude is lower than true altitude.

Sensitive Altimeters

54. [E17/2/1] Fill in the blank:
As you progress along your route of flight you should update your altimeter setting to the nearest source within ____________ nautical miles of your position.

55. [E17/Figure 29]
If an altimeter setting is not available before flight, to which altitude should the pilot adjust the altimeter?
A. The elevation of the nearest airport corrected to mean sea level.
B. The elevation of the departure area.
C. Pressure altitude corrected for nonstandard temperature.

56. [E17/2/1]
Which of the following is an example of a time where you might use *pressure altitude*?
A. When calculating how to lean the mixture.
B. When using an airplane's performance chart.
C. When you're planning to fly over a congested area.

Pressure Altitude

57. [E18/1/1]
What is pressure altitude?
A. The indicated altitude corrected for position and installation error.
B. The altitude indicated when the barometric pressure scale is set to 29.92.
C. The indicated altitude corrected for nonstandard temperature and pressure.

58. [E17-18, All]
Under which condition will pressure altitude be equal to true altitude?
A. When the atmospheric pressure is 29.92" Hg.
B. When standard atmospheric conditions exist.
C. When indicated altitude is equal to the pressure altitude.

59. [E18/2/1]
When you set a barometric pressure value of 29.92 into the Kollsman window, the altimeter
A. always reads true altitude.
B. reads pressure altitude.
C. density altitude.

60. [E18/Figure 30]
If it is necessary to change the setting in the altimeter's Kollsman window from 29.00 to 29.85, what change occurs?
A. 85 foot increase in indicated altitude.
B. 850 foot increase in density altitude.
C. 850 foot increase in indicated altitude.

61. [E18/Figure 30]
If it is necessary to change the setting in the altimeter's Kollsman window from 30.30 to 30.00, what change occurs?
A. 30 foot decrease in indicated altitude.
B. 30 foot increase in density altitude.
C. 300 foot decrease in indicated altitude.

62. [E18/Figure 30]
If it is necessary to change the setting in the altimeter's Kollsman window from 29.35 to 29.85, what change occurs?
A. 50 foot increase in indicated altitude.
B. 50 foot increase in density altitude.
C. 500 foot increase in indicated altitude.

63. [E17-18, All]
When true altitude and pressure altitude are equal, what conditions exist?
A. An atmospheric pressure of 29.92" Hg.
B. Standard atmospheric conditions.
C. Normal weather conditions.

64. [E18/3/2]
Pressure altitude is the height above a _____, which is nothing more than a fancy phrase for _____ reference point. This reference point is what the engineer's altimeter would have read if temperature and pressure at sea level were 59°F and 29.92" Hg.
A. reference point, true altitude
B. standard day plane, a real
C. standard datum plane, an imaginary

65. **[E19/Figure 32]** Fill in the indicated altitudes:

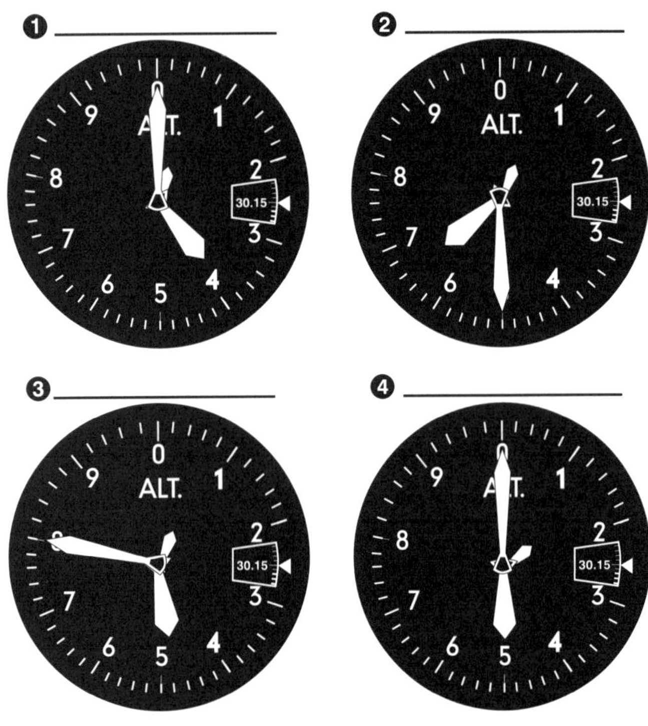

66. **[E19/Figure 32]** Refer to the figures in question **65.** Which altimeter(s) indicate(s) more than 10,000 feet?
A. 1, 2, and 3.
B. 2 and 4 only.
C. 4 only.

The Vertical Speed Indicator (VSI)

67. **[E20/1/4]**
The VSI's needle swings upward or downward reflecting the airplane's rate of climb or descent. The VSI is calibrated to read in _____.
A. feet per minute
B. miles per hour
C. knots per minute

68. **[E20/Figure 34]** Label the VSI's components:

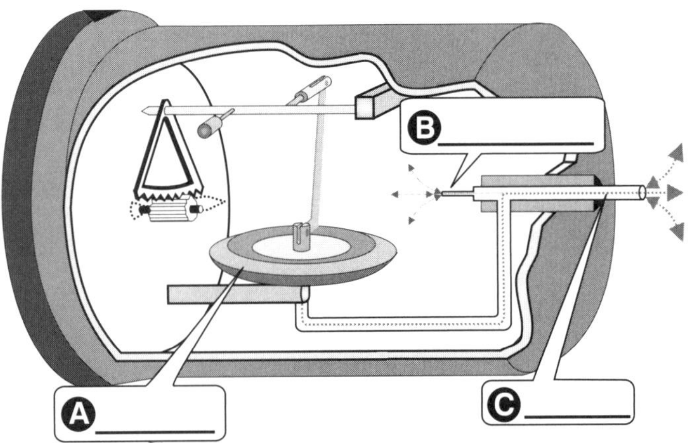

The Alternate Static Source

69. **[E21/2/1]**
A plugged static port will prevent any static air pressure change, causing the altimeter to freeze at its last indication and the VSI to read _____, regardless of altitude change.
A. a climb
B. zero
C. a descent

70. **[E21/3/1]**
If the primary static source becomes plugged, the _____ can be opened. The _____ now becomes the alternate source of static air pressure.
A. alternate static source, cabin
B. primary static source, cabin
C. pitot tube, cabin

The Gyroscopic Instruments

The Attitude Indicator

71. **[E22/1/2]**
The attitude indicator helps you determine the airplane's _____ and _____ condition.
A. yaw, turn
B. pitch, bank
C. roll, speed

72. **[E22/2/2]**
The attitude indicator's symbolic wings are attached to the instrument's case, while the airplane _____ about a horizon card that's attached to a stabilized gyro.
A. vibrates
B. rotates
C. remains stationary

73. **[E23/Figure 39]**
How should a pilot determine the direction of bank from an attitude indicator such as the one illustrated in question 76?
A. By the direction of deflection of the banking scale.
B. By the direction of deflection of the horizon bar.
C. By the relationship of the miniature airplane to the deflected horizon bar.

74. **[E24/1/2]**
The attitude indicator shows
A. climbs and descents.
B. pitch and yaw.
C. bank and pitch.

75. **[E25/1/1]**
The attitude indicator is able to display the airplane's attitude because of a gyroscopic principle known as
A. torque.
B. rigidity in space.
C. precession.

76. [E22/Figure 38] Label the attitude indicator's components.

THE ATTITUDE INDICATOR

A ________ B ________ C ________ D ________ E ________ F ________ G ________ H ________

77. [E23/Figure 39] Label the attitude indicators below by indicating the amount and direction of bank and pitch:

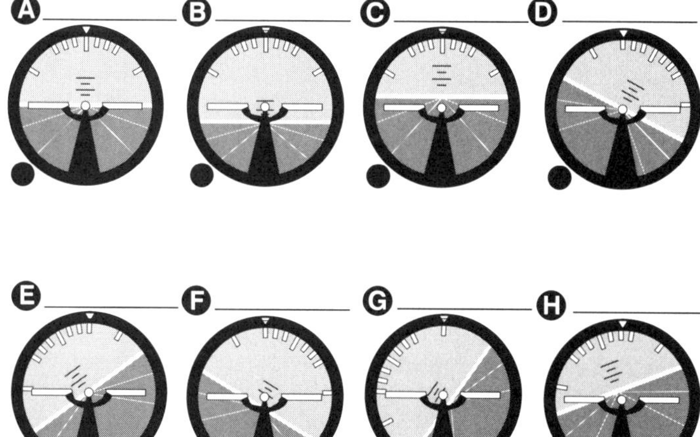

78. [E25/2/3]
Most light airplane attitude gyros are powered by a _____ pump.
A. vacuum
B. water
C. fuel

79. [E26/Figure 47]
The proper adjustment to make on the attitude indicator during level flight is to align the
A. horizon bar to the level-flight indication.
B. horizon bar to the miniature airplane.
C. miniature airplane to the horizon bar.

The Heading Indicator

80. [E27/2/2]
To receive accurate indications during flight from a heading indicator, the instrument must be
A. set prior to flight on a known heading.
B. calibrated on a compass rose at regular intervals.
C. periodically realigned with the magnetic compass as the gyro precesses.

81. [E27/2/1]
The heading indicator must be periodically reset to a known heading because of something known as
A. gyroscopic drift.
B. acceleration errors.
C. turning errors.

82. [E27/Figure 50] Label the parts of the heading indicator:

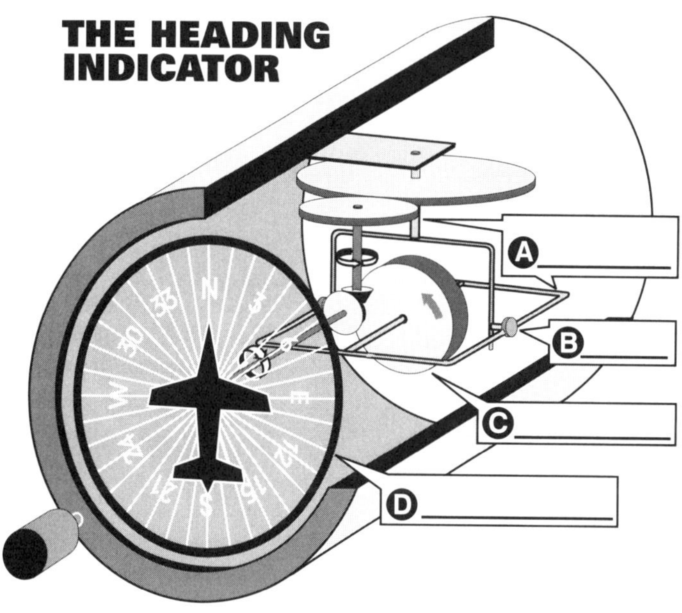

83. [E28/1/3]
If you're instructed to turn to a particular heading, simply look for the number on the instrument and turn in the _____ direction toward it.
A. longest
B. most scenic
C. shortest

The Turn Coordinator

84. [E29/1/1]
The turn coordinator is actually a _____ instrument that provides information on your airplane's direction of _____, rate of heading change and whether the airplane is _____ or _____ in the turns.
A. magnetic, turn, slipping, skidding
B. gyro, roll, slipping, skidding
C. gyro, pitch, climbing, descending

85. [E29/2/2]
Unlike the spinning gyros in the attitude indicator and heading indicator, the turn coordinator's gyro is usually spun by _____.
A. electricity
B. vacuum power
C. air pressure

86. [29/2/2]
The turn coordinator's gyro is electrically powered to keep at least one gyro instrument operating during a rare failure of the airplane's _____.
A. gear
B. mechanical fuel pump
C. vacuum pump

87. [E29/Figure 54] Label the parts of the turn coordinator:

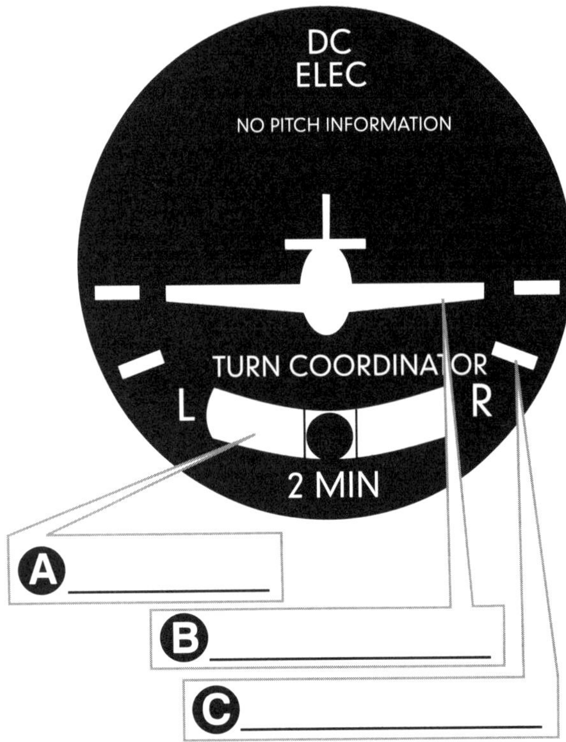

88. [E30/1/2]
Even though it may appear to, the turn coordinator doesn't show _____ angle. Don't be fooled by this. Only direction of _____ and _____ are derivable from the turn coordinator.
A. pitch, climb descent
B. bank, pitch, yaw
C. bank, roll or yaw, rate of turn

89. [E30/Figure 57]
A turn coordinator provides an indication of the
A. movement of the aircraft about the yaw and roll axes.
B. angle of bank up to but not exceeding 30 degrees.
C. attitude of the aircraft with reference to the longitudinal axis.

90. [E29/Figure 55] Label the parts of the turn coordinator:

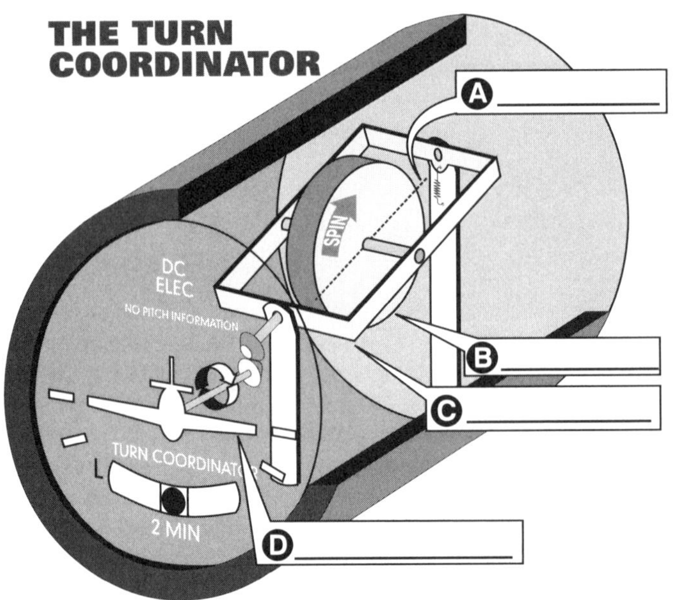

The Magnetic Compass Acceleration and Deceleration Error

91. [E31/2/3]
A magnetic compass responds to that phenomenon of the earth's magnetic pole, otherwise known as its _____.
A. flux magneto
B. magnetic field
C. magnetic dip

92. [E31/Figure 60] Label the parts of the compass:

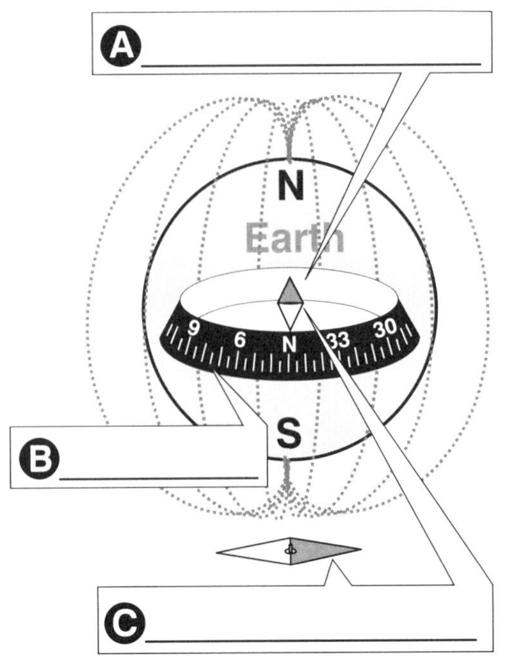

93. [E32/2/1]
One side of the needle is called the _____ and it always points towards the earth's magnetic north pole.
A. north-seeking end
B. south-seeking end
C. east-west seeking end

94. [E33/2/1]
Just like skis following dipping terrain, the magnetic compass needle wants to tilt downward with the magnetic field. This is called _____.
A. magnetic droop
B. magnetic dip
C. magnetic tilt

95. [E34/1/2]
In the northern hemisphere, the magnetic compass will normally indicate a turn toward the south when
A. a left turn is entered from an east heading.
B. a right turn is entered from a west heading.
C. the aircraft is decelerated while on a west heading.

96. [E34/1/2]
In the northern hemisphere, if an aircraft is accelerated or decelerated, the magnetic compass will normally indicate
A. a momentary turn.
B. correctly when on a north or south heading.
C. a turn toward the south.

97. [E34/1/2]
In the northern hemisphere, the magnetic compass will normally indicate a turn toward the north when
A. a left turn is entered from an east heading.
B. a right turn is entered from a west heading.
C. the aircraft is accelerated while on a west heading.

98. [E34/3/2]
In the northern hemisphere, a magnetic compass will normally initially indicate a turn toward the east if
A. an aircraft is decelerated while on a south heading.
B. an aircraft is accelerated while on a north heading.
C. a left turn is entered from a north heading.

Northerly Turn Errors

99. [E35/1/2]
In the northern hemisphere, a magnetic compass will normally initially indicate a turn toward the west if
A. a left turn is entered from a north heading.
B. a right turn is entered from a north heading.
C. an aircraft is accelerated while on a north heading.

100. [E35/1/3]
During flight, when are the indications of a magnetic compass accurate?
A. In straight-and-level unaccelerated flight.
B. As long as the airspeed is constant.
C. During turns if the bank does not exceed 18 degrees.

Chapter Five Answers

1. A/airspeed indicator, B/attitude indicator, C/altimeter, D/vertical speed indicator, E/heading indicator, F/turn coordinator
2. A/expandable capsule, B/ pitot tube, C/static line (or source)
3. A
4. B
5. A
6. C
7. C
8. B
9. A
10. C
11. B
12. A
13. B
14. C
15. A
16. B
17. C
18. C
19. C
20. B
21. B
22. C
23. C
24. C
25. C
26. C
27. C
28. A
29. A
30. B
31. A
32. B
33. C
34. C
35. C
36. A
37. A
38. B
39. C
40. A/expandable capsule, B/static line (or source)
41. A
42. B
43. A
44. C
45. A
46. B
47. C
48. C
49. C
50. B
51. A
52. C
53. A
54. 100
55. B
56. B
57. B
58. B
59. B
60. C
61. C
62. C
63. B
64. C
65. 1/4,000 feet, 2/6,500 feet, 3/4,800 feet, 4/15,000 feet
66. C
67. A
68. A/expandable capsule, B/calibrated leak, C/static source
69. B
70. A
71. B
72. B
73. C
74. C
75. B
76. A/vertical indicator, B/10 degree bank lines, C/60 degree bank line, D/horizon line, E/artificial airplane wing, F/adjustment knob, G/sky pointer, H/degree pitch line
77. A/straight & level, B/pitch up, C/pitch down, D/left turn, 30 degree bank, E/right turn, 30 degree bank F/pitch up, left turn, 30 degree bank, G/pitch up, right turn, 60 degree bank H/pitch down, right turn, 20 degree bank
78. A
79. C
80. C
81. A
82. A/gimbal, B/swivel point, C/gyro, D/face card
83. C
84. B
85. A
86. C
87. A/inclinometer, B/rate of turn needle, C/standard rate turn index
88. C
89. A
90. A/gimbal roll axis, B/gyro, C/gimbal, D/turn needle
91. B
92. A/north seeking end, B/compass card, C/compass needle
93. A
94. B
95. C
96. B
97. C
98. C
99. B
100. A

Note: To ensure that you have the most current answers to these questions, please check the *Book & Slide Updates* section at Rod Machado's web site: www.rodmachado.com

Chapter Six

Federal Aviation Regulations: How FAR Can We Go?

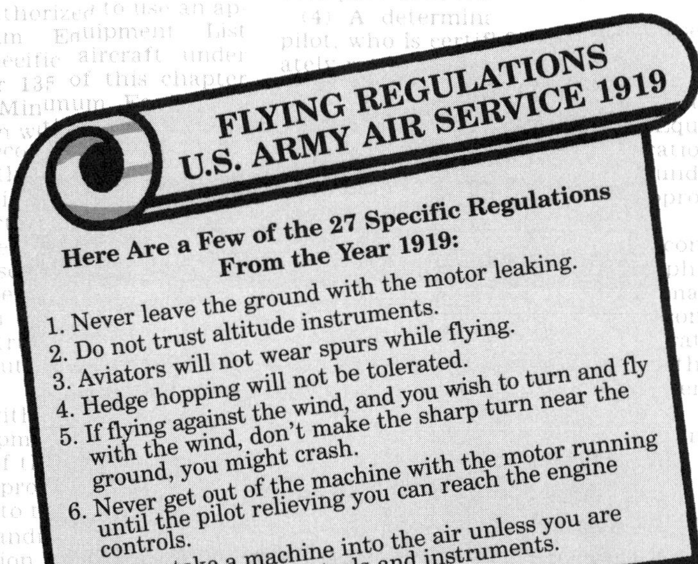

Definitions

Category

1. [F2/1/1]
For the purpose of pilot certification, the term *category of aircraft* represents something flyable having _____ operating characteristics.
A. redundant
B. dissimilar
C. similar

2. [F2/1/1]
With respect to the certification of airmen, which are all a category of aircraft?
A. Gyroplane, helicopter, airship, free balloon.
B. Airplane, rotorcraft, glider, lighter-than-air, powered-lift.
C. Single-engine land and sea, multi-engine land and sea.

Class

3. [F2/2/2]
With respect to certification of airmen, a class is a subdivision of _____.
A. a category
B. the number of engines
C. the category of airplane only

4. [F2/2/2]
With respect to the certification of airmen, which is a class of aircraft?
A. Airplane, rotorcraft, glider, lighter-than-air.
B. Single-engine land and sea, multi-engine land and sea.
C. Lighter-than-air, airship, hot air balloon, gas balloon.

5. [F2/3/2]
With respect to the certification of aircraft, which is a class of aircraft?
A. Airplane, rotocraft, glider, balloon.
B. Normal, utility, aerobatic, limited.
C. Transport, restricted, provisional.

6. [F2/2/2]
Airplanes are made up of which of the following classifications or classes:
A. single-engine land, multi-engine land, single-engine retractable, single-engine fixed gear.
B. single-engine fixed gear, single-engine retractable, multi-engine fixed gear and multi-engine retractable.
C. single-engine land, single-engine sea, multi-engine land and multi-engine sea.

7. [F3/1/1]
When pilots refer to the make and model of airplane they fly, which of the following are they referring to:
A. A36 Beechcraft Bonanza, a Cessna 152 or a Piper Malibu.
B. single-engine land or single-engine sea.
C. airplane, gliders, rotorcraft, powered-lift.

Type Ratings

8. [F3/2/3]
The pilot in command is required to hold a type rating in which aircraft?
A. Aircraft operated under an authorization issued by the Administrator.
B. Aircraft having a gross weight of more than 12,500 pounds.
C. Aircraft involved in ferry flights, training flights, or test flights.

Visual Flight Rules (VFR)

9. [F3/3/2]
The basic premise of *VFR flight* is to_____ other aircraft.
A. see and avoid
B. see and trade paint with
C. see and be seen by

10. [F3/3/3]
Private pilots without an _____ rating are only allowed to fly under visual flight rules.
A. instrument
B. airline transport
C. into-a-cloud

Instrument Flight Rules (IFR)

11. [F4/1/1]
An instrument rating allows you to fly to your destination while in the _____ and land under low _____ conditions.
A. rain, humidity
B. clouds, ceiling
C. clouds, visibility

Night

12. [F4/2/2]
The definition of night time is
A. sunset to sunrise.
B. 1 hour after sunset to 1 hour before sunrise.
C. the time between the end of evening civil twilight and the beginning of morning civil twilight.

Pilot In Command (PIC)

13. [F4/2/3]
The person who has final authority and responsibility for the operation and safety of the flight has been designated as PIC before or during the flight, and holds the appropriate category, class and type rating, if appropriate, for the conduct of the flight is known as the_____.
A. pilot in command
B. aircraft dispatcher
C. flight crewmember

PART 61

FAR 61.3 Requirements for Certificates, Ratings and Authorizations

14. [F5/1/1]
To act as pilot in command or as a required flight crewmember (copilot or flight engineer, for example) on an aircraft, you must have your _____ in your _____ or readily accessible in the _____.
A. pilot certificate, personal possession, vehicle you drove to arrive at the airport
B. pilot certificate, personal possession, aircraft
C. medical certificate, office or van, nearest FBO

15. [F5/1/1]
When must a current pilot certificate be in the pilot's personal possession?
A. When acting as a crew chief during launch and recovery.
B. Only when passengers are carried.
C. Any time when acting as pilot in command or as a required flight crewmember.

16. [F5/1/1]
Private pilots acting as pilot in command, or in any other capacity as a required pilot flight crewmember, must have in their personal possession while aboard the aircraft a current
A. logbook endorsement to show that a flight review has been satisfactorily accomplished.
B. medical certificate and an appropriate pilot certificate.
C. endorsement on the pilot certificate to show that a flight review has been satisfactorily accomplished.

17. [F5/1/1]
What document(s) must be in your personal possession while operating as pilot in command of an aircraft?
A. Certificates showing accomplishment of a checkout in the aircraft and a current flight review.
B. A pilot certificate with an endorsement showing accomplishment of an annual flight review and a pilot logbook showing recency of experience.
C. An appropriate pilot certificate and an appropriate current medical certificate.

18. [F5/1/2]
Each person who holds a pilot certificate or a medical certificate shall present it for inspection upon the request of the Administrator, an authorized representative of the National Transportation Safety Board, or any
A. authorized representative of the Department of Transportation.
B. person in a position of authority.
C. federal, state, or local law enforcement officer.

19. [F5/1/2]
Who must you present your pilot or medical certificate to for inspection if asked:
A. the FBO.
B. the IRS.
C. a National Transportation Safety Board representative.

61.15 Offenses Involving Alcohol or Drugs

20. [F5/2/2]
If you're convicted for almost any illegal drug activity there's a very good chance you'll have your pilot certificate _____ or _____.
A. cancelled, destroyed
B. suspended, revoked
C. denied, restrained

21. [F5/2/3]
Once a certificate is revoked, it _____ exist(s).
A. is doubtful that it
B. no longer
C. continues to

22. [F5/3/1]
After November 29, 1990, two or more motor vehicle actions (action taken against your driver's license because of a drug or alcohol problem) within three years of each other are grounds for _____ of a pilot certificate or rating.
A. transferral
B. renewal
C. suspension or revocation

23. [F5/3/2]
The regulations require that you report all drug and alcohol motor vehicle actions to the FAA within _____ days.
A. 60
B. 30
C. 120

The Bottle to Throttle Rule

24. [F5/3/3]
Regulations prohibit you from acting as pilot in command or as a required crewmember (copilot or flight engineer, for instance) for _____ hours after consuming alcohol.
A. 24
B. 12
C. 8

25. [F6/1/1]
Another part of this rule states that you may not act as PIC if you have blood alcohol content of _____ or more (by weight).
A. .08%
B. .04%
C. .02%

61.23 Duration of Medical Certificates

26. [F7/1/3]
As a private pilot, you are required to have at least a third-class medical certificate to act as pilot in command. A third-class medical is valid for _____ calendar months if you're under 40 years of age or _____ calendar months if you're 40 or over.
A. 24, 36
B. 36, 24
C. 12, 48

27. [F7/2/2]
A third-class medical certificate is issued on August 10, this year, to a 45 year-old pilot. To exercise the privileges of a private pilot certificate, the medical certificate will be valid until midnight on
A. August 10, 2 years later.
B. August 31, 2 years later.
C. August 31, 3 years later.

28. [F7/2/2]
A third-class medical certificate is issued to a 19 year-old pilot on August 10, this year. To exercise the privileges of a recreational or private pilot certificate, the medical certificate will expire at midnight on
A. August 10, 2 years later.
B. August 31, 2 years later.
C. August 31, 3 years later.

29. [F7/2/2]
For private pilot operations, a second-class medical certificate issued to a 47 year old pilot on July 15, this year, will expire at midnight on
A. July 15, 2 years later.
B. July 31, 1 year later.
C. July 31, 2 years later.

30. [F7/1/1]
For private pilot operations, a first-class medical certificate issued to a 47 year old pilot on October 21, this year, will expire at midnight on
A. October 21, 2 years later.
B. October 31, next year.
C. October 31, 2 years later.

Additional Training Requirements

High Performance and Complex Endorsements

31. [F7/2/6] Fill in the blanks:
An airplane having more than 200 horsepower is considered a _____________ _____________ airplane.

32. [F7/2/6] Fill in the blank:
An airplane having retractable gear, flaps, and a controllable propeller is considered a _____________ airplane.

33. [F8/1/1]
Before a person holding a private pilot certificate may act as pilot in command of a *high performance* airplane, that person must have
A. passed a flight test in that airplane from an FAA inspector.
B. an endorsement in that person's logbook that he/she is competent to act as pilot in command.
C. received flight instruction from an authorized flight instructor who then endorses that person's logbook.

34. [F8/1/2]
In order to act as pilot in command of a *complex* airplane, one requirement a pilot must meet is to have
A. made three solo takeoffs and landings in a high performance airplane.
B. received flight instruction in an airplane that has retractable landing gear, flaps, and a controllable propeller.
C. passed a flight test in a high performance airplane.

Pressurized Airplanes Capable of Operating at High Altitudes

35. [F8/3/1]
Before a person holding a private pilot certificate may act as pilot in command of a pressurized airplane capable of operating above 25,000 feet MSL, that person must have
A. passed a flight test in that airplane from an FAA inspector.
B. a logbook endorsement saying that he or she is competent to act as pilot in command.
C. received and logged ground training from an authorized instructor as well as flight training in a pressurized aircraft.

Tailwheel Airplanes

36. [F9/2/2]
If you want to act as PIC of a tailwheel airplane, you must have
A. made three solo takeoffs and landings in a tailwheel airplane.
B. received and logged flight training from an authorized instructor in a tailwheel airplane.
C. passed a flight test in a tailwheel airplane.

Flight Reviews

37. [F10/1/2]
To act as pilot in command of an aircraft carrying passengers, a pilot must show by logbook endorsement the satisfactory completion of a flight review or completion of a pilot proficiency check within the preceding
A. 6 calendar months.
B. 12 calendar months.
C. 24 calendar months.

38. [F10/1/4]
If you complete one or more phases of an FAA-sponsored _____ award program, this also counts as a flight review.
A. landing contest
B. pilot proficiency
C. safety seminar

39. [F10/1/5]
The flight review is applicable for _____ you're certified to fly.
A. the specific category of aircraft
B. the specific class of aircraft
C. anything

61.57 Recent Flight Experience: Pilot In Command

40. [F11/1/3]
To act as pilot in command of an aircraft carrying passengers, the pilot must have made at least three takeoffs and three landings in an aircraft of the same category, class, and, if a type rating is required, of the same type, within the preceding
A. 90 days.
B. 12 calendar months.
C. 24 calendar months.

41. [F11/1/3]
To act as pilot in command of an aircraft carrying passengers, the pilot must have made three takeoffs and three landings within the preceding 90 days in an aircraft of the same
A. make and model.
B. category and class, but not type.
C. category, class, and type, if a type rating is required.

42. [F11/1/7]
The currency requirements to act as pilot in command must be
A. to a full stop in a tricycle gear airplane.
B. completed as touch and goes in a tricycle gear airplane.
C. to a full stop in a tailwheel airplane.

43. [F11/1/7]
The takeoffs and landings required to meet the recency of experience requirements for carrying passengers in a tailwheel airplane
A. may be touch and go or full stop.
B. must be touch and go.
C. must be to a full stop.

Recent Experience at Night

44. [F11/2/3]
The three takeoffs and landings that are required to act as pilot in command at night must be done during the time period from
A. sunset to sunrise.
B. 1 hour after sunset to 1 hour before sunrise.
C. the end of evening civil twilight to the beginning of morning civil twilight.

45. [F11/2/3]
To meet the recency of experience requirements to act as pilot in command carrying passengers at night, a pilot must have made at least three takeoffs and three landings to a full stop at night within the preceding 90 days in
A. the same category and class of aircraft to be used.
B. the same type of aircraft to be used.
C. any aircraft.

46. [F11/2/4]
If recency of experience requirements for night flight are not met and official sunset is 1830, the latest time passengers may be carried is
A. 1829.
B. 1859.
C. 1929.

61.60 Change of Address

47. [F12/1/3]
If a certificated pilot changes permanent mailing address and fails to notify the FAA Airmen Certification Branch of the new address, the pilot is entitled to exercise the privileges of the pilot certificate for a period of only
A. 30 days after the date of the move.
B. 60 days after the date of the move.
C. 90 days after the date of the move.

FAR 61.87 Solo Requirements for Student Pilots

48. [F13/2/1]
Before you can solo, your instructor must give you a pre-solo written test that consists of at least
A. 30 questions.
B. 70 questions.
C. 20 questions.

49. [F13/2/3]
Once you've been soloed, can you continue to fly without being under the supervision of your instructor?
A. Yes, you paid so you should fly.
B. No. Your instructor is required to endorse your logbook for solo every 90 days.
C. No. Your instructor is required to endorse your student pilot certificate for solo every 90 days.

50. [F13/3/3]
Student pilot certificates are only good for ____ calendar months.
A. 36
B. 24
C. 12

FAR 61.89 General Limitations

51. [F14/1/3]
Can a student pilot carry passengers?
A. Yes, as long as the student remains in the pattern.
B. Never.
C. Yes, as long as they are at least student pilots themselves.

52. [F14/3/1]
Student pilots are not allowed to fly when flight or surface visibilities are less than _____ miles during the day or less than _____ miles at night.
A. three, five
B. five, one
C. five, five

53. [F14/3/1]
Student pilots are not allowed to fly without
A. visual reference to the surface.
B. visual reference to the airport.
C. a VOR on board.

FAR 61.93 Cross Country Flight Requirements

54. [F15/1/3]
As a student pilot, your instructor may allow you to practice solo takeoffs and landings at another airport without having a cross country endorsement as long as the airport is within _____ nautical miles of the departure airport at which instruction is received.
A. 50
B. 25
C. 10

55. [F16/1/2]
A cross country endorsement on your student pilot certificate allows you the additional privilege of making repeated solo cross country flights to an airport within _____ nautical miles of the point of departure.
A. 50
B. 25
C. 10

FAR 61.103 Private Pilot Requirements

56. [F16/2/4]
A minimum age of _____ is required to solo an airplane.
A. 16
B. 17
C. 14

57. [F16/2/5]
Private pilot certificates require a minimum age of _____ for airplanes.
A. 16
B. 17
C. 14

FAR 61.107 Flight Experience

58. [F17/1/2]
The FAA requires a minimum of ____ hours total flight time to be eligible for the private certificate. This must include a minimum of _____ hours of flight training (time with an instructor), and _____ hours solo flight training (time alone in the airplane).
A. 40, 20, 10
B. 40, 15, 15
C. 35, 25, 15

FAR 61.113 Private Pilot Privileges and Limitations: Pilot in Command

59. [F18/1/4]
Basically, a private pilot certificate allows you to do two things:
A. fly solo and carry paying passengers.
B. fly at night and for hire.
C. fly without supervision and carry passengers.

60. [F18/1/5]
What exception, if any, permits a private pilot to act as pilot in command of an aircraft carrying passengers who pay for the flight?
A. If the passengers pay all the operating expenses.
B. If a donation is made to a charitable organization for the flight.
C. There is no exception.

61. [F18/1/6]
Private pilots may, for compensation or hire, act as pilot in command of an aircraft in connection with any business or employment if the flight is only _____ to that business or employment and the aircraft does not carry passengers or property for _____.
A. casually connected, flight at night
B. incidental, compensation or hire
C. incidental, sightseeing purposes

62. [F18/2/4]
Regarding general privileges and limitations, a private pilot may
A. act as pilot in command of an aircraft carrying a passenger for compensation if the flight is in connection with a business or employment.
B. share the operating expenses of a flight with a passenger.
C. not be paid in any manner for the operating expenses of a flight.

63. [F18/2/4]
According to regulations pertaining to general privileges and limitations, a private pilot may
A. be paid for the operating expenses of a flight if at least three takeoffs and three landings were made by the pilot within the preceding 90 days.
B. share the operating expenses of a flight with the passengers.
C. not be paid in any manner for the operating expenses of a flight.

PART 91 GENERAL OPERATING AND FLIGHT RULES

FAR 91.3 Responsibility and Authority of the Pilot In Command

64. [F20/1/1]
The final authority as to the operation of an aircraft is the
A. Federal Aviation Administration.
B. pilot in command.
C. aircraft manufacturer.

65. [F20/1/2]
When experiencing an emergency requiring immediate action, FAR 91.3 allows you to deviate from any _____ to the extent required to meet that _____.
A. rule, emergency
B. flight path, clearance
C. emergency, rule

66. [F20/1/2]
If an in-flight emergency requires immediate action, the pilot in command may
A. deviate from the FAR's to the extent required to meet the emergency, but must submit a written report to the Administrator within 24 hours.
B. deviate from the FAR's to the extent required to meet that emergency.
C. not deviate from the FAR's unless prior to the deviation approval is granted by the Administrator.

67. [F20/3/1]
When must a pilot who deviates from a regulation during an emergency send a written report of that deviation to the Administrator?
A. Within 7 days.
B. Within 10 days.
C. Upon request.

68. [F20/3/1]
When would a pilot be required to submit a detailed report of an emergency which caused the pilot to deviate from an ATC clearance?
A. When requested by ATC.
B. Immediately.
C. Within 7 days.

FAR 91.7 Civil Aircraft Airworthiness

69. [F20/3/4]
Who is responsible for determining if an aircraft is in condition for safe flight?
A. A certificated aircraft mechanic.
B. The pilot in command.
C. The owner or operator.

FAR 91.9 Civil Aircraft Flight Manual, Markings and Placard Requirements.

70. [F21/2/3]
Where may an aircraft's operating limitations be found?
A. On the airworthiness certificate.
B. In the current, FAA-approved flight manual, approved manual material, markings, and placards, or any combination thereof.
C. In the aircraft airframe and engine logbooks.

FAR 91.15 Dropping Objects

71. [F21/3/3]
Under what conditions may objects be dropped from an aircraft?
A. Only in an emergency.
B. If precautions are taken to avoid injury or damage to persons or property on the surface.
C. If prior permission is received from the Federal Aviation Administration.

FAR 91.17 Alcohol or Drugs

72. [F22/1/1]
No person may attempt to act as a crewmember of a civil aircraft with
A. .008 percent by weight or more alcohol in the blood.
B. .004 percent by weight or more alcohol in the blood.
C. .04 percent by weight or more alcohol in the blood.

73. [F22/1/1]
A person may not act as a crewmember of a civil aircraft if alcoholic beverages have been consumed by that person within the preceding _____________ hours.
A. 8
B. 12
C. 24

74. [F22/1/4]
Under what condition, if any, may a pilot allow a person who is obviously under the influence of drugs to be carried aboard an aircraft?
A. In an emergency or if the person is a medical patient under proper care.
B. Only if the person does not have access to the cockpit or pilot's compartment.
C. Under no condition.

FAR 91.103 Preflight Action

75. [F22/2/1]
Which preflight action is specifically required of the pilot prior to each flight?
A. Check the aircraft logbooks for appropriate entries.
B. Become familiar with all available information concerning the flight.
C. Review wake turbulence avoidance procedures.

76. [F22/3/1] Fill in the blank:
Preflight action, as required for all flights away from the vicinity of an airport, shall include an ___________ course of action if the flight cannot be completed as planned.

77. [F22/3/3]
In addition to other preflight actions for a VFR flight away from the vicinity of the departure airport, regulations specifically require the pilot in command to
A. review traffic control light signal procedures.
B. check the accuracy of the navigation equipment and the emergency locator transmitter (ELT).
C. determine runway lengths at airports of intended use and the aircraft's takeoff and landing distance data.

78. [General Knowledge Question]
What special check should be made on an aircraft during preflight after it has been stored an extended period of time?
A. ELT batteries and operation.
B. Condensation in the fuel tanks.
C. Damage or obstructions caused by animals, birds, or insects.

FAR 91.105 Flight Crewmembers at Stations

79. [F23/1/4]
Flight crewmembers are required to keep their safety belts and shoulder harnesses fastened during
A. takeoffs and landings.
B. all flight conditions.
C. flight in turbulent air.

80. [F23/1/4]
Which best describes the flight conditions under which flight crewmembers are specifically required to keep their safety belts and shoulder harnesses fastened?
A. Safety belts during takeoff and landing; shoulder harnesses during takeoff and landing.
B. Safety belts during takeoff and landing; shoulder harnesses during takeoff and landing and while en route.
C. Safety belts during takeoff and landing and while enroute; shoulder harnesses during takeoff and landing.

81. [F23/1/6]
You're not required to stay in your seat with your safety belt fastened as a flight crewmember when
A. you must attend to your physiological needs.
B. you want to look out the passenger's rear window.
C. you're on the enroute portion of your flight.

FAR 91.107 Use of Safety Belts

82. [F23/2/3]
With certain exceptions, safety belts are required to be secured about passengers during
A. taxi, takeoffs, and landings.
B. all flight conditions.
C. flight in turbulent air.

83. [F23/1/4 & F23/2/3]
Safety belts are required to be properly secured about which persons in an aircraft and when?
A. Pilots only, during takeoffs and landings.
B. Passengers, during taxi, takeoffs, and landings only.
C. Each person on board the aircraft during the entire flight.

84. [F23/2/3]
With respect to passengers, what obligation, if any, does a pilot in command have concerning the use of safety belts?
A. The pilot in command must instruct the passengers to keep their safety belts fastened for the entire flight.
B. The pilot in command must brief the passengers on the use of safety belts and notify them to fasten their safety belts during taxi, takeoff, and landing.
C. The pilot in command has no obligation in regard to passengers' use of safety belts.

85. [F23/2/3]
One exception to the use of safety belts by passengers during taxi, takeoff and landing occurs when
A. the passenger has not yet reached his or her second birthday and is held by an adult sitting in an approved seat.
B. the air is smooth, thus no seat belts are required.
C. the pilot and his or her passengers wear parachutes.

FAR 91.111 Operating Near Other Aircraft

86. [F24/2/1] Fill in the blank:
No person may operate an aircraft so close to another aircraft as to create a _____________ hazard.

87. [F24/2/2]
No person may operate an aircraft in formation flight
A. over a densely populated area.
B. in Class D airspace under special VFR.
C. except by prior arrangement with the pilot in command of each aircraft.

FAR 91.113 Right of Way Rules: Except Water Ops

Distress

88. [F24/3/2]
Which aircraft has the right of way over all other air traffic?
A. A balloon.
B. An aircraft in distress.
C. An aircraft on final approach to land.

Converging

89. [F24/3/3]
What action is required when two aircraft of the same category converge, but not head-on?
A. The faster aircraft shall give way.
B. The aircraft on the left shall give way.
C. Each aircraft shall give way to the right.

90. [F25/1/1]
What action should the pilots of a glider and an airplane take if on a head-on collision course?
A. The airplane pilot should give way to the left.
B. The glider pilot should give way to the right.
C. Both pilots should give way to the right.

91. [F25/2/1]
An airplane and an airship are converging. If the airship is left of the airplane's position, which aircraft has the right of way?
A. The airship.
B. The airplane.
C. Each pilot should alter course to the right.

Aircraft Categories

92. [F25/1/3]
Which aircraft has the right of way over the other aircraft listed?
A. Airship.
B. Balloon.
C. Gyroplane.

93. [F25/2/1]
Which aircraft has the right of way over the other aircraft listed?
A. Gyroplane.
B. Airship.
C. Aircraft refueling other aircraft.

Overtaking

94. [F25/2/2]
When one aircraft is overtaking another, which has the right of way?
A. The overtaking aircraft.
B. The aircraft begin overtaken.
C. The aircraft passing to the right.

95. [F25/2/2]
When you're overtaking another aircraft you should
A. alter your course to the left and pass well clear of the slower aircraft.
B. alter your course to the right and pass well clear of the slower aircraft.
C. fly directly over the top and buzz the slower, inconsiderate aircraft.

Landing

96. [F25/2/4]
When two or more aircraft are approaching an airport for the purpose of landing, the right of way belongs to the aircraft
A. that has the other to its right.
B. that is the least maneuverable.
C. at the lower altitude, but it shall not take advantage of this rule to cut in front of or to overtake another.

FAR 91.115 Right of Way Rules: Water Operations

97. [F26/1/2]
A seaplane and a motorboat are on crossing courses. If the motorboat is to the left of the seaplane, which has the right of way?
A. The motorboat.
B. The seaplane.
C. Both should alter course to the right.

FAR 91.117 Aircraft Speed

98. [F26/1/3] Fill in the blank:
Unless otherwise authorized, ______________ knots is the maximum indicated airspeed at which a person may operate an aircraft below 10,000 feet MSL.

99. [F26/2/3]
Unless otherwise authorized, the maximum indicated airspeed at which aircraft may be flown when at or below 2,500 feet AGL and within four nautical miles of the primary airport of Class C airspace is
A. 200 knots.
B. 230 knots.
C. 250 knots.

100. [F27/1/1]
When flying in the airspace underlying Class B airspace, the maximum speed authorized is
A. 200 knots.
B. 230 knots.
C. 250 knots.

101. [F27/1/1 & FIG-33, Page-F27]
When flying in a VFR corridor designated through Class B airspace, the maximum speed authorized is
A. 180 knots.
B. 200 knots.
C. 250 knots.

FAR 91.119 Minimum Safe Altitudes

102. [F27/3/3]
Except when necessary for takeoff or landing, what is the minimum safe altitude for a pilot to operate an aircraft anywhere?
A. An altitude allowing, if a power unit fails, an emergency landing without undue hazard to persons or property on the surface.
B. An altitude of 500 feet above the surface and no closer than 500 feet to any person, vessel, vehicle, or structure.
C. An altitude of 500 feet above the highest obstacle within a horizontal radius of 1,000 feet.

103. [F27/3/3]
Except when necessary for takeoff or landing, what is the minimum safe altitude required for a pilot to operate an aircraft over other than a congested area?
A. An altitude allowing, if a power unit fails, an emergency landing without undue hazard to persons or property on the surface.
B. An altitude of 500 feet AGL, except over open water or a sparsely populated area, which requires 500 feet from any person, vessel, vehicle, or structure.
C. An altitude of 500 feet above the highest obstacle within a horizontal radius of 1,000 feet.

MINIMUM ALTITUDES WHEN OPERATING OVER A CONGESTED AREA

104. [F27/3/5] Fill in the blanks:
Referring to the figure above, if the airplane shown is flying over a congested area, it must fly at a minimum altitude of ________ on its altimeter when operating within ________ feet of the building whose top is 920 feet MSL.

105. [F27/3/5]
Except when necessary for takeoff or landing, what is the minimum safe altitude required for a pilot to operate an aircraft over congested areas?
A. An altitude of 1,000 feet above any person, vessel, vehicle, or structure.
B. An altitude of 500 feet above the highest obstacle within a horizontal radius of 1,000 feet of the aircraft.
C. An altitude of 1,000 feet above the highest obstacle within a horizontal radius of 2,000 feet of the aircraft.

106. [F28/2/2]
Except when taking off or landing, the minimum safe altitude required by the regulations for a pilot to operate an aircraft over other than a congested area is an
A. altitude of 500 feet AGL, except over open water or a sparsely populated area, which requires 500 feet from any person, vessel, vehicle, or structure.
B. altitude allowing, if a power unit fails, an emergency landing without undue hazard to persons or property on the surface but never closer than 500 feet to a person.
C. altitude of 2,000 feet above the highest obstacle within a horizontal radius of 1,000 feet.

107. [F28/2/2]
Except when necessary for takeoff or landing, an aircraft may not be operated closer than what distance from any person, vessel, vehicle, or structure?
A. 500 feet.
B. 700 feet.
C. 1,000 feet.

FAR 91.121 Altimeter Settings

108. [F29/1/1 & E17/1/5]
Prior to takeoff, the altimeter should be set to which altitude or altimeter setting?
A. The current local altimeter setting, if available, or the departure airport elevation.
B. The corrected density altitude of the departure airport.
C. The corrected pressure altitude for the departure airport.

109. [E17/1/5]
If an altimeter setting is not available before flight, to which altitude should the pilot adjust the altimeter?
A. The elevation of the nearest airport corrected to mean sea level.
B. The elevation of the departure area.
C. Pressure altitude corrected for nonstandard temperature.

110. [F29/1/3]
At what altitude shall the altimeter be set to 29.92, when climbing to cruising flight level?
A. 14,500 feet MSL.
B. 18,000 feet MSL.
C. 24,000 feet MSL.

FAR 91.123 Compliance with ATC Clearances and Instructions

111. [F29/2/2]
When an ATC clearance has been obtained, no pilot in command may deviate from that clearance, unless that pilot obtains an amended clearance. The one exception to this regulation is
A. when the clearance states "at pilot's discretion."
B. an emergency.
C. if the clearance contains a restriction.

112. [F29/2/4]
An ATC clearance provides
A. priority over all other traffic.
B. adequate separation from all traffic.
C. authorization to proceed under specified traffic conditions in controlled airspace.

113. [F30/See *Emergency From an Emergency*]
What action, if any, is appropriate if the pilot deviates from an ATC instruction during an emergency and is given priority?
A. Take no special action since you are pilot in command.
B. File a detailed report within 48 hours to the chief of the appropriate ATC facility, if requested.
C. File a report with the FAA Administrator, as soon as possible.

FAR 91.125 ATC Light Signals

Light Signals in the Air

114. [F31/1/2]
If the aircraft's radio fails, what is the recommended procedure when landing at a controlled airport?
A. Observe the traffic flow, enter the pattern, and look for a light signal from the tower.
B. Enter a crosswind leg and rock the wings.
C. Flash the landing lights and cycle the landing gear.

115. [F31/1/2]
If your radio fails in flight under VFR conditions, you can
A. enter the traffic pattern after observing the flow of traffic.
B. enter the traffic pattern on the other side of the field so as not to disrupt traffic.
C. fly opposite the direction of traffic but avoid other airplanes.

116. [F31/1/2]
A steady green light signal directed from the control tower to an aircraft in flight is a signal that the pilot
A. is cleared to land.
B. should give way to other aircraft and continue circling.
C. should return for landing.

117. [F31/1/3]
A flashing green light signal directed from the control tower to an aircraft in flight is a signal that the pilot
A. is cleared to land.
B. should give way to other aircraft and continue circling.
C. should return for landing.

118. [F31/2/1]
If the control tower uses a light signal to direct a pilot to give way to other aircraft and continue circling, the light will be
A. flashing red.
B. steady red.
C. alternating red and green.

119. [F31/2/2]
A flashing red light signal directed from the control tower to an aircraft in flight is a signal that the pilot
A. is cleared to land.
B. should give way to other aircraft and continue circling.
C. should not land at that airport.

120. [F31/2/3]
An alternating red and green light signal directed from the control tower to an aircraft in flight is a signal to
A. hold position.
B. exercise extreme caution.
C. not land; the airport is unsafe.

121. [F31/2/3]
While on final approach for landing, an alternating green and red light followed by a flashing red light is received from the control tower. Under these circumstances, the pilot should
A. discontinue the approach, fly the same traffic pattern and approach again, and land.
B. exercise extreme caution and abandon the approach, realizing the airport is unsafe for landing.
C. abandon the approach, circle the airport to the right, and expect a flashing white light when the airport is safe for landing.

Light Signals on the Ground

122. [F31/3/4]
Which light signal from the control tower clears a pilot to taxi?
A. Flashing green.
B. Steady green.
C. Flashing white.

123. [F31/3/4]
Which light signal from the control tower clears a pilot for takeoff?
A. Flashing green.
B. Steady green.
C. Flashing white.

124. [F32/1/1]
Which light signal from the control tower indicates that a pilot should stop his or her taxi?
A. Flashing green.
B. Steady red.
C. Flashing white.

125. [F32/1/2]
A flashing red light signal from the control tower to a taxiing aircraft is an indication to
A. taxi at a faster speed.
B. taxi clear of the runway in use.
C. return to the starting point on the airport.

126. [F32/1/3]
A flashing white light signal from the control tower to a taxiing aircraft is an indication to
A. taxi at a faster speed.
B. taxi only on taxiways and not cross runways.
C. return to the starting point on the airport.

127. [F31/2/3]
An alternating red and green light signal directed from the control tower to a taxiing aircraft is a signal to
A. hold position.
B. exercise extreme caution.
C. not land; the airport is unsafe.

FAR 91.126 Operating on or in the Vicinity of an Airport in Class G Airspace

128. [F32/3/2]
Unless otherwise indicated, when approaching to land at an airport without an operating control tower in Class G airspace, each pilot of an airplane must
A. make all turns to the left.
B. make all turns to the right.
C. make all turns to give the passengers a good view.

FAR 91.127 Operations on or in the Vicinity of an Airport in Class E Airspace

129. [F33/1/3]
Unless otherwise indicated, when approaching to land at an airport without an operating control tower in Class E airspace, each pilot of an airplane must
A. make all turns to the left.
B. make all turns to the right.
C. make all turns to give the passengers a good view.

FAR 91.129 Operations in Class D Airspace

130. [F34/1/1] Fill in the blank:
Class D airspace generally extends upwards from the surface to ____________ feet above the ground.

131. [F34/1/1]
The lateral dimensions of Class D airspace are based on
A. the number of airports that lie within the Class D airspace.
B. 5 statute miles from the geographical center of the primary airport.
C. the instrument procedures for which the controlled airspace is established.

132. [F34/1/2]
Airspace at an airport with a part-time control tower is classified as Class D airspace only
A. when the weather minimums are below basic VFR.
B. when the associated control tower is in operation.
C. when the associated Flight Service Station is in operation.

133. [F34/1/2]
The purpose of a control tower is to provide information and instructions to
A. student pilots only.
B. all pilots except those flying balloons near the primary airport within that airspace.
C. airplanes taking off or landing at the primary airport within that airspace.

134. [F34/1/3]
The minimum equipment required to operate within Class D airspace is
A. a GPS receiver.
B. a two-way radio.
C. an interrociter.

135. [F34/1/3]
Unless otherwise authorized, two-way radio communication with Air Traffic Control is required for landings or takeoffs
A. at all tower controlled airports regardless of weather conditions.
B. at all tower controlled airports only when weather conditions are less than VFR.
C. at all tower controlled airports within Class D airspace only when weather conditions are less than VFR.

136. [F34/1/4]
In order to fly through Class D airspace or land at the primary airport, you must
A. establish two-way radio communication with the ATC facility responsible for that airspace.
B. establish two-way radio communication with any airplanes in that airspace.
C. establish two-way radio communication with the nearest Flight Service Station only.

137. [F34/2/4]
If another airport without an operating control tower lies within the boundaries of Class D airspace and you're interested in landing there, you must
A. fly directly to that airport without interfering in the primary airport's traffic pattern.
B. establish two-way radio communication only if you'll be overflying the primary airport.
C. establish two-way radio communication with the air traffic control tower at the primary airport before entering this airspace.

138. [F34/2/6]
If you're taking off from a non-tower satellite airport within Class D airspace do you still need to call the ATC facility for the primary airport?
A. No.
B. Yes.
C. Only if you want to.

139. [F35/1/3]
If you're approaching to land on a runway served by a visual approach slope indicator (VASI), you must
A. stay at or above the glideslope (until a lower altitude is necessary).
B. always remain 500 feet above any structure.
C. use the VASI only if it won't mess up a good landing.

140. [F35/2/2]
If instructed by ground control to taxi to Runway 9, the pilot may proceed
A. via taxiways and across runways to, but not onto Runway 9.
B. to the next intersecting runway where further clearance is required.
C. via taxiways and across runways to Runway 9, where an immediate takeoff may be made.

FAR 91.130 Operations in Class C Airspace

141. [F36/1/1]
The physical dimensions of Class C airspace consist of a lower section having a radius of _____ nautical miles and extending vertically from the ground to_____ feet above the airport elevation.
A. 5 & 2,500
B. 10 & 4,000
C. 5 & 4,000

Traffic Patterns

142. [F36/2/4]
If you're taking off or landing at a satellite airport within Class C airspace, you must
A. overfly the primary airport at or above 2,500 before entering the pattern at the satellite airport.
B. comply with the FAA arrival and departure traffic pattern established for that airport.
C. be squawking 1200 on your transponder only.

Communications - Arrival or Through Flight

143. [F36/2/5]
If you want to land at the primary airport in Class C airspace, or if you just want to fly through that airspace, you must
A. establish and maintain two-way radio communication with the facility providing air traffic services prior to entering that airspace.
B. establish and maintain two-way radio communication with the FSS nearest the primary airport.
C. establish two-way radio communication and, once established, it's not necessary to maintain communication with ATC.

144. [F36/2/5]
Two-way radio communication must be established with the Air Traffic Control facility having jurisdiction over the area prior to entering which class airspace?
A. Class C.
B. Class E.
C. Class G.

Communications - Departing Flight

145. [F36/3/2]
If you're departing the primary airport within Class C airspace, you must
A. establish and maintain two-way radio communication with the radar controller prior to departure.
B. establish transponder reception with the tower prior to departure.
C. establish and maintain two-way radio communication with the tower prior to departure.

146. [F36/3/2]
If you're departing a nontower satellite airport lying within the surface boundaries of Class C airspace, you must
A. establish and maintain two-way radio communication with the FSS having jurisdiction over that airspace.
B. establish and maintain two-way radio communication with the approach or departure control facility having jurisdiction over that airspace.
C. establish and maintain two-way radio communication with any ATC facility.

Equipment Requirements

147. [F36/3/3]
The minimum equipment required to operate within Class C airspace is
A. a transponder with altitude reporting capability.
B. a two-way radio with altitude reporting capability.
C. a two-way radio and a transponder with altitude reporting capability.

FAR 91.131 Operations in Class B Airspace

148. [F37/1/1]
Class B airspace usually extends up to
A. 2,500 feet MSL.
B. 4,000 feet MSL.
C. 10,000 feet MSL.

149. [F37/2/2 & F43/1/5]
With certain exceptions, all aircraft within 30 miles of a Class B primary airport from the surface upward to 10,000 feet MSL must be equipped with
A. an operable VOR or TACAN receiver and an ADF receiver.
B. instruments and equipment required for IFR operations.
C. an operable transponder having either Mode S or 4096-code capability with Mode C automatic altitude reporting capability.

150. [F37/1/1]
Class B airspace usually extends up to _____ feet MSL and often has _____ dimensions of many miles.
A. 2,500, vertical
B. 4,000, horizontal
C. 10,000, horizontal

Operating Rules

151. [F37/1/3]
Before you can enter Class B airspace you'll need _____ from the ATC facility having jurisdiction over that airspace.
A. a clearance
B. to establish and receive two-way radio communication
C. a transponder code

Pilot Requirements

152. [F37/1/4]
What minimum pilot certification is required for operation within Class B airspace?
A. Private pilot certificate or student pilot certificate with appropriate logbook endorsements.
B. Commercial pilot certificate.
C. Private pilot certificate with an instrument rating.

Equipment Required

153. [F37/2/2]
The minimum equipment required to operate within Class B airspace is a
A. a transponder with altitude encoding capability and, for IFR operations, a functioning VOR receiver.
B. two-way radio with all the frequencies needed for communication with ATC, a transponder with altitude encoding capability and, for IFR operations, a functioning VOR receiver.
C. two-way radio with all the frequencies needed for communication with ATC and a functioning VOR receiver.

FAR 91.133 Restricted and Prohibited Areas

154. [F38/1/1]
Under what condition, if any, may pilots fly through a restricted area?
A. When flying on airways with an ATC clearance.
B. With the controlling agency's authorization.
C. Regulations do not allow this.

FAR 91.135 Operations in Class A Airspace

155. [F38/2/3 & I4/1/2]
In which type of airspace are VFR flights prohibited?
A. Class A.
B. Class B.
C. Class C.

FAR 91.151 Fuel Requirements for Flight in VFR

156. [F38/2/4]
What is the specific fuel requirement for flight under VFR during daylight hours in an airplane?
A. Enough to complete the flight at normal cruising speed with adverse wind conditions.
B. Enough to fly to the first point of intended landing and to fly after that for 30 minutes at normal cruising speed.
C. Enough to fly to the first point of intended landing and to fly after that for 45 minutes at normal cruising speed.

157. [F38/2/4]
What is the specific fuel requirement for flight under VFR at night in an airplane?
A. Enough to complete the flight at normal cruising speed with adverse wind conditions.
B. Enough to fly to the first point of intended landing and to fly after that for 30 minutes at normal cruising speed.
C. Enough to fly to the first point of intended landing and to fly after that for 45 minutes at normal cruising speed.

Note: All questions dealing with VFR weather minimums (FAR 91.155 & 91.157) are covered in Chapter 9

FAR 91.159 VFR Cruising Altitude or Flight Level

158. [F39/1/2]
VFR cruising altitudes pertain only to VFR flights operating more than _____ feet above the surface,
A. 2,500
B. 4,000
C. 3,000

159. [F39/1/2]
What VFR cruising altitude is acceptable when operating on a Victor Airway while on a magnetic course of 175 degrees?
A. 4,500 feet.
B. 5,000 feet.
C. 5,500 feet.

160. [F39/1/2]
What cruising altitude is appropriate for a VFR flight on a magnetic course of 135 degrees?
A. Even thousands.
B. Even thousands plus 500 feet.
C. Odd thousands plus 500 feet.

161. [F39/1/2]
What VFR cruising altitude is appropriate when flying above 3,000 feet AGL on a magnetic course of 185 degrees?
A. 4,000 feet.
B. 4,500 feet.
C. 5,000 feet.

162. [F39/1/2]
Each person operating an aircraft at a VFR cruising altitude shall maintain an odd-thousand plus 500-foot altitude while on a
A. magnetic heading of 0 degrees through 179 degrees.
B. magnetic course of 0 degrees through 179 degrees.
C. true course of 0 degrees through 179 degrees.

FAR 91.203 Civil Aircraft: Certifications Required

163. [F40/1/2]
In addition to a valid airworthiness certificate, what documents or records must be aboard an aircraft during flight?
A. Aircraft engine and airframe logbooks, and owner's manual.
B. Radio operator's permit, and repair and alteration forms.
C. Operating limitations and registration certificate.

164. [F40/2/1]
How long does the airworthiness certificate of an aircraft remain valid?
A. As long as the aircraft has a current registration certificate.
B. Indefinitely, unless the aircraft suffers major damage.
C. As long as the aircraft is maintained and operated as required by Federal Aviation Regulations.

165. [F40/All]
In regard to preflighting an aircraft, what is the minimum expected of a pilot prior to every flight?
A. Fill the fuel tanks regardless of the distance or time you plan to fly.
B. Let the mechanic do the walkaround inspection.
C. Check the required documents aboard the aircraft.

166. [F40/2/2]
The airworthiness certificate must be displayed where it's visible to _____.
A. passengers.
B. crew members
C. an FAA inspector

FAR 91.207 Emergency Locator Transmitters

167. [F41/1/1]
When activated, an emergency locator transmitter (ELT) transmits on
A. 118.0 and 118.8 MHz.
B. 121.5 and 243.0 MHz.
C. 123.0 and 119.0 MHz.

168. [F41/1/1]
Which procedure is recommended to ensure that the emergency locator transmitter (ELT) has not been activated?
A. Turn off the aircraft ELT after landing.
B. Ask the airport tower if they are receiving an ELT signal.
C. Monitor 121.5 before engine shutdown.

169. [F41/1/2]
When must batteries in an emergency locator transmitter (ELT) be replaced or recharged, if rechargeable?
A. After any inadvertent activation of the ELT.
B. When the ELT has been in use for more than 1 cumulative hour.
C. When the ELT can no longer be heard over the airplane's communication radio receiver.

170. [F41/1/2]
When are non-rechargeable batteries of an emergency locator transmitter (ELT) required to be replaced?
A. Every 24 months.
B. When 50 percent of their useful life expires.
C. At the time of each 100-hour or annual inspection.

171. [F41/1/2]
When must the battery in an emergency locator transmitter (ELT) be replaced (or recharged if the battery is rechargeable)?
A. After half the battery's useful life.
B. During each annual and 100 hour inspection.
C. Every 24 calendar months.

172. [F41/2/1]
When may an emergency locator transmitter (ELT) be tested?
A. Anytime.
B. At 15 and 45 minutes past the hour.
C. During the first five minutes after the hour.

173. [F41/2/2]
An emergency locator transmitter (ELT) must be inspected within _____ calendar months after the last inspection.
A. 12
B. 24
C. 36

FAR 91.209 Aircraft Lights

174. [F41/2/3] Fill in the blanks:
Official night time for airplanes (in regards to the operation of its lights) is from ______________ to ______________.

175. [F41/2/3]
Except in Alaska, during what time period should lighted position lights be displayed on an aircraft?
A. End of evening civil twilight to the beginning of morning civil twilight.
B. 1 hour after sunset to 1 hour before sunrise.
C. Sunset to sunrise.

176. [F41/3/3]
Airplanes with an anticollision light system must have these lights _____ when the airplane is in operation.
A. on at all times
B. on from sunrise to sunset
C. off

177. [F41/3/3]
Pilots are allowed to turn off the anticollision lighting system
A. if the pilot in command determines, because of the operating conditions, it is unsafe.
B. if the passengers determine, because of the operating conditions, it is unsafe.
C. if the pilot in command simply feels like doing so in retaliation to being oppressed by the man.

FAR 91.211 Use of Supplemental Oxygen

178. [F42/3/2]
Unless each occupant is provided with supplemental oxygen, no person may operate a civil aircraft of U.S. registry above a maximum cabin pressure altitude of
A. 12,500 feet MSL.
B. 14,000 feet MSL.
C. 15,000 feet MSL.

179. [F42/2/2]
When operating an aircraft at cabin pressure altitudes above 12,500 feet MSL up to and including 14,000 feet MSL, supplemental oxygen shall be used during
A. the entire flight time at those altitudes.
B. that flight time in excess of 10 minutes at those altitudes.
C. that flight time in excess of 30 minutes at those altitudes.

180. [F42/2/2]
When operating an aircraft at cabin pressure altitudes above 14,000 feet MSL, supplemental oxygen shall be used by the PIC (and the required minimum flight crew) for
A. the entire flight time at those altitudes.
B. that flight time in excess of 10 minutes at those altitudes.
C. that flight time in excess of 30 minutes at those altitudes.

FAR 91.215 ATC Transponder and Altitude Reporting Equipment and Use

181. [F43/1/4&5]
An operable 4096-code transponder with an encoding altimeter is required in which airspace?
A. Class A, Class B (and within 30 miles of the Class B primary airport), and Class C.
B. Class D and Class E (below 10,000 feet MSL).
C. Class D and Class G (below 10,000 feet MSL).

182. [F43/2/4]
If your transponder or its Mode C capability fails in flight or fails at some intermediate stop on a cross country flight, it's possible that ATC can provide an on-the-spot waiver if you
A. call the Center or approach controller with your request if you intend to fly in any of the areas or airspace in which a transponder is required.
B. continue flying in any of the airspace requiring a transponder but just don't tell anyone.
C. call the nearest FSS and explain your problem to them.

183. [F43/2/1]
An operable 4096-code transponder with an encoding altimeter is required
A. in all airspace of the 48 contiguous states and the District of Columbia at and above 12,500 feet MSL, excluding the airspace at and below 1,500 feet above the surface.
B. in all airspace of the 48 contiguous states and the District of Columbia at and above 2.500 feet MSL, excluding the airspace at and below 1,200 feet above the surface.
C. in all airspace of the 48 contiguous states and the District of Columbia at and above 10,000 feet MSL, excluding the airspace at and below 2,500 feet above the surface.

184. [F43/2/4] Fill in the blank:
If you don't have a transponder, then ATC wants at least ____________ hour(s) notice before they approve flight in an area requiring a transponder.

FAR 91.303 Aerobatic Flight

185. [F44/1/2]
In which controlled airspace is acrobatic flight prohibited?
A. Class D airspace, Class E airspace designated for Federal Airways.
B. All Class E airspace below 1,500 feet AGL.
C. All Class G airspace.

186. [F44/1/2]
No person may operate an aircraft in aerobatic flight when the flight visibility is less than
A. 3 miles.
B. 5 miles.
C. 7 miles.

187. [F44/1/2]
What is the lowest altitude permitted for aerobatic flight?
A. 1,000 feet AGL.
B. 1,500 feet AGL.
C. 2,000 feet AGL.

188. [F44/1/2]
No person may operate an aircraft in aerobatic flight when
A. flight visibility is less than 5 miles.
B. over any congested area of a city, town, or settlement.
C. less than 2,500 feet AGL.

FAR 91.307 Parachutes and Parachuting

189. [F44/2/2]
With certain exceptions, when must each occupant of an aircraft wear an approved parachute?
A. When a door is removed from the aircraft to facilitate parachute jumpers.
B. When intentionally pitching the nose of the aircraft up or down 30 degrees or more.
C. When intentionally banking in excess of 30 degrees.

190. [F44/2/2]
A chair-type parachute must have been packed by a certificated and appropriately rated parachute rigger within the preceding
A. 60 days.
B. 120 days.
C. 90 days.

191. [F44/3/2]
An approved chair-type parachute may be carried in an aircraft for emergency use if it has been packed by an appropriately rated parachute rigger within the preceding
A. 120 days.
B. 180 days.
C. 365 days.

FAR 91.313 Restricted Category Civil Aircraft: Operating Limitations

192. [F45/1/2]
Which is normally prohibited when operating a restricted category civil aircraft?
A. Flight for compensation or hire when a passenger is being carried.
B. Flight over a sparsely populated area.
C. Flight within Class K airspace.

193. [F45/1/3]
Which is normally prohibited when operating a restricted category civil aircraft?
A. Flight under instrument flight rules.
B. Flight over a densely populated area.
C. Flight within Class D airspace.

FAR 91.319 Aircraft Having Experimental Certificates: Operating Limitations

194. [F45/3/3]
Unless otherwise specifically authorized, no person may operate an aircraft that has an experimental certificate
A. beneath the floor of Class B airspace.
B. over a densely populated area or on a congested airway.
C. from the primary airport within Class D airspace.

FAR 91.403 Aircraft Maintenance: General

195. [F46/1/1]
The responsibility for ensuring that an aircraft is maintained in an airworthy condition is primarily that of the
A. pilot in command.
B. owner or operator.
C. mechanic who performs the work.

FAR 91.407 Operations After Maintenance, Preventive Maintenance, Rebuilding or Alteration

196. [F46/1/4]
If an alteration or repair substantially affects an aircraft's operation in flight, that aircraft must be test flown by an appropriately rated pilot and approved for return to service prior to being operated
A. by any private pilot.
B. with passengers aboard.
C. for compensation or hire.

197. [F46/1/4]
Before passengers can be carried in an aircraft that has been altered in a manner that may have appreciably changed its flight characteristics, it must be flight tested by an appropriately-rated pilot who holds at least a
A. commercial pilot certificate with an instrument rating.
B. private pilot certificate.
C. commercial pilot certificate and a mechanic's certificate.

198. [F46/2/2] Fill in the blanks:
A _____ or _____ pilot may perform preventive maintenance on an aircraft and approve it for a return to service.

199. [F46/2/2]
Preventive maintenance has been performed on an aircraft. What paperwork is required?
A. A full, detailed description of the work done must be entered in the airframe logbook.
B. The date the work was completed, and the name of the person who did the work must be entered in the airframe and engine logbook.
C. The signature, certificate number, and kind of certificate held by the person approving the work and a description of the work must be entered in the aircraft maintenance records.

200. [F46/2/2]
Which operation would be described as preventive maintenance?
A. Servicing landing gear wheel bearings.
B. Alteration of main seat support brackets.
C. Engine adjustments to allow automotive gas to be used.

201. [F46/2/2]
Which operation would be described as preventive maintenance?
A. Repair of landing gear brace struts.
B. Replenishing hydraulic fluid.
C. Repair of portions of skin sheets by making additional seams.

202. [F46/2/2]
What regulation allows a private pilot to perform preventive maintenance?
A. 14 CFR Part 43.7.
B. 14 CFR Part 91.403.
C. 14 CFR Part 61.113.

203. [F46/2/2]
Who may perform preventive maintenance on an aircraft and approve it for return to service?
A..Student or recreational pilot.
B. Private or commercial pilot.
C. None of the above.

FAR 91.409 Inspections

204. [F46/2/3]
An aircraft's annual inspection was performed on July 12, this year. The next annual inspection will be due no later than
A. July 1, next year.
B. July 13, next year.
C. July 31, next year.

205. [F46/2/4]
What aircraft inspections are required for rental aircraft that are also used for flight instruction?
A. Annual and 100 hour inspections.
B. Biannual and 100 hour inspections.
C. Annual and 50 hour inspections.

206. [F46/2/4]
An aircraft had a 100 hour inspection when the tachometer read 1259.6. When is the next 100 hour inspection due?
A. 1349.6 hours.
B. 1359.6 hours.
C. 1369.6 hours.

207. [F47/1/1]
If necessary, the 100 hour time limit may be exceeded by not more than _____ hours to reach a place where the inspection can be done.
A. 5
B. 10
C. 25

208. [F47/1/1]
A 100 hour inspection was due at 3302.5 hours on the tachometer. The 100 hour inspection was actually done at 3309.5 hours. When is the next 100 hour inspection due?
A. 3312.5 hours.
B. 3402.5 hours.
C. 3409.5 hours.

FAR 91.413 ATC Transponder Tests and Inspections

209. [F47/1/2]
No person may use an ATC transponder unless it has been tested and inspected within at least the preceding
A. 6 calendar months.
B. 12 calendar months.
C. 24 calendar months.

210. [F47/1/2]
Maintenance records show the last transponder inspection was performed on September 1, 1993. The next inspection was due no later than
A. September 30, 1994.
B. September 1, 1995.
C. September 30, 1995.

211. [F47/2/1]
The responsibility for ensuring that maintenance personnel make the appropriate entries in the aircraft maintenance records indicating the aircraft has been approved for return to service lies with the
A. owner or operator.
B. pilot in command.
C. mechanic who performed the work.

FAR 91. 417 Maintenance Records

212. [F47/2/1]
Completion of an annual inspection and the return of the aircraft to service should always be indicated by
A. the relicensing date on the registration certificate.
B. an appropriate notation in the aircraft maintenance records.
C. an inspection sticker placed on the instrument panel that lists the annual inspection completion date.

213. [F47/2/1]
To determine the expiration date of the last annual aircraft inspection, a person should refer to the
A. airworthiness certificate.
B. registration certificate.
C. aircraft maintenance records.

214. [F47/2/2]
Which records or documents shall the owner or operator of an aircraft keep to show compliance with an applicable airworthiness directive?
A. Aircraft maintenance records.
B. Airworthiness certificate and pilot's operating handbook.
C. Airworthiness and registration certificates.

215. [F47/2/2]
What should an owner or operator know about airworthiness directives (ADs)?
A. They are for informational purposes only.
B. They are voluntary.
C. They are mandatory.

216. [F47/2/2]
May a pilot operate an aircraft that is not in compliance with an Airworthiness Directive (AD)?
A. Yes, if allowed by the AD.
B. Yes, under VFR conditions only.
C. Yes, ADs are only voluntary.

217. [F47/2/2]
Who is responsible for ensuring Airworthiness Directives (ADs) are complied with?
A. Mechanic with inspection authorization (IA).
B. Owner or operator.
C. Repair station.

218. [F40/All]
The airworthiness of an aircraft can be determined by a preflight inspection and a
A. statement from the owner that the aircraft is airworthy.
B. logbook endorsement from a flight instructor.
C. review of the maintenance records.

National Transportation Safety Board Regulations

219. [F48/2/2]
If an aircraft is involved in an accident that results in substantial damage to the aircraft, the nearest NTSB field office should be notified
A. immediately.
B. within 48 hours.
C. within 7 days.

220. [F48/2/3]
Which incident requires an immediate notification to the nearest NTSB field office?
A. A forced landing due to engine failure.
B. Landing gear damage due to a hard landing.
C. Flight control system malfunction or failure.

221. [F48/2/6]
Which incident would necessitate an immediate notification to the nearest NTSB field office?
A. An in-flight generator/alternator failure.
B. An in-flight fire.
C. An in-flight loss of VOR receiver capability.

222. [F48/2/9]
Which incident requires an immediate notification be made to the nearest NTSB field office?
A. An overdue aircraft that is believed to be involved in an accident.
B. An in-flight radio communications failure.
C. An in-flight generator or alternator failure.

NTSB 830.10 Preservation of Aircraft Wreckage, Mail, Cargo and Records

223. [F48/3/1]
May aircraft wreckage be moved prior to the time the NTSB takes custody?
A. Yes, but only if moved by a federal, state, or local law enforcement officer.
B. Yes, but only to protect the wreckage from further damage.
C. No, it may not be moved under any circumstances.

NTSB 830.15 Reports and Statements to Be Filed

224. [F48/3/2]
The operator of an aircraft that has been involved in an *accident* is required to submit a report to the nearest field office of the NTSB
A. within 7 days.
B. within 10 days.
C. when requested.

225. [F48/3/2]
The operator of an aircraft that has been involved in an *incident* is required to submit a report to the nearest field office of the NTSB
A. within 7 days.
B. within 10 days.
C. when requested.

226. [F52, Postflight Briefing #6-4]
To act as pilot in command of an aircraft towing a glider, a pilot is required to have made within the preceding 12 months
A. at least three flights as an observer in a glider being towed by an aircraft.
B. at least three flights in a powered glider.
C. at least three actual or simulated glider tows while accompanied by a qualified pilot.

227. [F52, Postflight Briefing #6-4]
A certificated private pilot may not act as pilot in command of an aircraft towing a glider unless there is entered in the pilot's logbook a minimum of
A. 100 hours of pilot flight time in any aircraft.
B. 100 hours of pilot in command time in the aircraft category, class and type (if required), that the pilot is using to tow a glider.
C. 200 hours of pilot in command time in the aircraft category, class and type (if required) that the pilot is using to tow a glider.

228. Bonus Question [General Knowledge]
How should an aircraft preflight inspection be accomplished for the first flight of the day?
A. Quick walk around with a check of gas and oil.
B. Any sequence as determined by the pilot in command.
C. Thorough and systematic means recommended by the manufacturer.

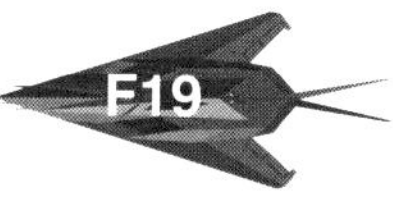

Chapter Six Answers

1. C
2. B
3. A
4. B
5. A
6. C
7 .A
8. B
9. A
10. A
11. C
12. C
13. A
14. B
15. C
16. B
17. C
18. C
19. C
20. B
21. B
22. C
23. A
24. C
25. B
26. B
27. B
28. C
29. C
30. C
31. high performance
32. complex
33. C
34. B
35. C
36. B
37. C
38. B
39. C
40. A
41. C
42. C
43. C
44. B
45. A
46. C
47. A
48. C
49. B
50. B
51. B
52. A
53. A
54. B
55. A
56. A
57. B
58. A
59. C
60. B
61. B
62. B
63. B
64. B
65. A
66. B
67. C
68. A
69. B
70. B
71. B
72. C
73. A
74. A
75. B
76. alternate
77. C
78. C
79. A
80. C
81. A
82. A
83. B
84. B
85. A
86. collision
87. C
88. B
89. B
90. C
91. A
92. B
93. C
94. B
95. B
96. C
97. B
98. 250
99. A
100. A
101. B
102. A
103. B
104. 1,920, 2,000
105. C
106. A
107. A
108. A
109. B
110. B
111. B
112. C
113. B
114. A
115. A
116. A
117. C
118. B
119. C
120. B
121. B
122. A
123. B
124. B
125. B
126. C
127. B
128. A
129. A
130. 2,500
131. B
132. B
133. C
134. B
135. A
136. A
137. C
138. B
139. A
140. A
141. C
142. B
143. A
144. A
145. C
146. B
147. C
148. C
149. C
150. C
151. A
152. A
153. B
154. B
155. A
156. B
157. C
158. C
159. C
160. C
161. B
162. B
163. C
164. C
165. C
166. A
167. B
168. C
169. B
170. B
171. A
172. C
173. A
174. sunset, sunrise
175. C
176. A
177. A
178. C
179. C
180. A
181. A
182. A
183. C
184. one
185. A
186. A
187. B
188. B
189. B
190. B
191. A
192. A
193. B
194. B
195. B
196. B
197. B
198. private, commercial
199. C
200. A
201. B
202. A
203. B
204. C
205. A
206. B
207. B
208. B
209. C
210. C
211. A
212. B
213. C
214. A
215. C
216. A
217. B
218. C
219. A
220. C
221. B
222. A
223. B
224. B
225. C
226. C
227. B
228. C

Note: To ensure that you have the most current answers to these questions, please check the *Book & Slide Updates* section at Rod Machado's web site: www.rodmachado.com

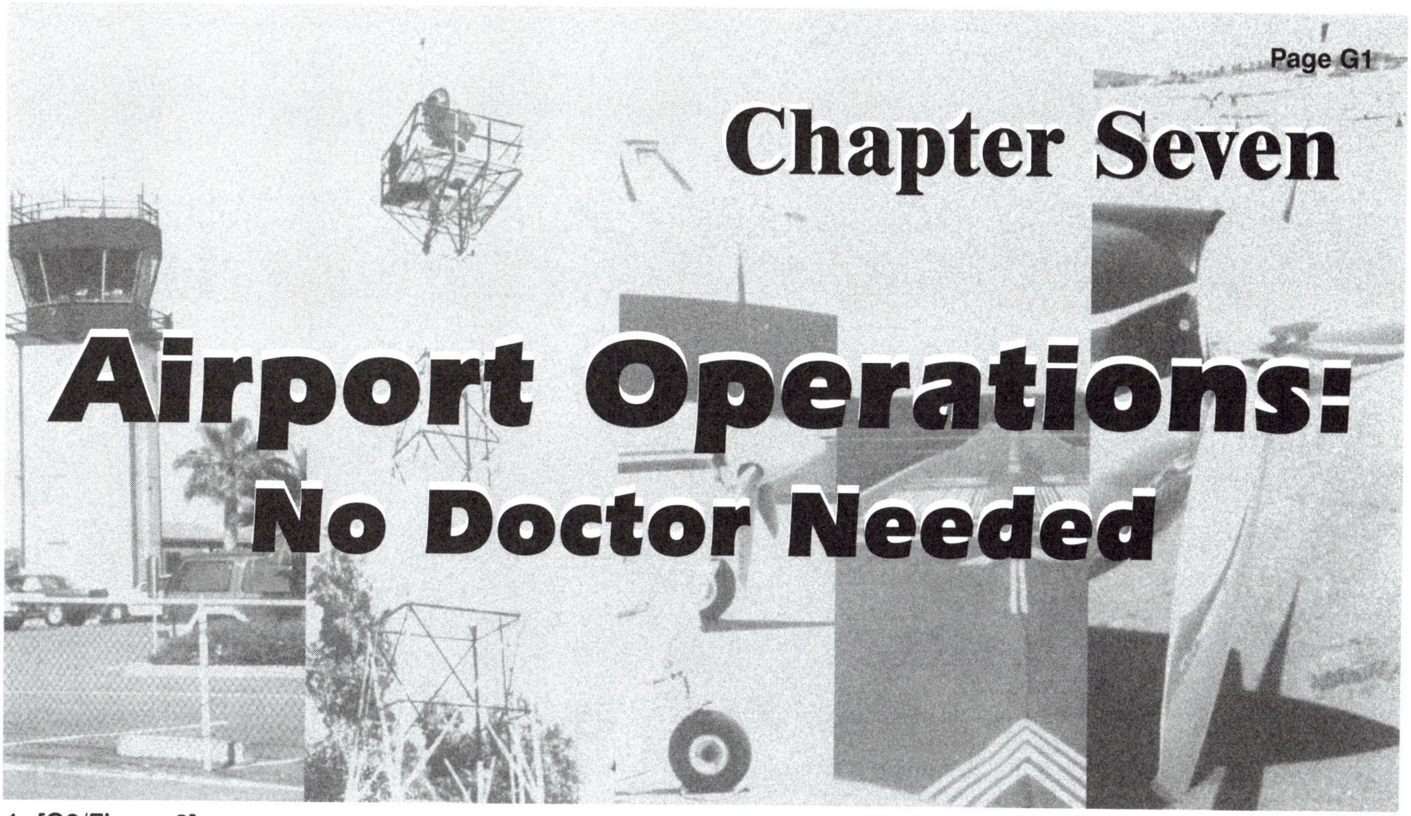

Chapter Seven

Airport Operations: No Doctor Needed

1. [G2/Figure 2]
The numbers 9 and 27 on a runway indicate that the runway is oriented approximately
A. 009 degrees and 027 degrees true.
B. 090 degrees and 270 degrees true.
C. 090 degrees and 270 degrees magnetic.

Runway Lighting

2. [G2/2/5]
Green lights announce the runway's
A. threshold.
B. midpoint.
C. last 2000 feet of usable surface.

3. [G2/Figure 3] Fill in the blanks:

BASIC RUNWAY LIGHTING

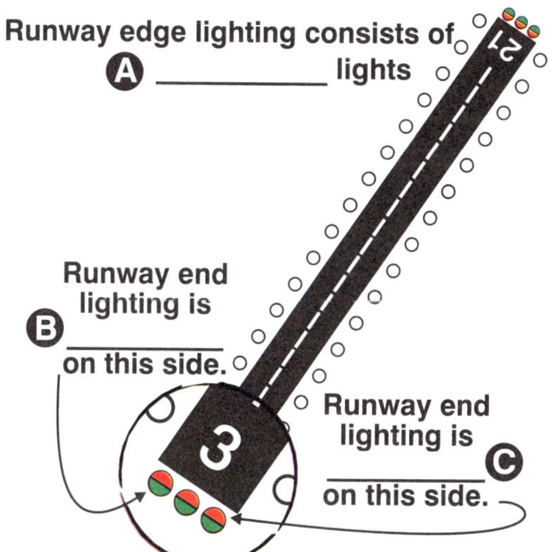

4. [G2/2/5]
The far end of the active runway is lit in
A. red.
B. green.
C. yellow.

Taxiway Markings

5. [G4/1/1]
Airport taxiway edge lights are identified at night by
A. white directional lights.
B. blue omnidirectional lights.
C. alternate red and green lights.

6. [G4/Figure 6] Fill in the blanks:

TAXIWAY MARKINGS

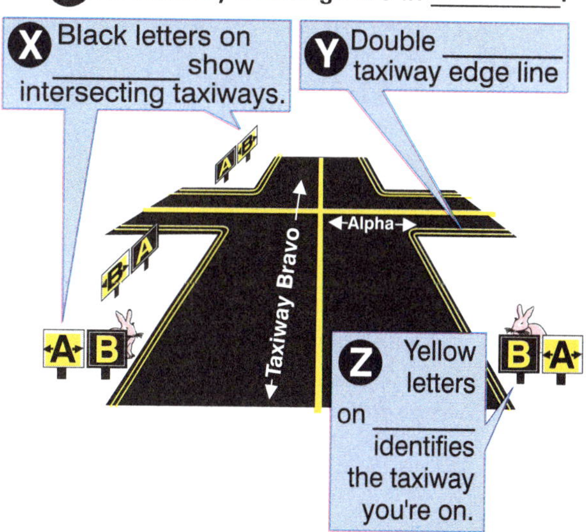

7. [G4/Figure 8] Fill in the blanks:

RUNWAY MARKINGS

A White numbers on ________ indicate mandatory hold points for all tower controlled airports. They indicate that you're about to taxi onto a runway (possibly an active one!).

K 30-12

30-12 K

B Runway hold markings consist of four ________ lines: two are solid & two are dashed.

C Solid double yellow lines require a ________ to cross at a controlled airport.

D If the broken double yellow lines are on your side, you should ________ them and enter the taxiway thus moving clear of the runway area.

---

8. [G5/Figure 9] Fill in the blanks:

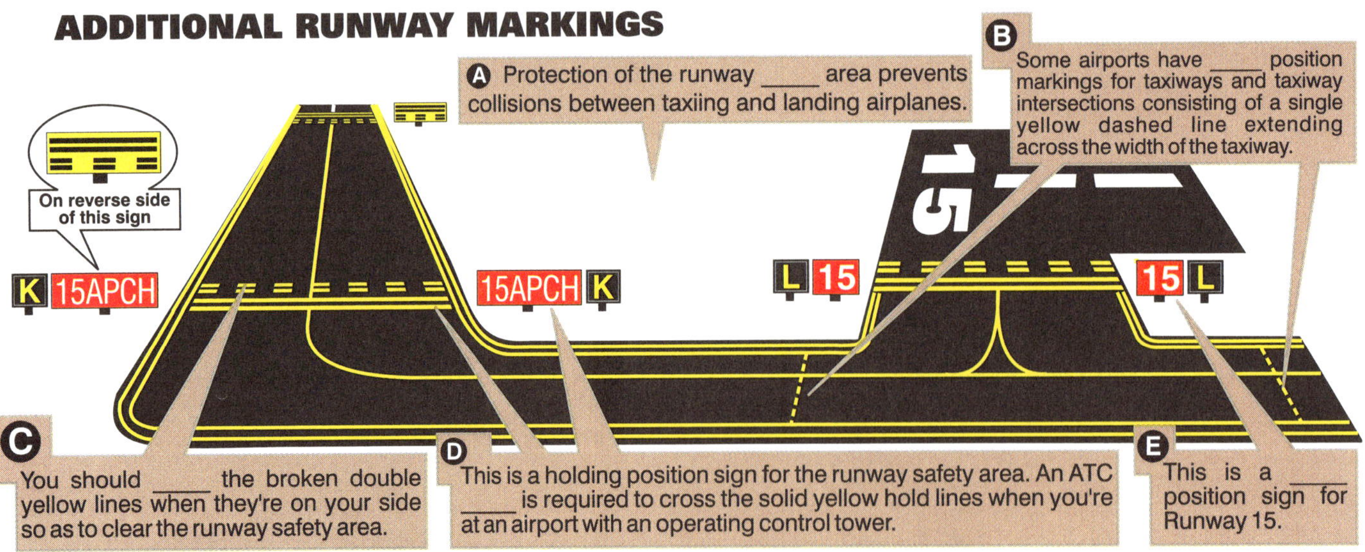

Additional Runway Markings

9. [G6/1/1] (Refer to figure in the top left hand corner of the next page.) The chevrons that appear on the end of Runway 24 indicate that the area

A. may be used only for taxiing.
B. is unusable for taxiing, takeoff, and landing.
C. cannot be used for landing, but may be used for taxiing and takeoff.

10. [G6/1/2] (Refer to figure in the top left hand corner of the next page.) That portion of the runway identified by the letter B may be used for

A. landing.
B. taxiing and takeoff.
C. taxiing and landing.

RUNWAY SURFACE MARKINGS

11. [G6/1/1&2] (Refer to the figure above)
According to the airport diagram, which statement is true?
A. Runway 24 is equipped at position A with emergency arresting gear to provide a means of stopping military aircraft.
B. Takeoffs may be started at position B on Runway 31, and the landing portion of this runway begins at the displaced threshold.
C. The takeoff and landing portion of Runway 6 begins at position A.

12. [G6/1/1] (Refer to the figure above)
What is the difference between area A and area B on the airport depicted?
A. "A" may be used for taxi and takeoff; "B" may be used only as an overrun.
B. "A" may be used for all operations except heavy aircraft landings; "B" may be used only as an overrun.
C. "A" may not be used at all; "B" may be used for all operations except landings.

13. [G6/Figure 10] (Refer to the figure above)
The large white X markings indicate a
A. stabilized area.
B. multiple heliport.
C. closed runway.

Airport Beacons

14. [G6/3/2]
Civilian airports with runway lights may be identified by a
A. flashing green and white rotating beacon.
B. flashing yellow light.
C. blue lighted square landing light.

15. [G6/3/2]
A lighted heliport may be identified by a
A. green, yellow, and white rotating beacon.
B. flashing yellow light.
C. blue lighted square landing light.

16. [G6/3/2]
A military air station can be identified by a rotating beacon that emits
A. white and green alternating flashes.
B. two quick, white flashes between green flashes.
C. green, yellow, and white flashes.

17. [G6/3/2]
How can a military airport be identified at night?
A. Alternate white and green light flashes.
B. Dual peaked (two quick) white flashes between green flashes.
C. White flashing lights with steady green at the same location.

18. [G7/1/1]
An airport's rotating beacon operated during daylight hours indicates
A. there are obstructions on the airport.
B. that weather at the airport located within surface-based controlled airspace is below basic VFR weather minimums.
C. the Air Traffic Control tower is not in operation.

The Traffic Pattern

19. [G8/Figure 13B] Fill in the blanks:

THE TRAFFIC PATTERN

_______ point (for turn to base leg)
_______ leg
_______ leg
Wind direction
A B C D E F G
30
12
_______ leg
_______ approach
_______ leg (parallel & offset)
_______ leg

20. [G8/1/3]
Airplane takeoffs are made into the wind, and the takeoff flight path is thus called the ________.
A. departure leg.
B. downwind leg.
C. opposite leg.

21. [G8/2/1]
If you're remaining in the pattern, a turn (generally a left turn) to the _____ will be made when the airplane is beyond the departure end of the runway and within _____ feet of the traffic pattern altitude.
A. departure leg, 500
B. crosswind leg, 300
C. perpendicular part, 1,000

22. [G8/2/2]
As the airplane continues its climb, another 90 degree turn is made. This places the airplane parallel to the runway, going opposite to the direction in which it will land. This is called the _____ because your direction is with the wind.
A. pattern leg
B. upwind leg
C. downwind leg

23. [G8/3/1]
Expect traffic pattern altitudes to range from _____ feet above the airport elevation, typically averaging about 1,000 feet AGL. The downwind leg is flown approximately _____ mile from the landing runway.
A. 600 to 1,500, 1/2 to one
B. 200 to 2,000, 1/10 to 1/5
C. 1,000 to 2,000, 3/8 to 1/2

24. [G8/3/2]
You continue downwind until passing a point abeam the beginning of the landing threshold of the runway. Then it's another 90 degree turn and you're on _____. From here you make one more 90 degree turn, onto _____.
A. base leg, final approach
B. final approach, base leg
C. base leg, the upwind leg

25. [G8/3/2]
The _____ is flown parallel to the runway in the direction of landing. It's often used during go-arounds or overflights to avoid departing traffic.
A. departure leg
B. sidestep leg
C. upwind leg

Crabbing in the Pattern

26. [G10/Figures 18 & 19]
The wind in traffic pattern A is from the ____; the wind in traffic pattern B is from the _____.

CROSSWIND CORRECTION

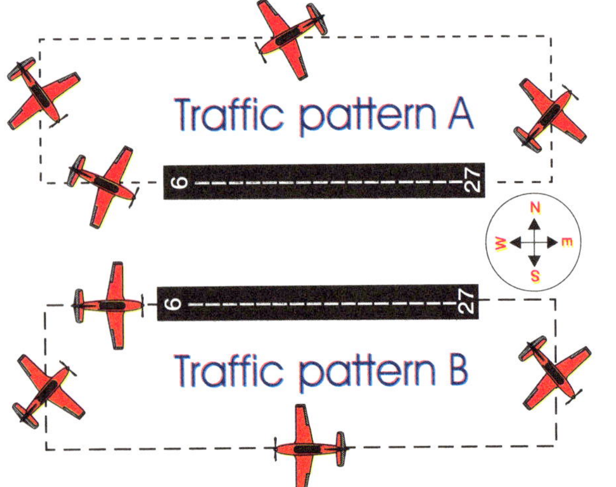

Entering the Traffic Pattern

27. [G10/3/4]
Before entering the traffic pattern at uncontrolled airports, you are expected to
A. observe the flow of traffic and conform to the traffic pattern in use.
B. observe traffic and maneuver any way as long as you don't cause a traffic conflict.
C. fly a traffic pattern that requires the least amount of maneuvering.

28. [G11/1/2]
The runway in use is normally determined by
A. noise abatement concerns.
B. wind direction.
C. direction of incoming traffic.

The Segmented Circle

29. [G12/Figure 21]
The traffic pattern indicated in the segmented circle shown in the figure below has been arranged to avoid flights over an area to the
A. south of the airport.
B. north of the airport.
C. northeast of the airport.

THE SEGMENTED CIRCLE

9 27

30. [G13/Figure 22]
(Refer to figure in the top left hand corner of the next page.) Which runway and traffic pattern should be used as indicated by the wind cone in the segmented circle (assume landing on a runway most nearly into the wind)?
A. Right hand traffic on Runway 20.
B. Right hand traffic on Runway 17.
C. Left hand traffic on Runway 2.

31. [G13/Figure 22]
(Refer to figure in the top left hand corner of the next page.) If you're landing on Runway 35 you should
A. expect a crosswind from the east.
B. expect a crosswind from the west.
C. not expect a crosswind on Runway 35.

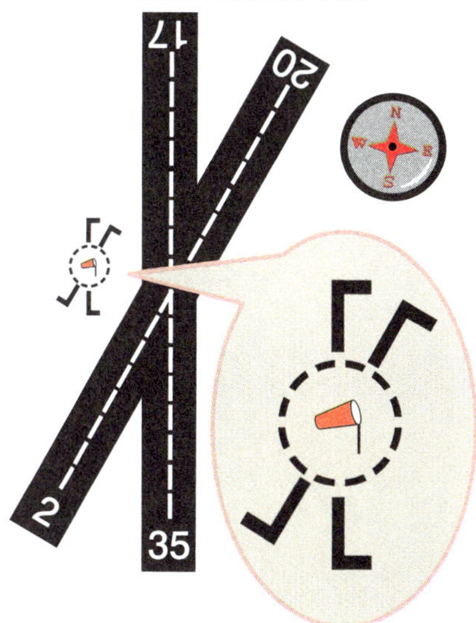

32. [G13/Figure 22] (Referring to the figure above) The segmented circle indicates that the airport traffic is
A. left-hand for Runway 35 and right-hand for Runway 17.
B. left-hand for Runway 17 and right-hand for Runway 35.
C. right-hand for Runway 2 and left-hand for Runway 20.

33. [G13/Figure 22] (Referring to the figure above) The segmented circle indicates that a landing on Runway 20 will be with a
A. right-quartering headwind.
B. left-quartering headwind.
C. left-quartering tailwind.

34. [G13/Figure 22] (Referring to the figure above) The segmented circle indicates that a landing on Runway 2 will be with a
A. left-quartering headwind.
B. left-quartering headwind.
C. right-quartering headwind.

Wind and Landing-Direction Indicators

35. [G14/Figure 24] Fill in the blanks: Based on the landing and wind direction indicators below, indicators ____________, ____________ and ____________ indicate a wind from the east.

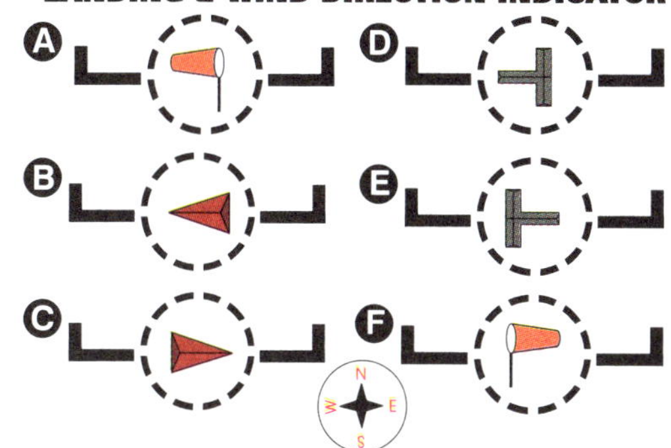

36. [G14/Figure 24]
Based on the landing and wind direction indicators below, indicators ____ and ____ indicate a wind from the south.

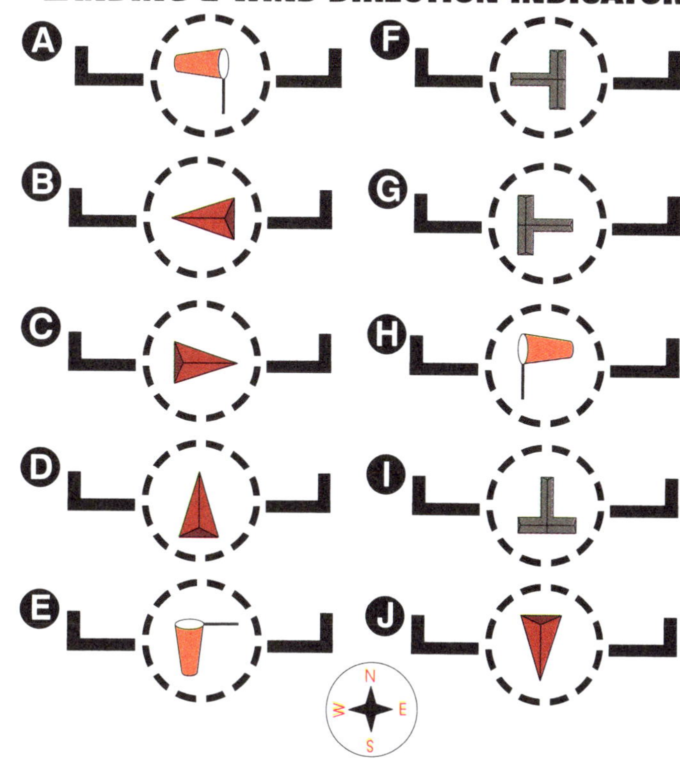

The 45 Degree Entry Point

37. [G15/1/4]
The preferred angle of entry into the downwind leg is to
A. enter at a 45 degree bank to the downwind leg.
B. enter at a 45 degree angle to the downwind leg.
C. enter any way you want onto the downwind, just don't go against the flow of traffic.

38. [G15/3/2]
Which is the correct traffic pattern departure procedure to use at a noncontrolled airport?
A. Depart in any direction consistent with safety, after crossing the airport boundary.
B. Make all turns to the left.
C. Comply with any FAA traffic pattern established for the airport.

CTAF (Common Traffic Advisory Frequency)

39. [G16/3/2]
At many uncontrolled airports a frequency known as the ____ is available for pilots to communicate their intentions and receive information during takeoff or landing.
A. guard frequency (GF)
B. tower frequency (TF)
C. common traffic advisory frequency (CTAF)

40. [G16/3/2]
The Common Traffic Advisory Frequency (CTAF) may be a:
A. CB frequency.
B. unicom, multicom, tower or FSS frequency.
C. FAA frequency.

Using Unicom and Multicom for Information

41. [G17/Figure 30]
What is the recommended communication procedure when inbound to land at Shafter-Minter Airport in the figure below?
A. Broadcast intentions when 10 miles out on the CTAF/MULTICOM frequency, 122.9 MHz.
B. Contact UNICOM when 10 miles out on 122.8 MHz.
C. Circle the airport in a left turn prior to entering traffic.

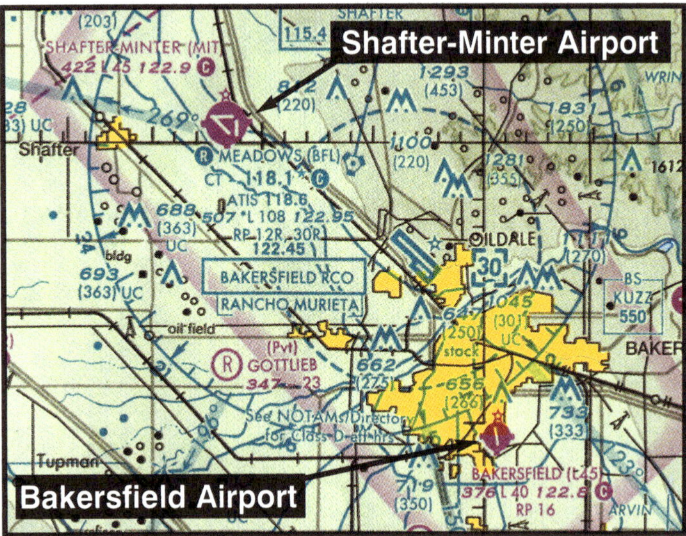

42. [G17/Figure 30]
What is the recommended communication procedure when inbound to land at Bakersfield airport above?
A. Broadcast intentions when 10 miles out on the CTAF/MULTICOM frequency, 122.9 MHz.
B. Contact UNICOM when 10 miles out on 122.8 MHz.
C. Circle the airport in a left turn prior to entering traffic.

43. [G18/1/5]
What is the recommended communication procedure when inbound to land at Paramount Farming airport below?
A. Transmit intentions on 122.75 or 122.85 MHz when 10 miles out and give position reports in the traffic pattern.
B. Contact UNICOM when 10 miles out on 122.8 MHz.
C. Circle the airport in a left turn prior to entering traffic.

44. [G18/2/1]
Prior to entering an Airport Advisory Area, a pilot should
A. monitor ATIS for weather and traffic advisories.
B. contact approach control for vectors to the traffic pattern.
C. contact the local FSS for airport and traffic advisories.

Finding Out What's Common

45. [G17/See *Finding Out What's Common*]
At a non privately-owned airport with no tower and no unicom frequency being used as the CTAF, you should tune your radio to _____. This frequency is referred to as multicom, because it's shared by many pilots at many places.
A. 122.9 MHz.
B. 123.0 MHz.
C. 123.6 MHz.

Automatic Terminal Information Service (ATIS)

46. [G18/3/2]
Automatic Terminal Information Service (ATIS) is the continuous broadcast of recorded information concerning
A. pilots of radar-identified aircraft whose aircraft is in dangerous proximity to terrain or to an obstruction.
B. non-essential information to reduce frequency congestion.
C. non-control information in selected high-activity terminal areas.

47. [G19/1/2]
When approaching an airport with an ATIS, you should listen to the ATIS broadcast when you're approximately ___ miles from the airport.
A. 25
B. 350
C. 10

48. [G19/1/2 & Figure 32]
The control tower frequency for Meadows airport (below) is
A. 118.1 MHz.
B. 118.6 MHz.
C. 122.95 MHz.

49. [G17/1/2 & Figure 32]
The CTAF frequency for Meadows airport (below) is
A. 118.1 MHz.
B. 118.6 MHz.
C. 122.95 MHz.

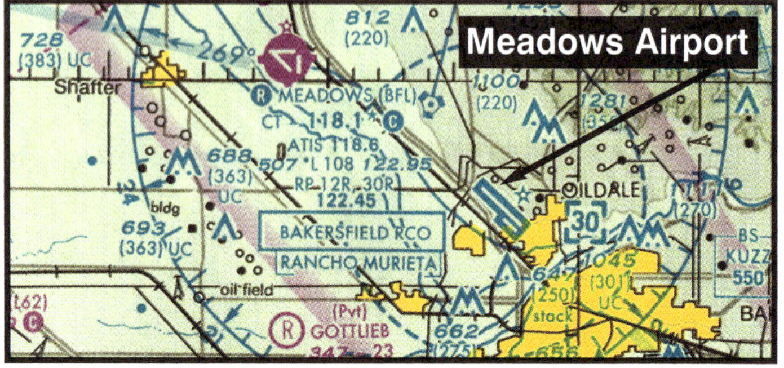

50. [G19/2/2]
ATIS broadcasts are updated when any _____ change occurs at the airport or upon the receipt of any official _____ weather information.
A. insignificant, new
B. significant, hourly or special
C. incredible, weekly

Pilot Control of Airport Lighting

51. [G20/3/1]
An asterisk (*) next to the lighting symbol (the "L") on the sectional chart may mean pilot control of airport lighting is _____ at that airport.
A. available, but lighting limitations may exist
B. never available
C. unusable

52. [G20/3/2]
To set the high intensity runway lights on medium intensity, the pilot should click the microphone seven times, then click it
A. one time.
B. three times.
C. five times.

53. [G20/3/2]
Once activated, pilot controlled lights will stay on for ____ minutes.
A. 20
B. 10
C. 15

54. [G21/2/3]
At some airports there are two small stroboscopic lights near the runway's threshold. These are called _____.
A. Jimmy Hendrix lights
B. Runway End Identifier Lights
C. the Approach Lights System

Visual Approach Slope Indicator (VASI)

55. [G22/1/2]
If, while on final approach to a runway equipped with a standard 2-bar VASI, you see red over white, then the aircraft is
A. above the glideslope.
B. below the glideslope.
C. on the glideslope.

56. [G22/3/2]
When approaching to land on a runway served by a VASI, the pilot shall
A. maintain an altitude that captures the glideslope at least 2 miles downwind from the runway threshold.
B. maintain an altitude at or above the glideslope.
C. remain on the glideslope & land between the two-light bar.

57. [G22/3/2]
Each pilot of an aircraft approaching to land on a runway served by a VASI shall
A. maintain a 3 degree glide to the runway.
B. maintain an altitude at or above the glideslope.
C. stay high until the runway can be reached in a power-off landing.

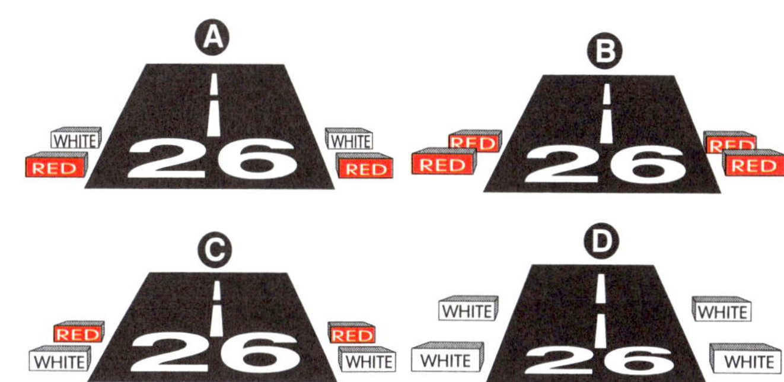

58. [G22/1/2]
VASI lights as shown by illustration D (above) indicate that the airplane is
A. off course to the left.
B. above the glideslope.
C. below the glideslope.

59. [G22/1/2]
VASI lights as shown by illustration B (above) indicate that the airplane is
A. below the glideslope.
B. on the glideslope.
C. above the glideslope.

60. [G22/1/2]
VASI lights as shown by illustration C (above) indicate that the airplane is
A. below the glideslope.
B. on the glideslope.
C. above the glideslope.

Precision Approach Path Indicator (PAPI)

61. [G23/1/2]
A below-glideslope indication from a precision approach path indicator (PAPI) is
A. four red lights.
B. three white lights and one red light.
C. two white lights and two red lights.

62. [G23/1/2]
A slightly high glideslope indication from a PAPI is
A. four white lights.
B. three white lights and one red light.
C. two white lights and two red lights.

Tricolor VASI

63. [G24/1/1]
An on-glideslope indication from a tri-color VASI is
A. a white light signal.
B. a green light signal.
C. an amber light signal.

64. [G24/1/1]
An above-glideslope indication from a tri-color VASI is
A. a white light signal.
B. a green light signal.
C. an amber light signal.

65. [G24/1/1]
A below-glideslope indication from a tri-color VASI is a
A. red light signal.
B. pink light signal.
C. green light signal.

Pulsating VASI Systems

66. [G24/2/1]
A below-glideslope indication from a pulsating approach slope indicator is a
A. pulsating white light.
B. steady white light.
C. pulsating red light.

Wake Turbulence

67. [G24/2/4 & G25/1/2]
Wingtip vortices are created only when an aircraft is
A. operating at high airspeeds.
B. heavily loaded.
C. developing lift.

68. [G24/3/3]
The greatest vortex strength occurs when the generating aircraft is
A. light, dirty, and fast.
B. heavy, dirty, and fast.
C. heavy, clean, and slow.

69. [G25/2/2 & G25/See Figure 44]
When landing behind a large aircraft, the pilot should avoid wake turbulence by staying
A. above the large aircraft's final approach path and landing beyond the large aircraft's touchdown point.
B. below the large aircraft's final approach path and landing before the large aircraft's touchdown point.
C. above the large aircraft's final approach path and landing before the large aircraft's touchdown point.

70. [G25/2/2 & G25/See Figure 44]
When departing behind a heavy aircraft, the pilot should avoid wake turbulence by maneuvering the aircraft
A. below and downwind from the heavy aircraft.
B. above and upwind from the heavy aircraft.
C. below and upwind from the heavy aircraft.

71. [G25/3/2]
When taking off or landing at an airport where heavy aircraft are operating, one should be particularly alert to the hazards of wingtip vortices because this turbulence tends to
A. rise from a crossing runway into the takeoff or landing path.
B. rise into the traffic pattern area surrounding the airport.
C. sink into the flightpath of aircraft operating below the aircraft generating the turbulence.

72. [G25/3/2]
Wingtip vortices created by large aircraft tend to
A. sink below the aircraft generating turbulence.
B. rise into the traffic pattern.
C. rise into the takeoff or landing path of a crossing runway.

73. [G26/See Figure 46]
The wind condition that requires maximum caution when avoiding wake turbulence on landing is a
A. light, quartering headwind.
B. light, quartering tailwind.
C. strong headwind.

ATC Wake Turbulence Separation Requirements

74. [G27/1/1]
For the purposes of wake turbulence separation minima, ATC classifies aircraft as _____ .
A. small, medium and large
B. heavy, large and small
C. big, bad and ugly

75. [G27/1/2]
Heavy aircraft are aircraft capable of takeoff weights of more than _____ pounds regardless of their weight during any particular phase of flight.
A. 255,000
B. 300,000
C. 12,500

76. [G27/1/2]
Large aircraft are aircraft of more than _____ pounds maximum certified takeoff weight up to _____ pounds.
A. 41,000, 255,000
B. 300,000, 500,000
C. 12,500, 14,000

77. [G27/1/2]
Small aircraft are aircraft of _____ pounds or less maximum certified takeoff weight.
A. 41,000
B. 300,000
C. 500,000

Taxiing in Crosswind Conditions

78. [G27/3/3]
Which aileron positions should a pilot generally use when taxiing in strong quartering headwinds?
A. Aileron up on the side from which the wind is blowing.
B. Aileron down on the side from which the wind is blowing.
C. Ailerons neutral.

79. [G27/3/3]
How should the flight controls be held while taxiing a tricycle-gear airplane into a left quartering headwind?
A. Left aileron up, elevator neutral.
B. Left aileron down, elevator neutral.
C. Left aileron up, elevator down.

80. [G27/3/3]
How should the flight controls be held while taxiing a tailwheel airplane into a right quartering headwind?
A. Right aileron up, elevator up.
B. Right aileron down, elevator neutral.
C. Right aileron up, elevator down.

81. [G28/1/2]
When taxiing with strong quartering tailwinds, which aileron positions should be used?
A. Aileron down on the downwind side.
B. Ailerons neutral.
C. Aileron down on the side from which the wind is blowing.

82. [G28/1/2]
Which wind condition would be most critical when taxiing a nosewheel equipped high wing airplane?
A. Quartering tailwind.
B. Direct crosswind.
C. Quartering headwind.

83. [G28/1/2]
How should the flight controls be held while taxiing a tricycle-gear airplane with a left quartering tailwind?
A. Left aileron up, elevator neutral.
B. Left aileron down, elevator down.
C. Left aileron up, elevator down.

84. [G28/1/2]
How should the flight controls be held while taxiing a tailwheel airplane with a left quartering tailwind?
A. Left aileron up, elevator neutral.
B. Left aileron down, elevator neutral.
C. Left aileron down, elevator down.

Postflight Briefing 7-1: Land & Hold Short Operations

85. [G28/See *Land and Hold Short Operations*]
A LAHSO clearance doesn't preclude a rejected landing (a go-around). If this is necessary, then go around, but maintain _____ from other traffic and promptly _____ the controller.
A. eyeball distance, give a thumbs up to
B. safe separation, notify
C. 1/2 mile, notify

86. [G28/See *Land and Hold Short Operations*]
It's very important that you give the controller a _____ of the LAHSO clearance. This lets the controller know that you know what's expected of you.
A. partial readback (call sign only)
B. written copy
C. full readback

87. [G28/See *Land and Hold Short Operations*]
Who should not participate in the Land and Hold Short Operations (LAHSO) program?
A. Recreational pilots only.
B. Student pilots.
C. Military pilots.

88. [G28/See *Land and Hold Short Operations*]
Who has final authority to accept or decline any land and hold short (LAHSO) clearance?
A. Owner/operator.
B. Pilot-in-command.
C. Second-in-command.

89. [G28/See *Land and Hold Short Operations*]
When should pilots decline a land and hold short (LAHSO) clearance?
A. When it will compromise safety.
B. Only when the tower operator concurs.
C. Pilots can not decline clearance.

90. [G28/See *Land and Hold Short Operations*]
Where is the "Available Landing Distance" (ALD) data published for an airport that utilizes Land and Hold Short Operations (LAHSO) published?
A. Aeronautical Information Manual (AIM).
B. 14 CFR Part 91, General Operating and Flight Rules.
C. Airport/Facility Directory (A/FD).

91. [G28/See *Land and Hold Short Operations*]
What is the minimum visibility for a pilot to receive a land and hold short (LAHSO) clearance?
A. 3 nautical miles.
B. 1 statute mile.
C. 3 statute miles.

Chapter Seven Answers

1. C
2. A
3. A/white, B/green, C/red
4. A
5. B
6. W/yellow, X/yellow, Y/yellow, Z/black
7. A/red, B/yellow, C/clearance, D/cross
8. A/safety, B/yellow, C/cross, D/clearance, E/holding
9. B
10. B
11. B
12. C
13. C
14. A
15. A
16. B
17. B
18. B
19. A/departure, B/crosswind, C/downwind, D/decision, E/base, F/final, G/upwind
20. A
21. B
22. C
23. A
24. A
25. C
26. southeast, east
27. A
28. B
29. A
30. C
31. A
32. B
33. C
34. C
35. A, C, D
36. I & J
37. B
38. C
39. C
40. B
41. A
42. B
43. A
44. C
45. A
46. C
47. A
48. A
49. A
50. B
51. A
52. C
53. C
54. B
55. C
56. B
57. B
58. B
59. A
60. B
61. A
62. B
63. B
64. C
65. A
66. C
67. C
68. C
69. A
70. B
71. C
72. A
73. B
74. B
75. A
76. A
77. A
78. A
79. A
80. A
81. C
82. A
83. B
84. C
85. B
86. C
87. B
88. B
89. A
90. C
91. C

Note: To ensure that you have the most current answers to these questions, please check the *Book & Slide Updates* section at Rod Machado's web site: www.rodmachado.com

Chapter Eight

Radio Operations: Aviation Spoken Here

Radio Technique

1. [H2/1/1]
Using the radio is no great mystery. When transmitting, hold the radio close to your _____.
A. mouth
B. ear
C. tongue

VHF Transmissions

2. [H3/1/4]
Aviation transmissions are usually on the _____ portion of the radio spectrum.
A. very high frequency (VHF)
B. ultra high frequency (UHF)
C. very long frequency (VLF)

3. [H3/2/1]
The downside of VHF is that the range is limited by _____.
A. static
B. lunar interference
C. line of sight

4. [H3/2/2]
The Federal Communications Commission (FCC) assigns frequencies ranging from _____ megahertz (MHz) to _____ MHz for aviation use.
A. 200, 850
B. 118.0, 135.975
C. 119.7, 149.325

Talking the Talk

5. [H5/3/4]
The correct method of stating 4,500 feet MSL to ATC is
A. "FOUR THOUSAND FIVE HUNDRED."
B. "FOUR POINT FIVE."
C. "FORTY-FIVE HUNDRED FEET MSL."

6. [H5/3/5]
The correct method of stating 10,500 feet MSL to ATC is
A. "TEN THOUSAND, FIVE HUNDRED FEET."
B. "TEN POINT FIVE."
C. "ONE ZERO THOUSAND, FIVE HUNDRED."

7. [H5/See *Frequent Flyer Frequencies*] Fill in the blanks with the items, services or facilities covered by the following frequencies:

118.0 to 135.975 MHz - ______________________________

121.5 MHz - ______________________________

121.6 to 121.9 MHz - ______________________________

122.0 MHz - ______________________________

122.1 MHz - ______________________________

122.2 MHz - ______________________________

122.7 MHz - ______________________________

122.725 MHz - ______________________________

122.75 MHz - ______________________________

122.8 MHz - ______________________________

122.85 MHz - ______________________________

122.9 MHz - ______________________________

122.925 MHz - ______________________________

122.95 MHz - ______________________________

122.975 MHz - ______________________________

123.0 MHz - ______________________________

123.025 MHz - ______________________________

123.05 & 123.075 MHz - ______________________

Controlled Airports

8. [H7/1/3]
After landing at a tower controlled airport, when should the pilot contact ground control?
A. When advised by the tower to do so.
B. Prior to turning off the runway.
C. After reaching a taxiway that leads directly to the parking area.

Control Tower Communications

9. [H7/Figure 7]
The control tower frequency for General Fox Field (below) is
A. 122.95 MHz.
B. 120.3 MHz.
C. 133.875 MHz.

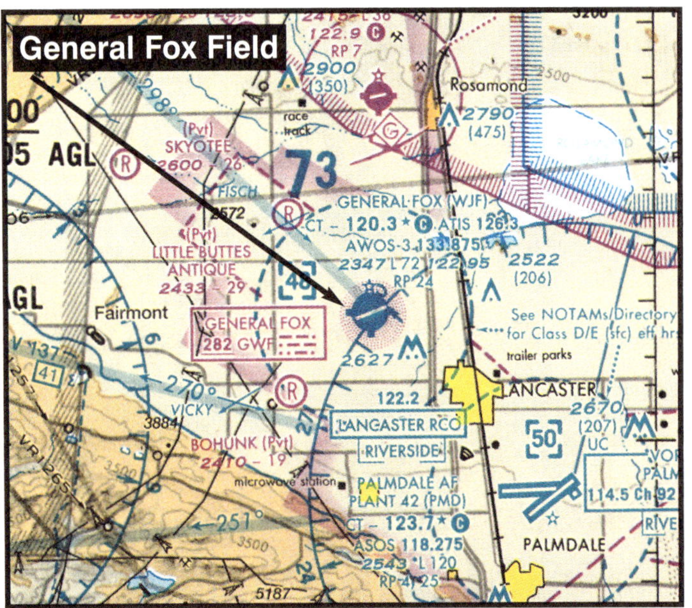

10. [G3 Insert & H8/1/3]
When flying Cessna N2132B, the proper phraseology for initial contact with Tulane FSS is
A. "TULANE RADIO, CESSNA TWO ONE THREE TWO BRAVO, RECEIVING MOTOWN VORTAC, OVER."
B. "TULANE STATION, CESSNA TWO ONE THREE TWO BEE, RECEIVING MOTOWN VORTAC, OVER."
C. "TULANE FLIGHT SERVICE STATION, CESSNA NOVEMBER THREE TWO BRAVO, RECEIVING MOTOWN VORTAC, OVER."

Flight Service Station Frequencies

11. [H8/2/1]
A heavy-lined VOR frequency box around the Riverside VOR frequency box (shown below) indicates what two frequencies are available and monitored?
A. 122.0 MHz & 121.5 MHz
B. 122.95 MHz & 121.5 MHz
C. 122.2 MHz & 121.5 MHz

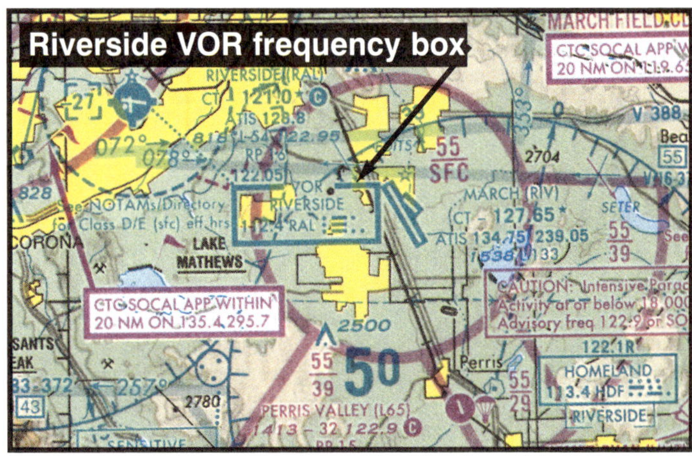

12. [H8/Figure 10]
What frequencies are available for communicating with the Riverside FSS (refer to the figure below)?
A. 122.05 MHz, 122.0 MHz, 121.5 MHz
B. 112.4 MHz, 122.2 MHz, 121.5 MHz
C. 122.05 MHz, 122.2 MHz, 121.5 MHz

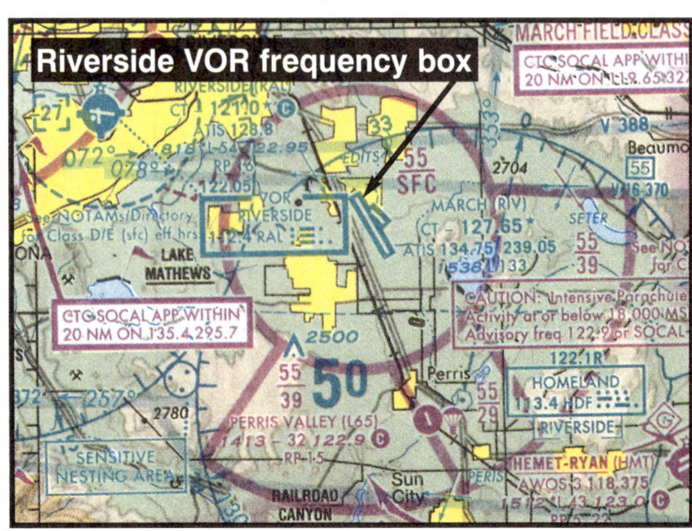

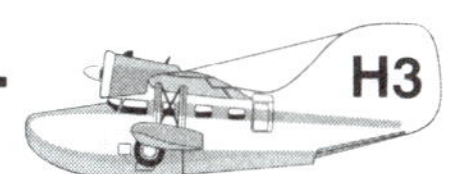

13. [H9/1/1]
Referring to the figure below, on what frequency could you contact Riverside FSS if you're in the vicinity of Hector VOR?
A. Transmit on 122.1 MHz, listen on 112.7 MHz.
B. Transmit on 110.2 MHz, listen on 122.1 MHz.
C. Transmit on Channel 39 MHz, listen on 110.2 MHz.

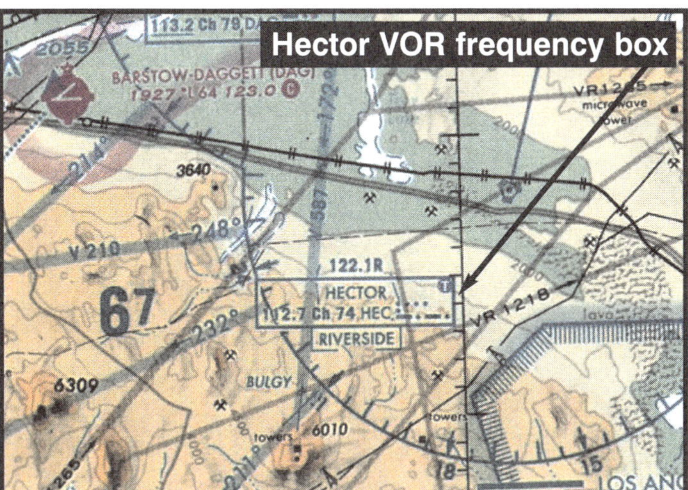

14. [H9/Figure 13]
A line underneath the Daggett VOR frequency as shown below indicates no _____ capability exists on that frequency.
A. Morse code
B. sound
C. voice

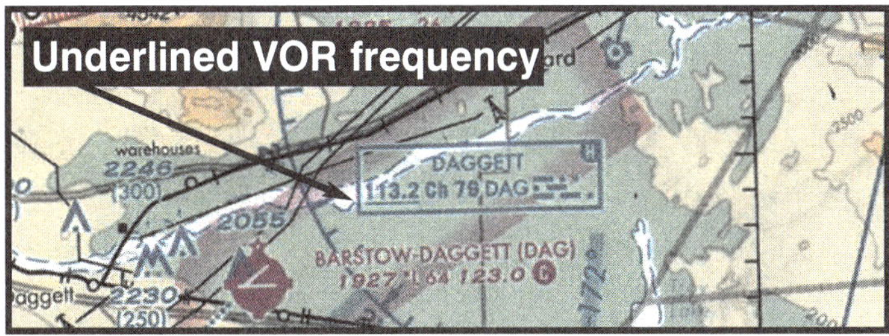

15. [H10/Figure 15]
Referring to the figure below, what frequency could you use to contact the Riverside FSS when flying in the vicinity of Barstow-Daggett airport?
A. 113.2
B. 122.2
C. 123.0

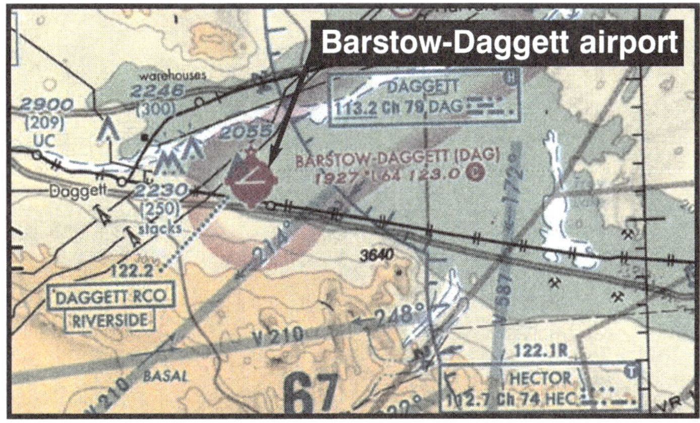

HIWAS and TWEB

16. [H10/2/3]
Referring to the figure below, on what frequency can a pilot receive Hazardous Inflight Weather Advisory Service (HIWAS) in the vicinity of Daggett VOR?
A. 122.7 MHz
B. 113.2 MHz
C. 123.0 MHz

17. [H10/3/2]
HIWAS is a continuous broadcast of in-flight _____ advisories.
A. FAA inspector location
B. weather
C. Notam

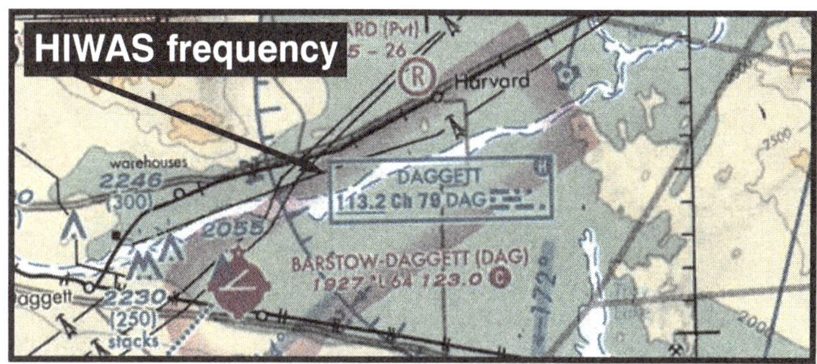

18. [H11/1/3]
TWEB's are recorded, repeating broadcasts over selected low frequency or VOR navigational stations. These broadcasts contain _____ information for routes at or near those navigational aids.
A. Notam
B. weather
C. ramp check

19. [H11/1/3]
Referring to the figure below, on what frequency can a pilot receive a TWEB broadcast in the vicinity of Hector VOR?
A. 112.7 MHz
B. 123.0 MHz
C. 122.2 MHz

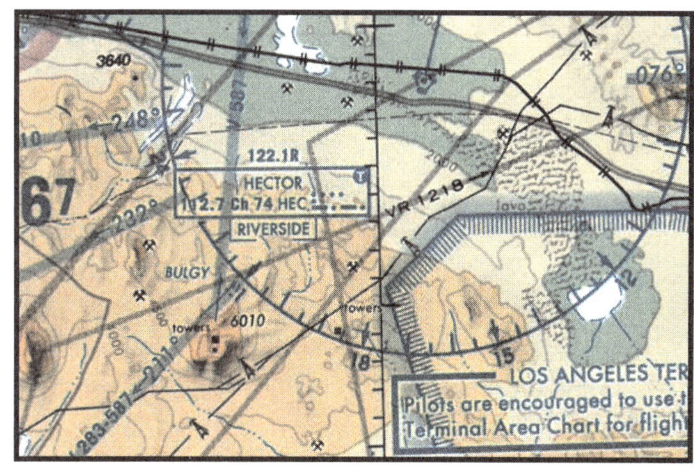

The Emergency Frequency

20. [H11/2/1]
The international emergency frequency is _____.
A. 121.5 MHz
B. 122.5 MHz
C. 122.2 MHz

The Airport/Facility Directory

EAGLE CO REGIONAL (EGE) 4 W UTC−7(−6DT) N39°38.55′ W106°55.06′ **DENVER**
6535 B S4 **FUEL** 100, 100LL, JET A1, JET A1 + OX 1, 3 ARFF Index C **H–2C, L–5D, 6E, 8F**
RWY 07–25: H8000X150 (ASPH—GRVL) S–60, D–115 MIRL **IAP**
RWY 07: REIL. Tree. Rgt tfc. 0.9% up. **RWY 25:** MALSF. PAPI(P4L)—GA 3.0° TCH 45′. 1.1% down.
AIRPORT REMARKS: Attended 1400–0200Z‡. CLOSED to unscheduled air carrier operations with more than 30 passenger seats except PPR call arpt manager 970–524–9490. High unmarked terrain all quadrants. Ngt ops discouraged to pilots unfamiliar with arpt. Rwy 07 mountain top 10:1 clearance 12000′ from thld 1500′ left of rwy centerline extended. Recommend all acft departing Rwy 25 initiate a left turn as soon as altitude and safety permit to avoid high terrain. Extensive military helicopter training operations surface to 1000′ AGL within 25 NM radius Eagle Co Arpt 1330–0500Z‡. No snow removal at nights. Rwy 25 PAPI only visible to 6° left of centerline due to terrain. After 0300Z‡ ACTIVATE MALSF Rwy 25 MIRL Rwy 07–25, PAPI Rwy 25 and REIL Rwy 07—CTAF.
WEATHER DATA SOURCES: AWOS–3 135.575 (970) 524–7386.
COMMUNICATIONS: CTAF 118.2 **UNICOM** 122.95
DENVER FSS (DEN) TF 1–800–WX–BRIEF. NOTAM FILE EGE.
RCO 122.2 (DENVER FSS)
DENVER CENTER APP/DEP CON 134.5
TOWER 118.2 FCT (1400–0200Z‡) VFR only. **GND CON** 121.8
AIRSPACE: CLASS D svc 1400–0200Z‡ other times CLASS E.
RADIO AIDS TO NAVIGATION: NOTAM FILE DEN.
SNOW (L) VORW/DME 109.2 SXW Chan 29 N39°37.77′ W106°59.47′ 065° 3.5 NM to fld. 8060/12E. Unmonitored.
ILS/DME 110.1 I–EGE Chan 38 Rwy 25 (LOC only). LOC/DME unmonitored.

21. [H12/Figure 20]
Traffic patterns in effect at Eagle County airport are
A. to the right on Runway 7 and Runway 25.
B. to the left on Runway 7 and to the left on Runway 25.
C. to the right on Runway 7 and to the left on Runway 25.

22. [H12/Figure 20]
Where is Eagle County airport located with relation to the city?
A. west approximately 4 miles.
B. east approximately 4 miles.
C. east approximately 10 miles.

23. [H12/Figure 20 G17/1/2]
What is the recommended communications procedure for landing at Eagle County airport during the hours when the tower is not in operation?
A. monitor airport traffic and announce your position and intentions on 118.2 MHz.
B. contact UNICOM on 122.95 MHz for traffic advisories.
C. monitor ATIS for airport conditions, then announce your position on 122.95 MHz.

Radar and the ATC System

24. [H13/1/2]
Transponders allow airplanes to be identified on radar via the _____ radar beacon system.
A. secondary
B. rotating
C. primary

Transponders

25. [H13/2/3]
Unless otherwise authorized, if flying a transponder equipped aircraft, a pilot should squawk which VFR code?
A. 1200
B. 7600
C. 7700

26. [H14/1/2 & I3/Figure 1]
When operating under VFR below 18,000 feet MSL, unless otherwise authorized, what transponder code should be selected?
A. 1200
B. 7600
C. 7700

27. [H14/2/3]
If you have an emergency, what code should you squawk?
A. 1200
B. 7600
C. 7700

28. [H15/2/2]
If you are being hijacked in your Cessna 152, what code should you squawk?
A. 7500
B. 7600
C. 7700

29. [H16/1/1]
If you have a loss of radio communication, what code should you squawk?
A. 1200
B. 7600
C. 7700

30. [H16/1/2]
When making routine transponder code changes, pilots should avoid inadvertent selection of which codes?
A. 0700, 1700, 7000
B. 1200, 1500, 7000
C. 7500, 7600, 7700

31. [H15/See *Airborne Cowboy: Riding a DF Steer*]
The letters VHF/DF appearing in the *Airport/Facility Directory* for a certain airport indicate that
A. this airport is designated as an airport of entry.
B. the Flight Service Station has equipment with which to determine your direction from the station.
C. this airport has a direct-line phone to the Flight Service Station.

32. [H15/See *Airborne Cowboy: Riding a DF Steer*]
To use VHF/DF facilities for assistance in locating an aircraft's position, the aircraft must have a
A. VHF transmitter and receiver.
B. 4096-code transponder.
C. VOR receiver and DME.

33. [H15/See *Airborne Cowboy: Riding a DF Steer*]
When using a DF steer, the FSS specialist will have you _____ your microphone for a given length of time.
A. elevate
B. key (depress the transmit key)
C. mouth

Radar Services for Pilots

34. [H16/3/3]
An ATC radar facility issues the following advisory to a pilot flying on a heading of 090 degrees:

"TRAFFIC 3 O'CLOCK, 2 MILES, WESTBOUND..."

Where should the pilot look for this traffic?
A. East.
B. South.
C. West.

35. [H16/3/3]
An ATC radar facility issues the following advisory to a pilot flying on a heading of 360 degrees:

"TRAFFIC 10 O'CLOCK, 2 MILES, SOUTHBOUND..."

Where should the pilot look for this traffic?
A. Northwest.
B. Northeast.
C. Southwest.

36. [H16/3/3]
An ATC radar facility issues the following advisory to a pilot during a local flight:

"TRAFFIC 2 O'CLOCK, 5 MILES, NORTHBOUND..."

Where should the pilot look for this traffic?
A. Between directly ahead and 90 degrees to the left.
B. Between directly behind and 90 degrees to the right.
C. Between directly ahead and 90 degrees to the right.

37. [H16/3/3]
An ATC radar facility issues the following advisory to a pilot flying north in a calm wind:

"TRAFFIC 9 O'CLOCK, 2 MILES, SOUTHBOUND..."

Where should the pilot look for this traffic?
A. South.
B. North.
C. West.

Radar Assistance to VFR Aircraft Basic Radar Service

38. [H17/2/3]
Basic radar service in the terminal radar program is best described as
A. safety alerts, traffic advisories and limited vectoring to VFR aircraft.
B. mandatory radar service provided by the Automated Radar Terminal System (ARTS) program.
C. wind shear warning at participating airports.

39. [H17/2/4]
From whom should a departing VFR aircraft request radar traffic information during ground operations?
A. Clearance delivery.
B. Tower, just before takeoff.
C. Ground control, on initial contact.

Radar Services for Pilots

40. [H18/3/2]
TRSA Service provides
A. IFR separation (1,000 feet vertical and 3 miles lateral) between all aircraft.
B. warning to pilots when their aircraft are in unsafe proximity to terrain, obstructions, or other aircraft.
C. sequencing and separation for participating VFR aircraft.

41. [H19/3/1]
If Air Traffic Control advises that radar service is terminated when the pilot is departing Class C airspace, the transponder should be set to code
A. 0000
B. 1200
C. 4096

Class C Service

42. [H19/3/2]
Class C service provides _____ service, _____ between all IFR and VFR aircraft, and _____ of the VFR arrivals to the primary airport .
A. vector, sightseeing, sequencing
B. basic radar, 1,000 foot separation, sequencing
C. basic radar, separation, sequencing

Class B Service

43. [H19/3/3]
Class B service provides _____ service, _____ between IFR and VFR aircraft, and _____ of VFR arrivals to the primary airport.
A. vector, sightseeing, sequencing
B. basic radar, 1,000 foot separation, sequencing
C. basic radar, separation, sequencing

Clearance Delivery

44. [H20/1/1]
What publication could you look at to determine what stage of radar service is available at an airport?
A. Airport/Facility Directory.
B. Aeronautical Information Manual.
C. Federal Aviation Regulations.

45. [H20/3/1]
Who would you contact if you wanted to obtain a clearance to depart the primary airport lying within Class B airspace?
A. Ground Control.
B. Flight Service Station.
C. Clearance Delivery.

Chapter Eight Answers

1. A
2. A
3. C
4. B
5. A
6. C
7.
118.0 to 135.975 MHz - Air Traffic Control general frequency range
121.5 MHz - Emergency frequency and ELT's
121.6 to 121.9 MHz - Ground Control at tower airports
122.0 MHz - FSS Enroute Flight Advisory Service (Flight Watch)
122.1 MHz - FSS receive-only and transmit other frequency
122.2 MHz - FSS Common enroute frequency
122.7 MHz - Unicom at airports without an operating control tower
122.725 MHz - Unicom at airports without an operating control tower
122.75 MHz - Air-to-air communications and private airports not open to public
122.8 MHz - Unicom at airports without an operating control tower
122.85 MHz - Air to air; private airports not open to public
122.9 MHz - Multicom; activities of a temporary, seasonal, emergency nature or search and rescue, as well as airports with no tower, FSS or Unicom
122.925 MHz - Multicom, forestry management and fire suppression, fish and game management and protection and environmental monitoring and protection
122.95 MHz - Unicom at airports with a control tower or FSS on field
122.975 MHz - Unicom at airports without an operating control tower
123.0 MHz - Unicom at airports without an operating control tower
123.025 MHz - Air-to-air communication general aviation helicopters
123.05 & 123.075 MHz - Unicom at airports without an operating control tower
8. A
9. B
10. A
11. C
12. C
13. A
14. C
15. B
16. B
17. B
18. B
19. A
20. A
21. C
22. A
23. A
24. A
25. A
26. A
27. C
28. A
29. B
30. C
31. B
32. A
33. B
34. B
35. A
36. C
37. C
38. A
39. C
40. C
41. B
42. C
43. C
44. A
45. C

Note: To ensure that you have the most current answers to these questions, please check the *Book & Slide Updates* section at Rod Machado's web site: www.rodmachado.com

Chapter Nine

Airspace: The Wild Blue, Green & Red Yonder

Eye See

1. [I2/1/2]
Any time you are flying, you will be operating under one of two primary sets of rules: ________ or_________.
A. instrument flight rules (IFR), Part 121 rules
B. visual flight rules (VFR), instrument flight rules (IFR)
C. visual flight rules (VFR), part 141 flight rules

Controlled and Uncontrolled Airspace

2. [I3/1/1]
Two basic types of airspace exist in the United States: _____ and _____.
A. controlled, uncontrolled
B. Class E, uncontrolled
C. controlled, special use

The Big Picture

3. [I3/3/3]
Class A, B, C, D and E is _____ airspace. Class G is _____ airspace.
A. uncontrolled, purple
B. controlled, VFR only
C. controlled, uncontrolled

Class A Airspace

4. [I4/1/1]
Class A airspace begins at what altitude?
A. 14,500 feet MSL.
B. 18,000 feet MSL.
C. 29,920 feet MSL.

5. [I4/1/2]
In which type of airspace is VFR flight prohibited?
A. Class A.
B. Class B.
C. Class C.

6. [I4/1/3]
At what altitude shall the altimeter be set to 29.92, when climbing to cruising flight level?
A. 14,500 feet MSL.
B. 18,000 feet MSL.
C. 24,000 feet MSL.

Class E at and Above 10,000 Feet MSL

7. [I5/1/2]
The airspace lying directly below Class A airspace is Class _____ airspace.
A. E
B. F
C. G

8. [I5/1/3]
Class E airspace is _____ airspace.
A. uncontrolled
B. controlled
C. IFR only

9. [I5/3/1&2]
Class E airspace generally begins at _____ AGL and sometimes begins at a lower altitude of _____ AGL.
A. 700, 200
B. 14,500, 1,200
C. 1,200, 700

Class E at and Above 10,000 Feet MSL

10. [I6/1/1]
For VFR flight operations above 10,000 feet MSL and more than 1,200 feet AGL, the minimum horizontal distance from clouds required is
A. 1,000 feet.
B. 2,000 feet.
C. 1 mile.

11. [I6/1/1]
For VFR flight operations above 10,000 feet MSL and more than 1,200 feet AGL, the minimum required vertical distance above a cloud is
A. 1,000 feet.
B. 2,000 feet.
C. 1 mile.

12. [I6/1/1]
For VFR flight operations above 10,000 feet MSL and more than 1,200 feet AGL, the minimum required vertical distance below a cloud is
A. 1,000 feet.
B. 2,000 feet.
C. 1 mile.

13. [I6/1/1]
For VFR flight operations above 10,000 feet MSL and more than 1,200 feet AGL, the minimum required flight visibility is
A. 5 miles.
B. 2,000 feet.
C. 1 mile.

Class E Below 10,000 Feet MSL

14. [I7/2/1]
VFR flight in controlled airspace above 1,200 feet AGL and below 10,000 feet MSL requires a minimum visibility and vertical cloud clearance of
A. 3 miles, and 500 feet below or 1,000 feet above the clouds in controlled airspace.
B. 5 miles, and 1,000 feet below or 1,000 feet above the clouds at all altitudes.
C. 5 miles, and 1,000 feet below or 1,000 feet above the clouds only in Class A airspace.

15. [I7/2/1]
The minimum distance from clouds required for VFR operations on an airway (most airways begin at 1,200 feet AGL) below 10,000 feet MSL is
A. remain clear of clouds.
B. 500 feet below, 1,000 feet above, and 2,000 feet horizontally.
C. 500 feet above, 1,000 feet below, and 2,000 feet horizontally.

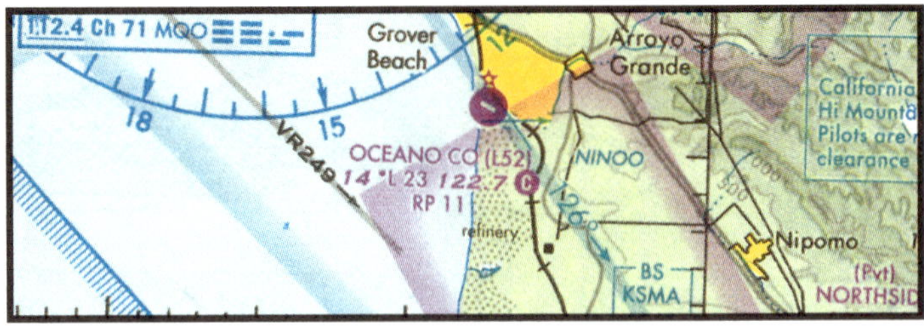

16. [I7/2/1]
What is the minimum flight visibility and cloud clearance requirement after departing Oceano airport (see above) at 2,000 feet AGL while heading in a northwesterly direction?
A. 3 miles, and 500 feet below or 1,000 feet above the clouds in controlled airspace.
B. 5 miles, and 1,000 feet below or 1,000 feet above the clouds at all altitudes.
C. 5 miles, and 1,000 feet below or 1,000 feet above the clouds only in Class A airspace.

17. [I7/2/1]
What minimum flight visibility is required for VFR flight operations on an airway below 10,000 feet MSL?
A. 1 mile.
B. 3 miles.
C. 4 miles.

18. [I7/2/1]
During operations within controlled airspace at altitudes of less than 1,200 feet AGL, the minimum horizontal distance from clouds requirement for VFR flight is
A. 1,000 feet.
B. 1,500 feet.
C. 2,000 feet.

Class E Airspace Starting at 700 Feet AGL

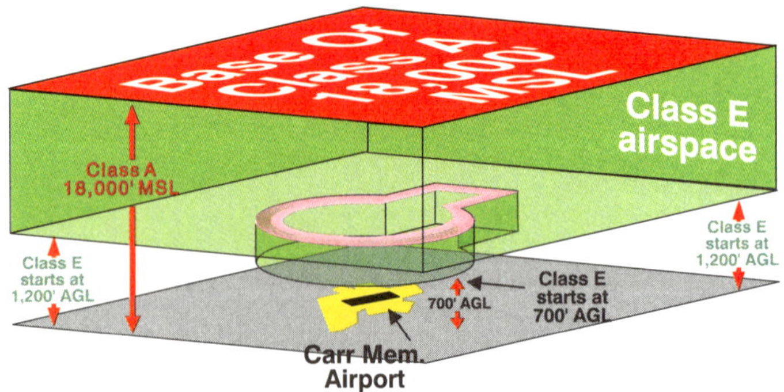

19. [I8/1/2]
The visibility and cloud clearance requirements to operate VFR during daylight hours in the vicinity of Carr Memorial airport (see the figure above) between 1,200 feet AGL and 10,000 feet MSL are
A. 1 mile and clear of clouds.
B. 1 mile and 1,000 feet above, 500 feet below, and 2,000 feet horizontally from clouds.
C. 3 miles and 1,000 feet above, 500 feet below, and 2,000 feet horizontally from clouds.

20. [I8/1/2]

The visibility and cloud clearance requirements to operate VFR during daylight hours over Carr Memorial airport (see figure bottom right previous page) at more than 700 feet AGL are

A. 1 mile and clear of clouds.
B. 1 mile and 1,000 feet above, 500 feet below, and 2,000 feet horizontally from each cloud.
C. 3 miles and 1,000 feet above, 500 feet below, and 2,000 feet horizontally from each cloud.

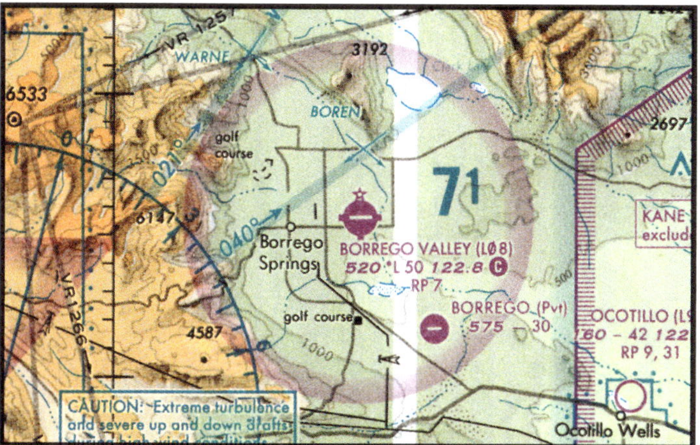

21. [I8/1/2]

The visibility and cloud clearance requirements to operate VFR during daylight hours over Borrego Valley airport (in the figure above) at more than 1,500 feet AGL are

A. 1 mile and clear of clouds.
B. 3 miles and 1,000 feet above, 500 feet below, and 2,000 feet horizontally from each cloud.
C. 1 mile and 1,000 feet above, 500 feet below, and 2,000 feet horizontally from each cloud.

22. [I9/1/3]

Referring to the figure above, the visibility and cloud clearance requirements to operate VFR during daylight hours in the Class E extension (point Z) of San Luis Obispo's airport at more than 700 feet AGL are

A. 1 mile and clear of clouds.
B. 1 mile and 1,000 feet above, 500 feet below, and 2,000 feet horizontally from each cloud.
C. 3 miles and 1,000 feet above, 500 feet below, and 2,000 feet horizontally from each cloud.

Additional Requirements in Surface-Based Controlled Airspace

23. [I9/1/5]

When operating at an airport having any type of surface-based controlled airspace established for it, the reported ground visibility at the airport must be at least _____ statute mile(s).

A. five
B. one
C. three

24. [I9/1/5]

If the ground visibility isn't reported in surface-based controlled airspace, then the flight visibility during takeoff, landing or when operating in the traffic pattern must be at least _____ statute miles.

A. three
B. five
C. one

25. [I9/1/6]

When operating at an airport having any type of surface-based controlled airspace established for it, the ceiling at the airport can be no lower than _____ AGL.

A. 1,000 feet
B. 3,000 feet
C. 500 feet

26. [I9/2/2]

For aviation purposes, ceiling is defined as the height above the earth's surface of the

A. lowest reported obscuration and the highest layer of clouds reported as overcast.
B. lowest broken or overcast layer or vertical visibility into an obscuration.
C. lowest layer of clouds reported as scattered, broken, or thin.

27. [I9/3/2]

To depart Desert Resorts airport in the figure shown above, what minimum visibility and ceiling must exist at the airport (assume Class E, surface-based airspace is active)?

A. 500 foot ceiling, 3 miles visibility.
B. 1,000 foot ceiling, 3 miles visibility.
C. 1,000 foot ceiling, 1 mile visibility.

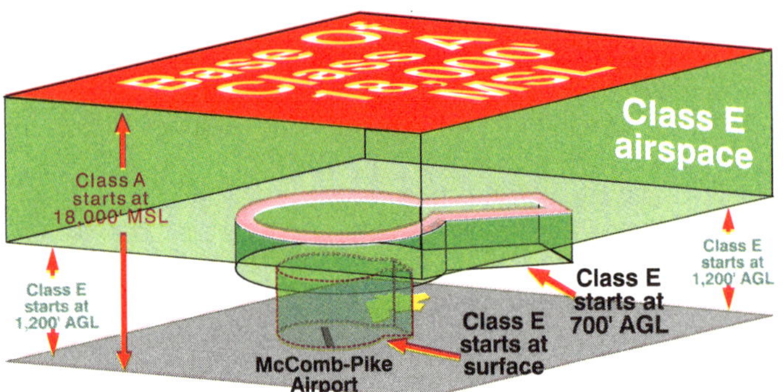

28. [I9/3/2]
To depart McComb-Pike airport in the figure shown above, what minimum ceiling must exist at the airport? (Assume Class E, surface-based airspace is active.)
A. 500 foot ceiling.
B. 1,000 foot ceiling.
C. 1,500 foot ceiling.

Special VFR Clearance

29. [I11/1/2]
A SVFR clearance allows you to operate below _____ feet MSL down to the surface, within the _____ boundaries of surface-based controlled airspace
A. 10,000, lateral
B. 1,200, lateral
C. 14,500, 10 mile

30. [I11/1/3]
To fly SVFR, a pilot must have:
A. a clearance from ATC.
B. a 1,000 foot ceiling.
C. five miles of flight visibility.

31. [I11/1/4]
The official weather observer's hours of operation normally coincide with the hours during which the surface-based controlled airspace exists. The actual hours that surface-based controlled airspace exists are depicted in the _____.
A. AIM (Aeronautical Information Manual)
B. A/FD (Airport/Facility Directory)
C. Federal Aviation Regulations

32. [I11/1/2, 4 & 5]
A special VFR clearance authorizes the pilot of an aircraft to operate VFR while within Class E (or Class B, C, and D) airspace when the visibility is
A. less than 1 mile and the ceiling is less than 1,000 feet.
B. at least 1 mile and the aircraft can remain clear of clouds.
C. at least 3 miles and the aircraft can remain clear of clouds.

33. [I11/1/5]
What is the minimum weather condition required for airplanes operating under special VFR in Class E (or Class B, C, and D) airspace?
A. 1 mile flight visibility.
B. 1 mile flight visibility and 1,000-foot ceiling.
C. 3 miles flight visibility and 1,000-foot ceiling.

Obtaining a SVFR Clearance

34. [I11/3/4 & I12/1/1]
What ATC facility should the pilot contact to receive a special VFR departure clearance in Class D airspace?
A. Automated Flight Service Station.
B. Air Traffic Control Tower.
C. Air Route Traffic Control Center.

35. [I12/2/3]
No person may operate an airplane within Class D airspace at night under special VFR unless the
A. flight can be conducted 500 feet below the clouds.
B. airplane is equipped for instrument flight.
C. flight visibility is at least 3 miles.

36. [I12/2/3]
What are the minimum requirements for airplane operations under special VFR in Class D airspace at night?
A. The airplane must be under radar surveillance at all times while in Class D airspace.
B. The airplane must be equipped for IFR with an altitude reporting transponder.
C. The pilot must be instrument rated, and the airplane must be IFR equipped.

Satellite Airports Lying Within the Primary Airport's Surface-Based Controlled Airspace

37. [I12/2/4]
Assume that you are operating at a satellite airport (Port Angeles in the figure above) located within another airport's surface-based controlled airspace (Fairchild's surface-based, Class E airspace above). In this instance, visibility is determined by pilots on the _____ but ceilings are determined by the _____ at the primary airport.
A. honor system, pilot
B. official weather observer, pilot
C. honor system, official weather observer

Class G Airspace

38. [I14/3/3]
What minimum visibility and clearance from clouds are required for VFR operations in Class G airspace at 700 feet AGL or lower during daylight hours?
A. 1 mile visibility and clear of clouds.
B. 1 mile visibility, 500 feet below, 1,000 feet above, and 2,000 feet horizontal clearance from clouds.
C. 3 miles visibility and clear of clouds.

39. [I14/3/3]
What minimum visibility and clearance from clouds are required in Class G airspace at 1,200 feet AGL or below during daylight hours?
A. 1 mile visibility and clear of clouds.
B. 3 miles visibility and clear of clouds.
C. 3 miles visibility, 500 feet below the clouds.

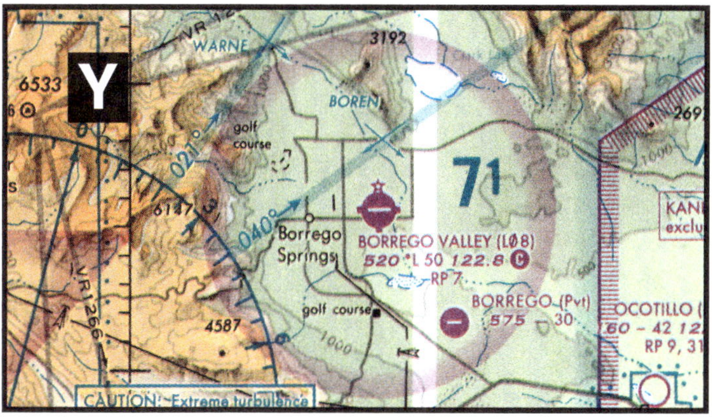

40. [I14/Figure 16]
Referring to the figure above, the airspace overlying Borrego Valley airport is uncontrolled from the surface to
A. 700 feet AGL.
B. 1,700 feet MSL.
C. 4,000 feet AGL.

41. [I14/Figure 17]
Referring to the figure above, the visibility and cloud clearance requirements to operate VFR during daylight hours directly over Borrego Valley airport at less than 700 feet AGL are
A. 1 mile and clear of clouds.
B. 1 mile and 1,000 feet above, 500 feet below, and 2,000 feet horizontally from each cloud.
C. 3 miles and 1,000 feet above, 500 feet below, and 2,000 feet horizontally from each cloud.

42. [I14/Figure 16]
Referring to the figure above, the airspace directly over position "Y" near Borrego Valley airport is uncontrolled from the surface to
A. 700 feet AGL.
B. 1,700 feet MSL.
C. 1,200 feet AGL.

43. [I15/1/2]
Referring to the figure above, if the flight visibility is only one statute mile, what is the maximum height AGL you can fly when departing Oceano airport to the northwest?
A. 1,500 feet AGL.
B. 1,200 feet AGL.
C. 700 feet AGL.

Night Operations in Class G Airspace at 1,200 Feet AGL and Below

44. [I15/2/2]
Outside controlled airspace, the minimum flight visibility requirement for VFR flight at and below 1,200 feet AGL at night is
A. 1 mile.
B. 3 miles.
C. 5 miles.

45. [I15/3/2]
One exception to nighttime minimum visibility of three miles for airplanes operating in Class G airspace occurs when operating in the _____.
A. vicinity of the airport
B. traffic pattern
C. view of a tower controller

46. [I15/3/2]
At night, in Class G airspace, if the flight visibility is less than three statute miles but not less than one statute mile during night hours, an airplane may be operated clear of clouds if it is flown in the airport traffic pattern within _____ of the runway.
A. two miles
B. gliding distance
C. one-half mile

Operations in Class G Airspace Above 1,200 Feet AGL

47. [I17/Figure 21]
Outside controlled airspace, the minimum flight visibility requirement for VFR flight above 1,200 feet AGL and below 10,000 feet MSL during daylight hours is
A. 1 mile.
B. 3 miles.
C. 5 miles.

48. [I17/Figure 21]
During operations at altitudes of more than 1,200 feet AGL and at or above 10,000 feet MSL, the minimum distance above clouds requirement for VFR flight is
A. 500 feet.
B. 1,000 feet.
C. 1,500 feet.

49. [I17/Figure 21]
During operations outside controlled airspace at altitudes of more than 1,200 feet AGL, but less than 10,000 feet MSL, the minimum distance below clouds requirement for VFR flight at night is
A. 500 feet.
B. 1,000 feet.
C. 1,500 feet.

50. [I17/Figure 21]
During operations within controlled airspace at altitudes of more than 1,200 feet AGL, but less than 10,000 feet MSL, the minimum distance above clouds requirement for VFR flight is
A. 500 feet.
B. 1,000 feet.
C. 1,500 feet.

51. [I17/Figure 21]
Outside controlled airspace, the minimum flight visibility requirement for VFR flight above 1,200 feet AGL and below 10,000 feet MSL during daylight hours is
A. 1 mile.
B. 3 miles.
C. 5 miles.

52. [I17/Figure 21]
During operations outside controlled airspace at altitudes of more than 1,200 feet AGL, but less than 10,000 feet MSL, the minimum flight visibility for VFR flight at night is
A. 1 mile.
B. 3 miles.
C. 5 miles.

53. [I17/Figure 21]
During operations outside controlled airspace at altitudes of more than 1,200 feet AGL, but less than 10,000 feet MSL, the minimum visibility and distance from clouds for VFR flight at night is
A. 1 mile and clear of clouds.
B. 1 mile and 1,000 feet above, 500 feet below, and 2,000 feet horizontally from each cloud.
C. 3 miles and 1,000 feet above, 500 feet below, and 2,000 feet horizontally from each cloud.

54. [I17/Figure 21]
During operations outside controlled airspace at altitudes of more than 1,200 feet AGL and above 10,000 feet MSL, the minimum visibility and distance from clouds for VFR flight at night is
A. 5 miles and clear of clouds.
B. 1 mile and 1,000 feet above, 500 feet below, and 2,000 feet horizontally from each cloud.
C. 5 miles and 1,000 feet above, 1,000 feet below, and 1 mile horizontally from each cloud.

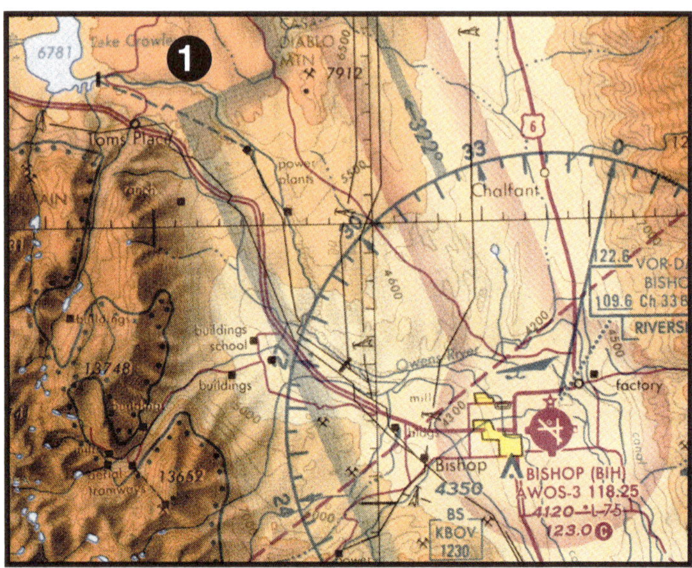

55. [I16/All]
Referring to the figure above, the airspace overlying Bishop Airport is
A. Class D airspace from the surface to the floor of the overlying Class E airspace.
B. Class E airspace from the surface to 1,200 feet MSL.
C. Class G airspace from the surface to 700 feet AGL.

56. [I16/All]
Referring to the figure above, identify the airspace to the upper left of Bishop (position 1) that exists from the surface to 14,500 feet MSL.
A. Class G airspace - surface to 14,500 feet MSL.
B. Class G airspace - surface to 3,500 feet MSL; Class E airspace - 3,500 feet MSL to 14,500 feet MSL.
C. Class G airspace - surface to 3,500 feet MSL; Class E airspace - 3,500 feet MSL to 10,000 feet MSL; Class G airspace - 10,000 feet MSL to 14,500 feet MSL.

Class D Airspace

57. [I19/1/2 & I18/2/2]
Airspace at an airport with a part-time control tower is classified as Class D airspace only
A. when the weather minimums are below basic VFR.
B. when the associated control tower is in operation.
C. when the associated Flight Service Station is in operation.

58. [I19/1/3]
Referring to the figure above, which point (1, 2, or 3) represents the borders of Class D surface-based airspace?
A. Point 3
B. Point 2
C. Point 1

59. [I19/1/3]
A blue segmented circle on a sectional chart depicts which class airspace?
A. Class B.
B. Class C.
C. Class D.

60. [I19/1/2]
The lateral dimensions of Class D airspace are based on
A. the number of airports that lie within the Class D airspace.
B. 5 statute miles from the geographical center of the primary airport.
C. the instrument procedures for which the controlled airspace is established.

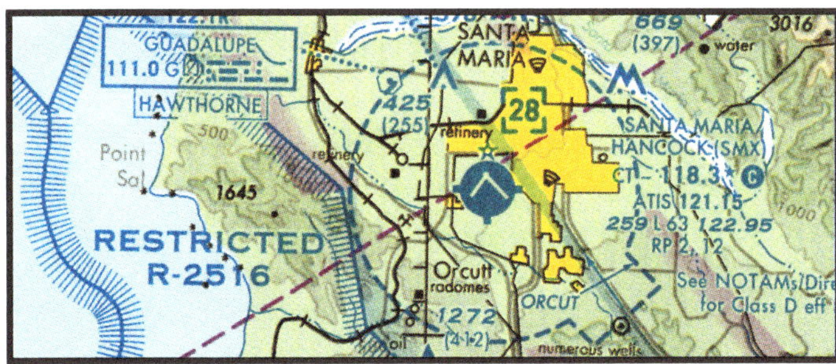

61. [I19/1/3 & I19/Figure 26]
Referring to the figure above, the airspace directly overlying Santa Maria airport is
A. Class B airspace to 10,000 feet MSL.
B. Class C airspace to 5,000 feet MSL.
C. Class D airspace to 2,800 feet MSL.

62. [I19/2/2]
Unless otherwise authorized, two-way radio communication with Air Traffic Control is required for landings or takeoffs
A. at all tower controlled airports regardless of weather conditions.
B. at all tower controlled airports only when weather conditions are less than VFR.
C. at all tower controlled airports within Class D airspace only when weather conditions are less than VFR.

Weather Minimums for Class D Airspace

63. [I20/3/1]
The basic VFR weather minimums for operating an aircraft within Class D airspace are
A. 500 foot ceiling and 1 mile visibility.
B. 1,000 foot ceiling and 3 miles visibility.
C. clear of clouds and 2 miles visibility.

64. [I20/3/1&2]
No person may take off or land an aircraft under basic VFR at an airport that lies within Class D airspace unless the
A. flight visibility at that airport is at least 1 mile.
B. ground visibility at that airport is at least 1 mile.
C. ground visibility at that airport is at least 3 miles.

65. [I20/3/1]
Normal VFR operations in Class D airspace with an operating control tower require the ceiling and visibility to be at least
A. 1,000 feet and 1 mile.
B. 1,000 feet and 3 miles.
C. 2,500 feet and 3 miles.

66. [I21/1/1]
During the day, you can usually tell if the primary airport is below basic VFR minimums by looking at the airport's _____.
A. anemometer
B. rotating beacon
C. tower

67. [I21/1/1]
An airport's rotating beacon operated during daylight hours indicates
A. there are obstructions on the airport.
B. that weather at the airport located in Class D airspace is below basic VFR weather minimums.
C. the Air Traffic Control tower is not in operation.

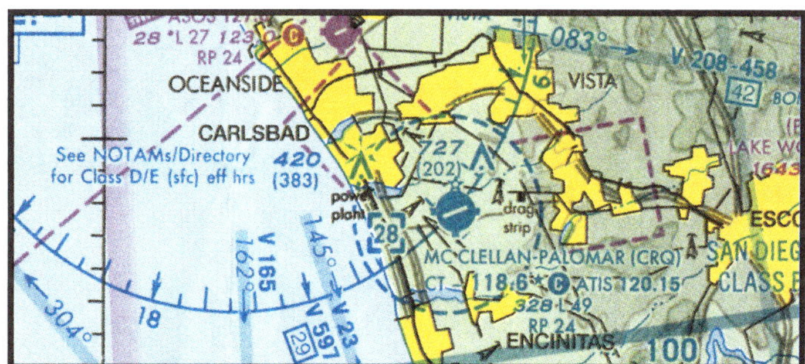

68. [I21/1/1]
Would a SVFR clearance from McClellan-Palomar airport to the northwest allow you to fly at 2,000 feet AGL with one mile visibility, then descend and land at Oceanside airport?
A. Yes.
B. No.
C. Only if you received a SVFR clearance to land at Oceanside airport.

Satellite Airports Within Class D Airspace

69. [I21/1/3]
A non-towered satellite airport, within the same Class D airspace as that designated for the primary airport, requires radio communication be established and maintained with the
A. satellite airport's Unicom.
B. associated Flight Service Station.
C. primary airport's control tower.

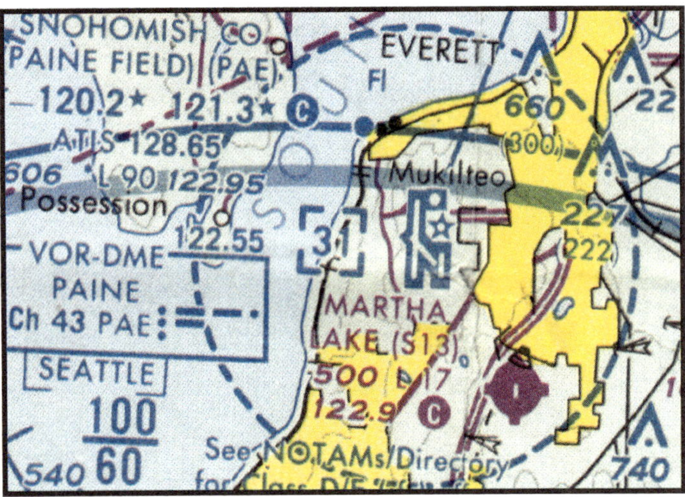

70. [I21/1/3]
When landing at Martha Lake airport in the figure above, you must establish and maintain communication with ATC on frequency
A. 122.55 MHz.
B. 122.9 MHz.
C. 120.2 MHz.

Class C Airspace

71. [I22/1/1]
An operating _____ (as in Class D airspace) as well as _____ approach control services are associated with the existence of Class C airspace.
A. airplane, radar
B. control tower, radar
C. radio, radar

72. [I22/1/2]
Class C airspace is geometrically shaped like two cylinders. Considering the entire structure, the surface-based inner cylinder extends upward to approximately _____ AGL and has a five nautical mile radius from the center of the _____ airport.
A. 4,000 feet, primary
B. 1,200 feet, primary
C. 1,200 feet, satellite

73. [I22/1/2]
The upper or outer cylinder of Class C airspace normally begins at _____ AGL and has a _____ nautical mile radius from the center of the primary airport. The upper limit of the top cylinder is generally found at _____ feet above the elevation of the primary airport.
A. 1,200 feet, 10, 4,000
B. 4,000 feet, 5, 10,000
C. 4,000 feet, 10, 10,000

74. [I22/1/2]
The vertical limit of Class C airspace above the primary airport is normally
A. 1,200 feet AGL.
B. 3,000 feet AGL.
C. 4,000 feet AGL.

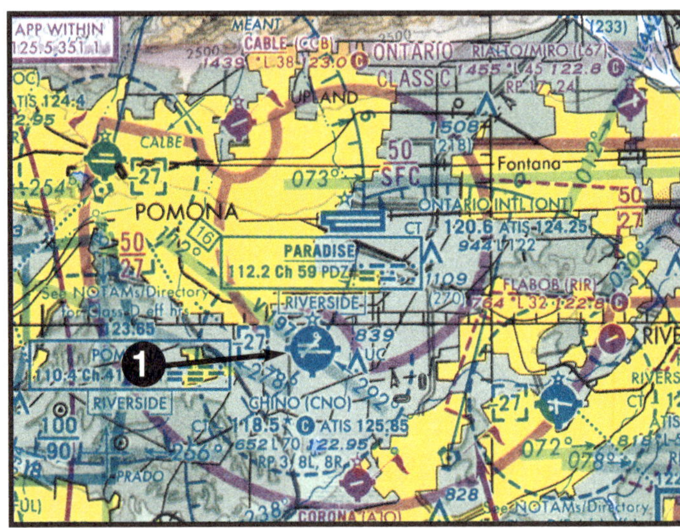

75. [I22/1/3]
Referring to the figure above, where does Class C airspace begin above Chino airport (Position 1)?
A. 2,700 feet MSL.
B. At the surface.
C. 5,000 feet MSL.

76. [I22/1/3]
Referring to the figure above, where does Class C airspace end above Chino airport?
A. 2,700 feet MSL.
B. At the surface.
C. 5,000 feet MSL.

Equipment Requirements to Operate Within Class C Airspace

77. [I23/1/2]
Two-way radio communication must be established with the Air Traffic Control facility having jurisdiction over the area prior to entering which class airspace?
A. Class C.
B. Class E.
C. Class G.

78. [I23/1/3]
All operations within Class C airspace must be in
A. accordance with instrument flight rules.
B. compliance with ATC clearances and instructions.
C. an aircraft equipped with a 4096-code transponder with Mode C encoding capability.

79. [I23/1/2 & I23/2/3]
Which initial action should a pilot take prior to entering Class C airspace?
A. Contact Approach Control on the appropriate frequency.
B. Contact the tower and request permission to enter.
C. Contact the FSS for traffic advisories.

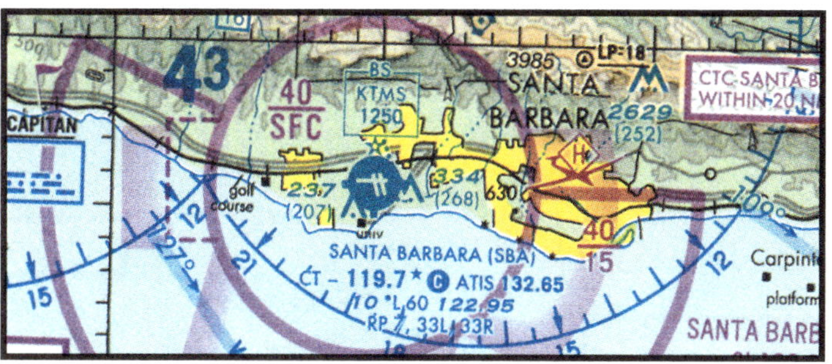

80. [I23/1/3]
Referring to the figure above, what minimum equipment is required to land and take off at Santa Barbara airport?
A. Mode C transponder and VOR receiver.
B. Mode C transponder and two-way radio.
C. Mode C transponder, VOR receiver, and DME.

81. [I23/1/2&3]
What minimum radio equipment is required for operation within Class C airspace?
A. Two-way radio communication equipment and a 4096-code transponder.
B. Two-way radio communication equipment, a 4096-code transponder, and DME.
C. Two-way radio communication equipment, a 4096-code transponder, and an encoding altimeter.

Class C Service

82. [I23/2/3]
The normal radius of the outer area of Class C airspace is
A. 5 nautical miles.
B. 15 nautical miles.
C. 20 nautical miles.

Satellite Airports Within Class C Airspace

83. [I23/3/2]
Under what condition may an aircraft operate from a satellite airport within Class C airspace?
A. The pilot must file a flight plan prior to departure.
B. The pilot must monitor ATC until clear of the Class C airspace.
C. The pilot must contact ATC as soon as practicable after takeoff.

Weather Minimums for Class C Airspace

84. [I24/1/3]
The weather minimums for Class C airspace are exactly the same as they are for Class _____ airspace and Class _____ airspace below _____ MSL.
A. B, A, 10,000 feet
B. D, E, 10,000 feet
C. D, E, 1,200 feet

Class B Airspace

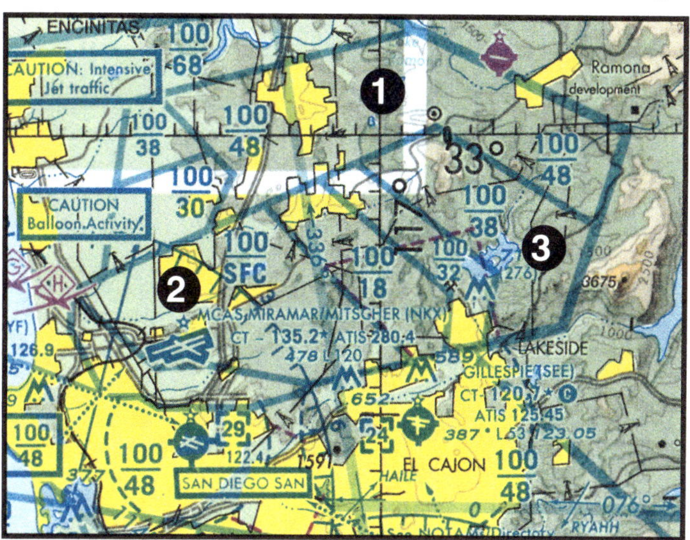

85. [I25/2/3]
Referring to the figure above, the floor of Class B airspace at position 1 is
A. at the surface.
B. 4,800 feet MSL.
C. 3,800 feet MSL.

86. [I25/2/3]
Referring to the figure above, the floor of Class B airspace at position 2 is
A. at the surface.
B. 4,800 feet MSL.
C. 3,800 feet MSL.

87. [I25/2/3]
Referring to the figure above, the floor of Class B airspace at position 3 is
A. at the surface.
B. 4,800 feet MSL.
C. 3,800 feet MSL.

88. [I25/2/3]
Referring to the figure above, the top of Class B airspace at position 1 is
A. 10,000 feet MSL.
B. 4,800 feet MSL.
C. 3,800 feet MSL.

89. [I25/2/2]
Typically, Class B airspace has a radius of 15 to 30 miles from the primary airport and it extends vertically from the surface to _____ MSL.
A. 12,000 feet
B. 5,000 feet
C. 10,000 feet

Requirements to Enter Class B Airspace

90. [I26/1/3]
What minimum pilot certification is required for operation within Class B airspace?
A. Airline transport pilot certificate.
B. Private pilot certificate or student pilot certificate with appropriate logbook endorsements.
C. Private pilot certificate with an instrument rating.

91. [I26/1/3]
The minimum pilot certification required for operation within Class B airspace?
A. Private pilot certificate or, student or recreational pilot certificate with appropriate logbook endorsements.
B. Commercial pilot certificate.
C. Flight instructor certificate only.

92. [I26/1/4]
While you need only establish and maintain radio communication for Class C or D airspace, you need to obtain a _____ to enter Class B airspace.
A. clearance
B. permission slip
C. transponder code

93. [I26/3/2]
What minimum radio equipment is required for VFR operation within Class B airspace?
A. Two-way radio communication equipment and a 4096-code transponder.
B. Two-way radio communication equipment, a 4096-code transponder, and an encoding altimeter.
C. Two-way radio communication equipment, a 4096-code transponder, an encoding altimeter, and a VOR or TACAN receiver.

94. [I26/3/2]
If for any reason the Class B airspace doesn't extend to 10,000 feet MSL, a transponder with Mode C capability is required when operating above the _____ and within the_____ of Class B airspace.
A. lateral boundaries, periphery
B. ceiling, lateral boundaries
C. ceiling, transmission range

95. [I27/1/1]
To operate within Class B airspace, _____ mile(s) visibility is required. However, you only need to remain _____ clouds instead of the typical 1,000'/500'/2,000' distance minimums.
A. three, clear of
B. five, clear of
C. one, 1 mile from all

Special VFR Within Class B Airspace

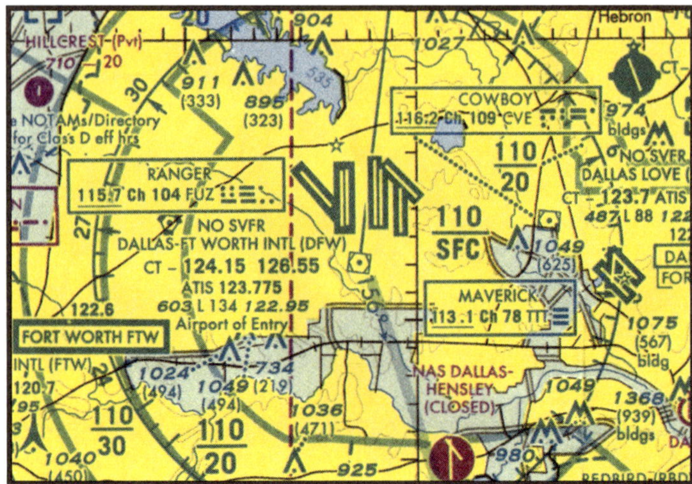

96. [I27/1/3]
Referring to the figure above, is fixed-wing Special VFR authorized at Dallas-Ft. Worth authorized?
A. Yes. SVFR is authorized at all airports.
B. Yes. SVFR is authorized as long as the pilot has a clearance.
C. No. SVFR is not authorized for fixed-wing aircraft at this airport.

Corridors and Circumnavigating Class B Airspace

97. [I27/2/2]
A speed limit of _____ knots exists underneath the lateral limits of Class B airspace or VFR corridors through Class B.
A. 250
B. 200
C. 180

Transponder and Mode C Within 30 NM of Certain airports

98. [I28/1/1]
With certain exceptions, all aircraft within 30 miles of a Class B primary airport from the surface upward to 10,000 feet MSL must be equipped with
A. an operable VOR or TACAN receiver and an ADF receiver.
B. instruments and equipment required for IFR operations.
C. an operable transponder having either Mode S or 4096-code capability with Mode C automatic altitude reporting capability.

99. [I28/1/1]
An operable 4096-code transponder with an encoding altimeter is required in which airspace?
A. Class A, Class B (and within 30 miles of the Class B primary airport), and Class C.
B. Class D and Class E (below 10,000 feet MSL).
C. Class D and Class G (below 10,000 feet MSL).

100. [I28/1/1]
An operable 4096-code transponder and Mode C encoding altimeter are required in
A. Class B airspace and within 30 miles of the Class B primary airport.
B. Class D airspace.
C. Class E airspace below 10,000 feet MSL.

Transponders and Mode C Above 10,000 Feet MSL

101. [I28/1/2]
A transponder with Mode C is also required in all the airspace of the 48 contiguous United States and the District of Columbia when operating at and above _____ feet MSL, excluding the airspace at and below _____ feet AGL.
A. 10,000, 2,500
B. 10,000, 4,000
C. 2,500, 1,200

Transponders in Controlled Airspace

102. [I28/2/2]
If your airplane has a transponder, the rules require that it be turned on (including the Mode C capability) any time you are operating in _____ airspace.
A. special use
B. uncontrolled
C. controlled

Transponder and Mode C Deviations

103. [I28/2/4]
If you have a transponder lacking Mode C capability, you can request a _____ to operate within airspace requiring Mode C at any time.
A. deviation
B. scout plane
C. clearance

Speed Restriction in Class C and D Airspace

104. [I28/3/3]
When any aircraft is within four nautical miles of the primary airport in Class C and D airspace and at or below 2,500 feet AGL, a _____ knot speed restriction applies.
A. 180
B. 200
C. 250

Terminal Radar Service Area

105. [I29/1/2]
Recall that in a TRSA, ATC provides _____ between all participating VFR aircraft and all IFR aircraft.
A. separation
B. sequencing and separation
C. 1,000 feet separation

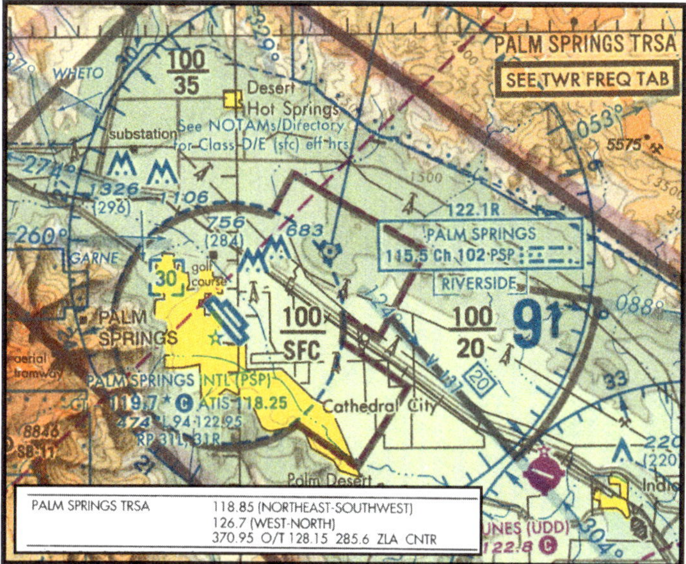

106. [I29/2/1]
Referring to the figure above, if you're approaching Palm Springs anywhere from west to north from the airport, you can contact Approach Control on _____.
A. 118.85 MHz
B. 126.7 MHz
C. 128.15 MHz

107. [I29/2/3]
If you happen to be talking to approach control (when approaching) or ground control (before departure) and don't want TRSA service, you should state,
A. "Not going to happen, man."
B. "Negative TRSA service."
C. "Negative radar service."

Special Use Airspace

108. [I30/2/3]
Prohibited areas are defined by _____ lines.
A. red dashed
B. red hatched
C. blue hatched

109. [I30/3/2]
Restricted areas restrict flights due to the unusual activities conducted within them. These areas often contain invisible hazards to aircraft such as the firing of _____, _____, _____.
A. artillery, aerial gunnery, guided missiles
B. artillery, lasers, rocks
C. bullets, rockets, gum wads

110. [I30/3/4]
Before you can enter or fly through a restricted area, you need permission from the _____.
A. administrator
B. FAA
C. controlling agency

Warning Areas

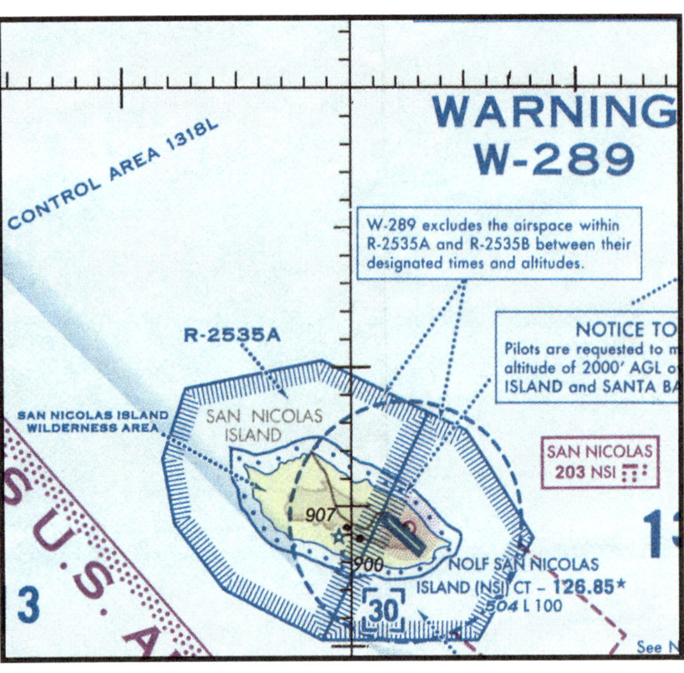

111. [I31/2/2 & I30/3/2]
Referring to the figure above, what hazards to aircraft may exist in warning areas such as Warning W-289?
A. Unusual, often invisible, hazards such as aerial gunnery or guided missiles over international waters.
B. High volume of pilot training or unusual type of aerial activity.
C. Heavy military aircraft traffic in the approach and departure area of the North Atlantic Control Area.

Alert Areas

112. [I32/1/1]
Responsibility for collision avoidance in an alert area rests with
A. the controlling agency.
B. all pilots.
C. Air Traffic Control.

Military Operations Areas

113. [I32/1/2 & I32/2/3]
What hazards to aircraft may exist in a MOA?
A. Unusual, often invisible, hazards to aircraft such as artillery firing.
B. High density military training activities.
C. Parachute jump operations.

114. [I32/3/3]
What action should a pilot take when operating under VFR in a Military Operations Area (MOA)?
A. Obtain a clearance from the controlling agency prior to entering the MOA.
B. Operate only on the airways that transverse the MOA.
C. Exercise extreme caution when military activity is being conducted.

Military Training Routes

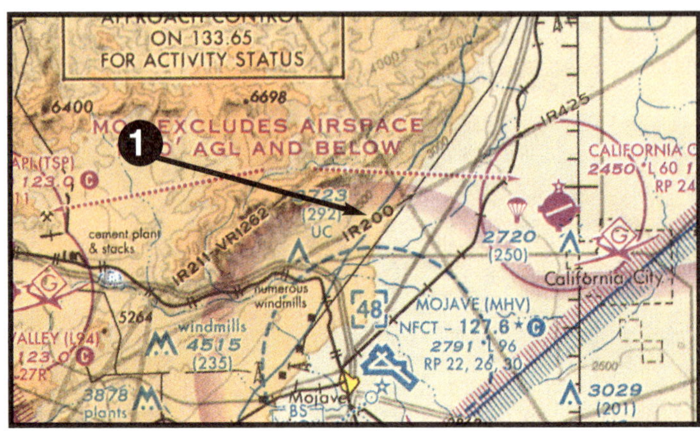

115. [I32/3/4]
Referring to the figure above, what type of military flight operations should a pilot expect along IR 200?
A. IFR training flights above 1,500 feet AGL at speeds in excess of 250 knots.
B. VFR training flights above 1,500 feet AGL at speeds less than 250 knots.
C. Instrument training flights below 1,500 feet AGL at speeds in excess of 150 knots.

Variable Floors of Class E Airspace

116. [I34/1/1]
The width of a federal airway from either side of the centerline is
A. 4 nautical miles.
B. 6 nautical miles.
C. 8 nautical miles.

117. [I34/1/2]
Unless otherwise specified, federal airways include that Class E airspace extending upward from
A. 700 feet above the surface up to and including 17,999 feet MSL.
B. 1,200 feet above the surface up to and including 17,999 feet MSL.
C. the surface up to and including 18,000 feet MSL.

Postflight Briefing 9-1: Variable Floors of Class E Airspace

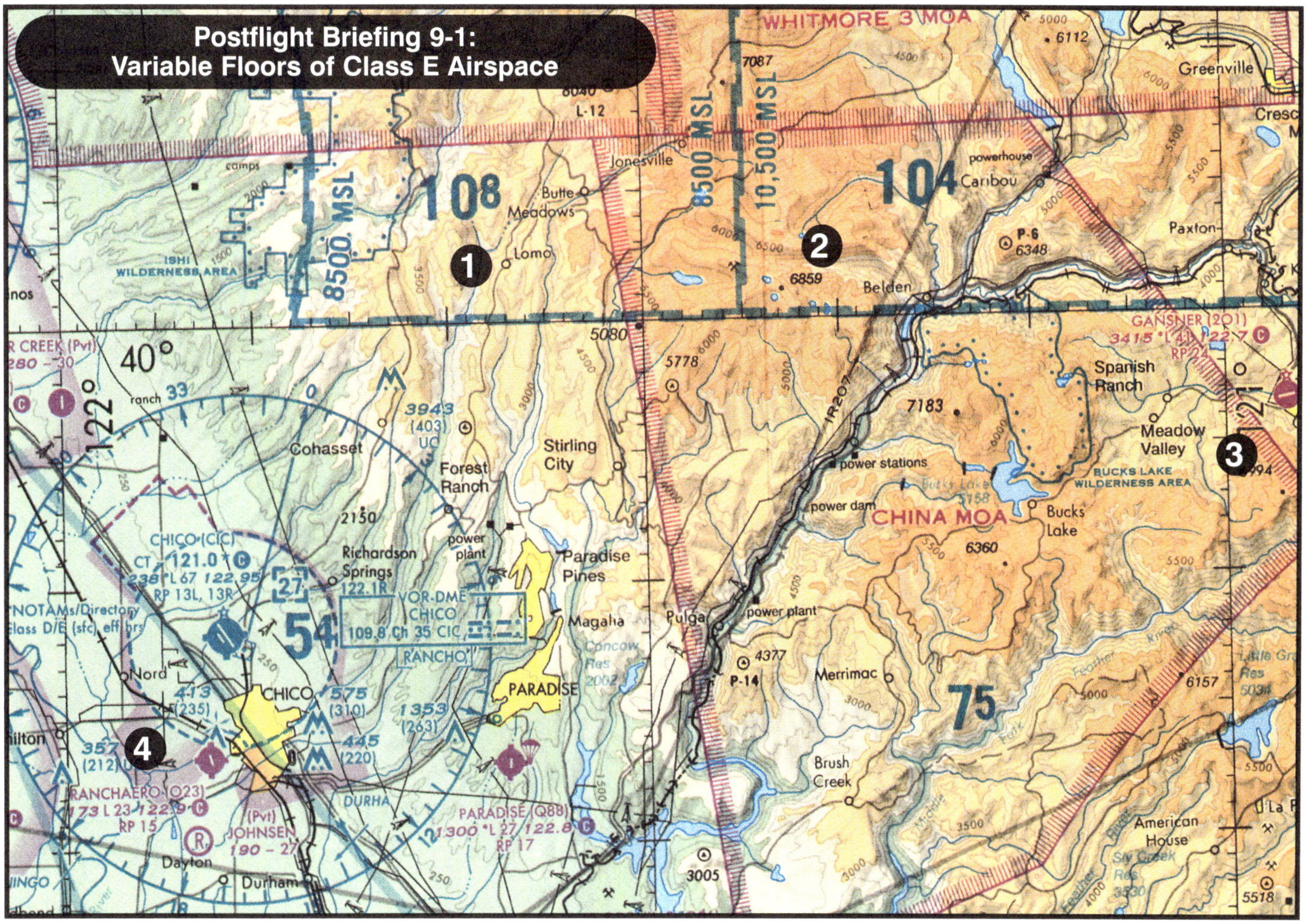

118. [All of page I34]
Referring to the figure above, the vertical limits of that portion of Class E airspace over position 1 are
A. 1,200 feet AGL to 8,500 feet MSL.
B. 8,500 feet MSL to 12,500 feet MSL.
C. 8,500 feet MSL to 17,999 feet MSL.

119. [All of page I34]
Referring to the figure above, the vertical limits of that portion of Class E airspace over position 2 are
A. 1,200 feet AGL to 8,500 feet MSL.
B. 10,500 feet MSL to 17,999 feet MSL.
C. 10,400 feet MSL to 17,999 feet MSL.

120. [All of page I34]
Referring to the figure above, the vertical limits of that portion of Class E airspace directly over Ganser airport (position 3) are
A. 1,200 feet AGL to 17,999 feet MSL.
B. 7,500 feet MSL to 12,500 feet MSL.
C. 7,500 feet MSL to 17,999 feet MSL.

121. [All of page I34]
Referring to the figure above, the vertical limits of that portion of Class E airspace over position 4 are
A. 700 feet AGL to 8,500 feet MSL.
B. 1,200 feet MSL to 14,500 feet MSL.
C. 700 feet AGL to 17,999 feet MSL.

Chapter Nine Answers

1. B
2. A
3. C
4. B
5. A
6. B
7. A
8. B
9. C
10. C
11. A
12. A
13. A
14. A
15. B
16. A
17. B
18. C
19. C
20. C
21. B
22. C
23. C
24. A
25. A
26. B
27. B
28. B
29. A
30. A
31. B
32. B
33. A
34. B
35. B
36. C
37. C
38. A
39. A
40. A
41. A
42. C
43. B
44. B
45. B
46. C
47. A
48. B
49. A
50. B
51. A
52. B
53. C
54. C
55. C
56. A
57. B
58. C
59. C
60. C
61. C
62. A
63. B
64. C
65. B
66. B
67. B
68. B
69. C
70. C
71. B
72. A
73. A
74. C
75. A
76. C
77. A
78. C
79. A
80. B
81. C
82. C
83. C
84. B
85. B
86. A
87. C
88. A
89. C
90. B
91. A
92. A
93. B
94. B
95. A
96. C
97. B
98. C
99. A
100. A
101. A
102. C
103. A
104. B
105. B
106. B
107. B
108. C
109. A
110. C
111. A
112. B
113. B
114. C
115. A
116. A
117. B
118. C
119. B
120. A
121. C

Note: To ensure that you have the most current answers to these questions, please check the *Book & Slide Updates* section at Rod Machado's web site: www.rodmachado.com

Chapter Ten

Aviation Maps: The Art of the Chart

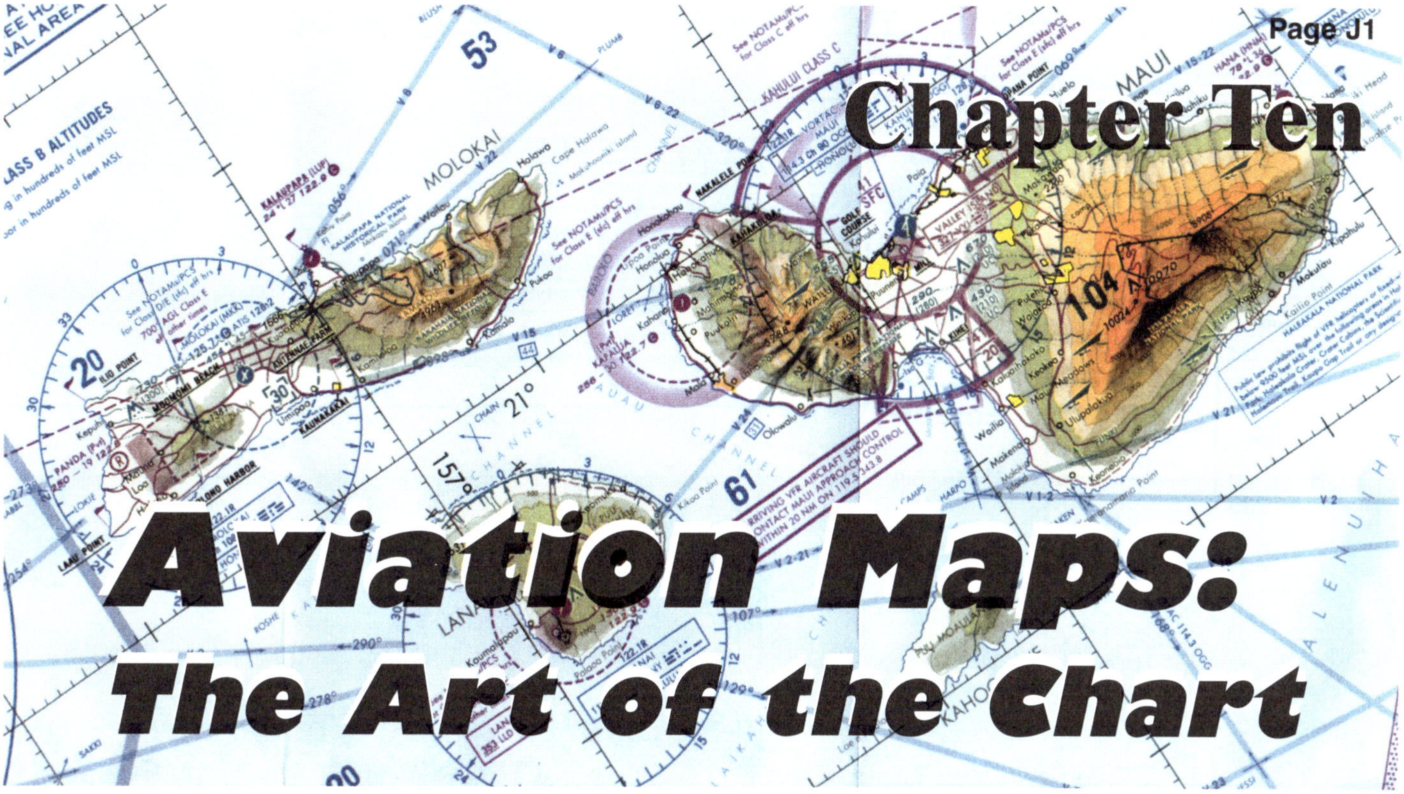

The Lambert Conformal Conic Projection

1. [J2/Figure 1]
When drawing lines on a Lambert Conformal Conic Projection, a straight line represents a
A rhumb line.
B. great circle route.
C. great rhumb line.

2. [J2/Figure 1]
On a Lambert Conformal Conic Projection, the two standard parallels represent the positions where
A. no distortion exists.
B. maximum distortion exists.
C. good things happen.

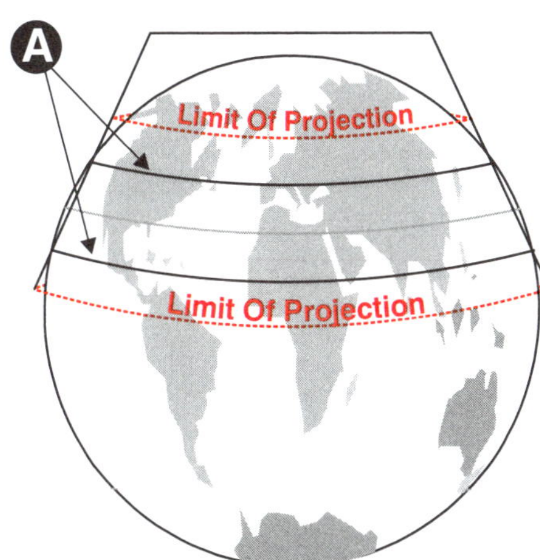

3. [J2/Figure 2] Fill in the blank:
Based on the figure above, the two lines identified by "A" in the Lambert Conformal Conic Projection are known as ______________ parallels.

HOUSTON
SECTIONAL AERONAUTICAL CHART
SCALE 1:500,000
Lambert Conformal Conic Projection Standard Parallels 25°20′ and 30°40′
Horizontal Datum: North American Datum of 1983 (World Geodetic System 1984)
Topographic data corrected to May 1995

4. [J2/Figure 2]
Based on the sectional chart excerpt above, what are the standard parallels on which this lambert conformal conic projection is based?
A. 25 degrees 20 minutes north latitude, 30 degrees 40 minutes north latitude.
B. 25 degrees 20 minutes west longitude, 30 degrees 40 minutes west longitude.
C. 1 degree north latitude & 500,000 degrees north latitude.

The Aeronautical Sectional Chart

5. [J2/1/1]
Sectional charts are valid for flight planning for
A. 12 months.
B. 6 months.
C. a lot of things.

6. [J2/3/2]
Changes on the sectional chart occurring prior to the next publication cycle can be found in the
A. FARs.
B. pilots operating handbook.
C. Airport/Facility Directory.

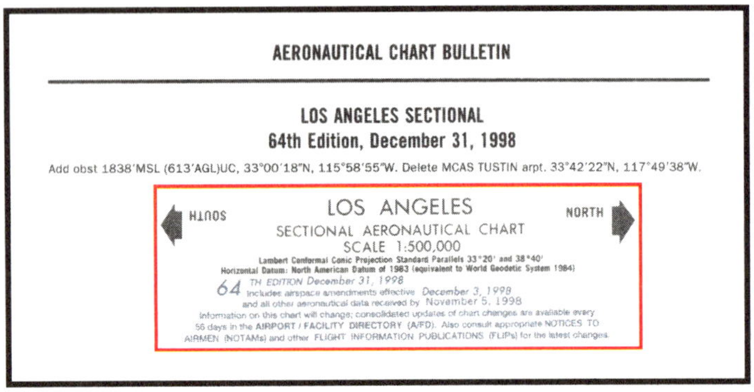

AERONAUTICAL CHART BULLETIN

LOS ANGELES SECTIONAL
64th Edition, December 31, 1998

Add obst 1838'MSL (613'AGL)UC, 33°00'18"N, 115°58'55"W. Delete MCAS TUSTIN arpt. 33°42'22"N, 117°49'38"W.

SOUTH — LOS ANGELES — NORTH
SECTIONAL AERONAUTICAL CHART
SCALE 1:500,000
Lambert Conformal Conic Projection Standard Parallels 33°20' and 38°40'
Horizontal Datum: North American Datum of 1983 (equivalent to World Geodetic System 1984)
64 TH EDITION December 31, 1998
Includes airspace amendments effective December 3, 1998
and all other aeronautical data received by November 5, 1998
Information on this chart will change; consolidated updates of chart changes are available every 56 days in the AIRPORT / FACILITY DIRECTORY (A/FD). Also consult appropriate NOTICES TO AIRMEN (NOTAMs) and other FLIGHT INFORMATION PUBLICATIONS (FLIPs) for the latest changes.

7. [J2/3/2]
According the Airport/Facility Directory for Los Angeles shown above, what change should you make to your Los Angeles sectional chart in order to make this chart as accurate as possible?
A. No change at all. The changes shown in the A/FD excerpt were already incorporated in this issue of the sectional chart.
B. No change. Just wait for the next issue of the sectional chart to show these changes.
C. Take your pen and mark the position of the 1,838 obstacle and make a note that MCAS airport is deleted.

World Aeronautical Charts (WAC)

8. [J4/1/2]
World Aeronautical Charts are valid for
A. 12 months.
B. 6 months.
C. until updated by Notam.

9. [J5/1/1]
World Aeronautical Charts have a scale of
A. 1 to 500,000.
B. 1 to 250,000.
C. 1 to 1,000,000.

VFR Terminal Area Charts.

10. [J5/1/1]
VFR Terminal Area Charts are good for
A. 12 months.
B. 6 months.
C. until updated by Notam.

11. [J5/1/1]
VFR Terminal Area Charts have a scale of
A. 1 to 500,000.
B. 1 to 250,000.
C. 1 to 1,000,000.

Topographical Information on a Sectional Chart

Relief

12. [J5/2/3]
Contour lines on a topographical chart join areas of
A. equal pressure.
B. equal density.
C. equal height.

13. [J6/1/1]
On a sectional chart, contour lines are commonly spaced at intervals of
A. 500 feet.
B. 100 feet.
C. 200 feet.

14. [J6/1/1]
Referring to the figure above, the contour lines are spaced at intervals of _____.
A. 500 feet
B. 100 feet
C. 200 feet

Color

15. [J6/2/1]
A specific color shown on a topographic chart doesn't precisely indicate the height of terrain, it indicates _____ heights within which terrain can be found in those areas.
A. specific
B. a random selection of
C. a range of

16. [J6/2/1]
The area of terrain identified by area "A" in the figure located in top left hand corner of the opposite page has terrain that varies from
A. 500 feet to 2,000 feet.
B. sea level to 1,000 feet.
C. sea level to 2,000 feet .

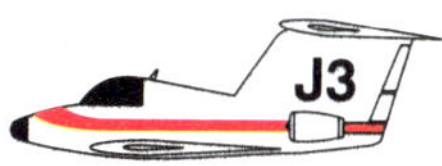

Spot Elevation Symbols

17. [J6/2/2]
Normally, spot elevations (shown as black dots) are chosen by mapmakers to indicate the _____ on a particular mountain range or ridge.
A. high point
B. low point
C. obstacle points

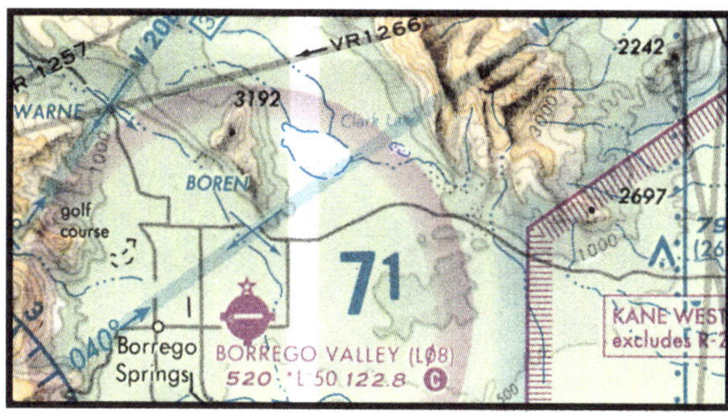

18. [J6/2/2]
What is the highest spot elevation shown in the sectional chart excerpt above?
A. 2,242 feet.
B. 2,697 feet.
C. 3,192 feet.

Spot Elevations Showing Highest Terrain

19. [J7/1/1]
A single spot elevation showing the highest terrain is found within the _____ bounded by lines of latitude and longitude.
A. quadrangles
B. biangles
C. triangles

20. [J7/1/1] Fill in the blank:
Referring to the figure in the upper right hand side of this page, the highest terrain for the quadrangle shown is _____________.

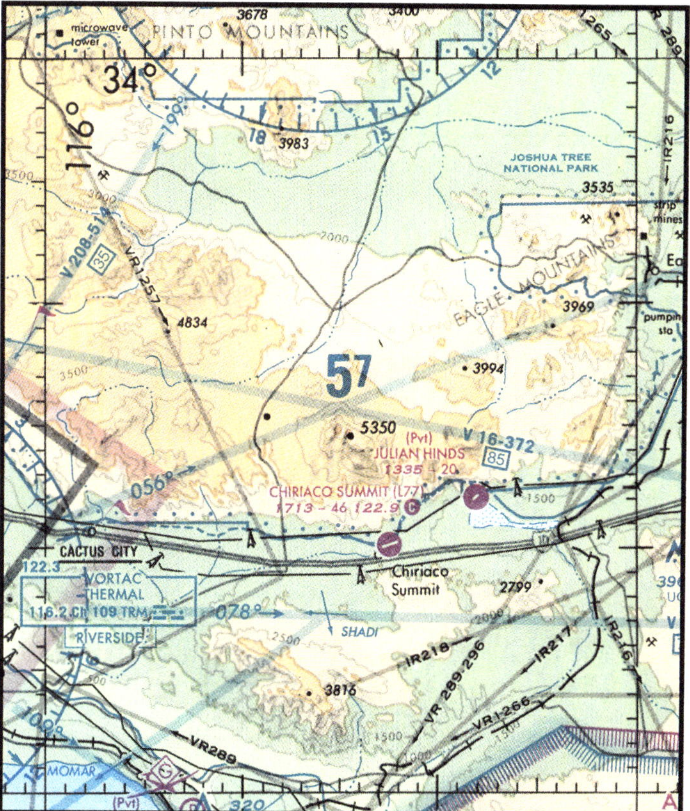

Maximum Elevation Figures (MEF)

21. [J7/1/2]
Maximum elevation figures (MEFs) represent the highest elevation of terrain and other obstacles (towers, trees, etc.) within _____.
A. any area on the chart
B. a quadrangle
C. a magenta bordered area

22. [J7/1/2]
The maximum elevation figure shown for the quadrangle in the figure above is
A. 5,350 feet.
B. 5,700 feet.
C. 3,944 feet.

Obstacles

23. [J8/1/1 & J8/2/1&2]
Referring to the figure to the right, what is the height of the obstacle approximately 4 nautical miles due north of Midway city?
A. 620 feet AGL.
B. 260 feet MSL.
C. 260 feet AGL.

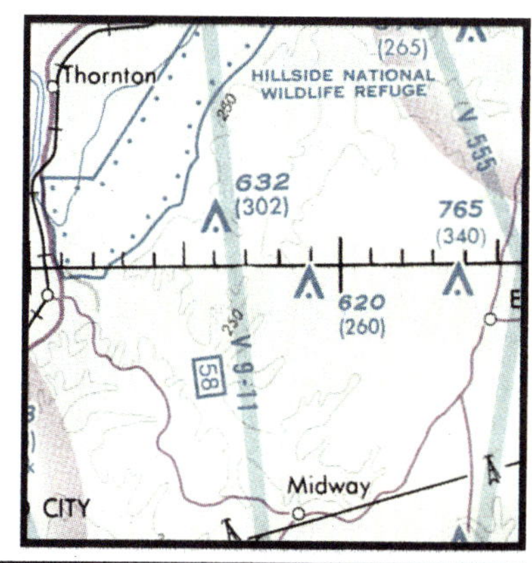

24. [J8/1/1 & J8/2/1&2] Referring to the figure on the right, what minimum altitude is required to fly over the lighted obstacle located just north of the city of Hamburg? (Assume that the entire area is a congested area.)
A. 1,483 feet MSL.
B. 1,483 feet AGL.
C. 1,323 feet MSL.

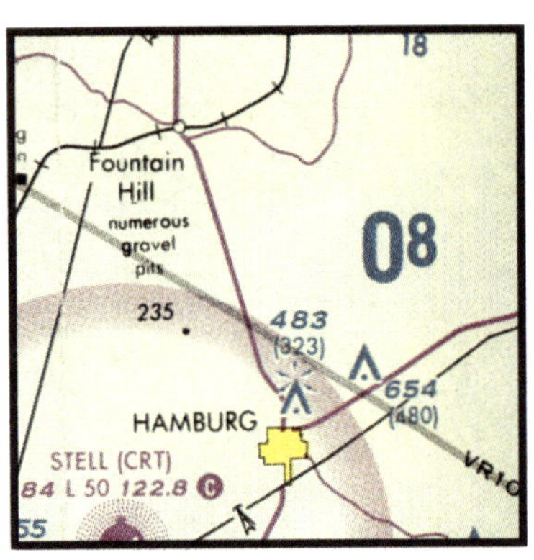

25. [J8/1/1 & J8/2/1&2] Referring to the figure on the right, the top of the obstacle approximately 3 miles southwest of the city of Lexington is
A. 579 feet AGL.
B. 265 feet MSL.
C. 579 feet MSL.

26. [J8/1/1 & J8/2/1&2] What minimum altitude is necessary to vertically clear the lighted obstacle on the southwest side of Hobbs airport by 500 feet?
A. 2,500 feet MSL.
B. 2,615 feet MSL.
C. 2,615 feet AGL.

Roads

27. [J8/2/3] What does arrow A point to in the figure to the right?
A. A railroad track.
B. A power transmission line.
C. A road.

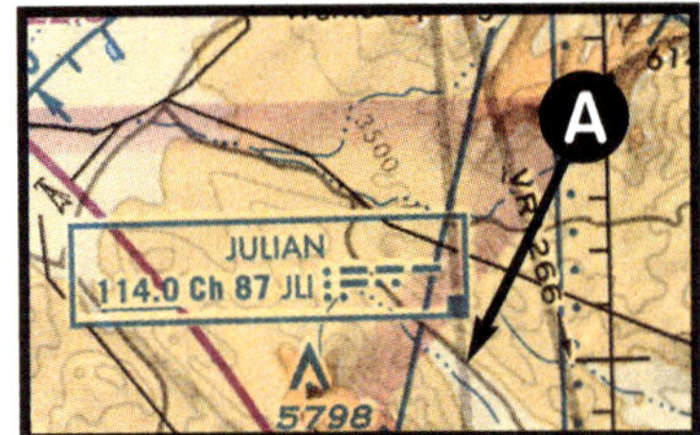

Railroad Tracks

28. [J8/2/3] Referring to the figure to the right, what does arrow A point to?
A. A railroad track.
B. A power transmission line.
C. A superhighway.

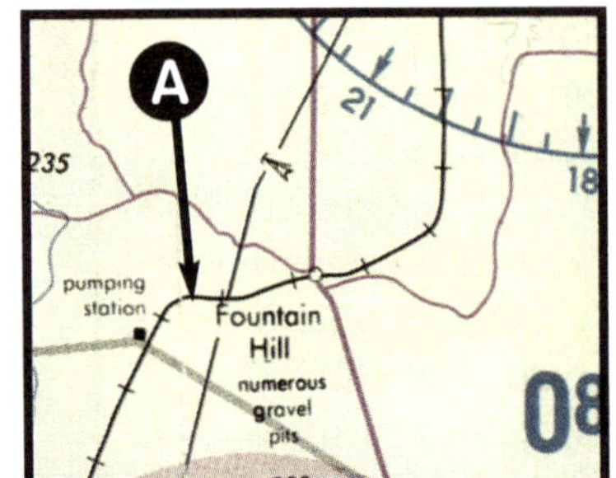

Wires

29. [J9/1/2] What does arrow A point to in the figure to the right?
A. Guy wires extending from radio or TV towers.
B. Power transmission lines.
C. A single-rail railroad.

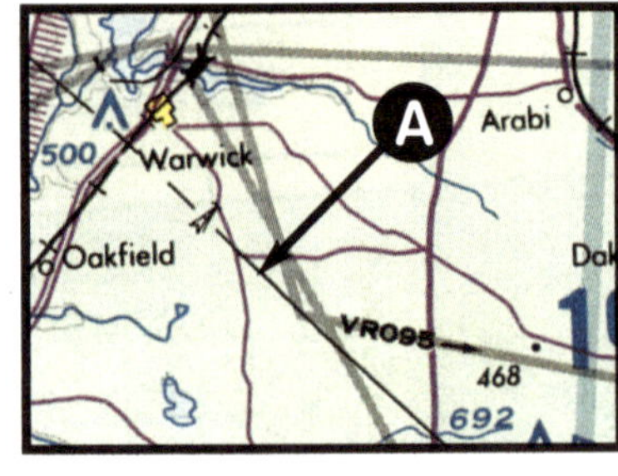

Airport

30. [J11/1/1] Airports are coded by colors on the map. Those airports colored in _____ don't have an air traffic control tower. Those shown in _____ have a tower (although it may not be in operation 24 hours a day—most aren't).
A. magenta, black
B. magenta, blue
C. blue, magenta

31. [J11/2/2] Normally, both the magenta and blue airport symbols are circles unless the airport has a hard surfaced runway greater than _____.
A. 5,000 feet
B. 10,000 feet
C. 8,000 feet

32. [J12/1/1] Any airport having a darkened circle with the runways in reverse-bold white has a _____ runway between 1,500 and 8,000 feet in length.
A. soft surfaced
B. hard surfaced
C. asphalt covered

33. [J12/Figure 35] Referring to the figure to the right, which public airports depicted have fuel?
A. Carson and Dayton Valley.
B. Douglas, Pinenut and Parker.
C. Douglas and Carson.

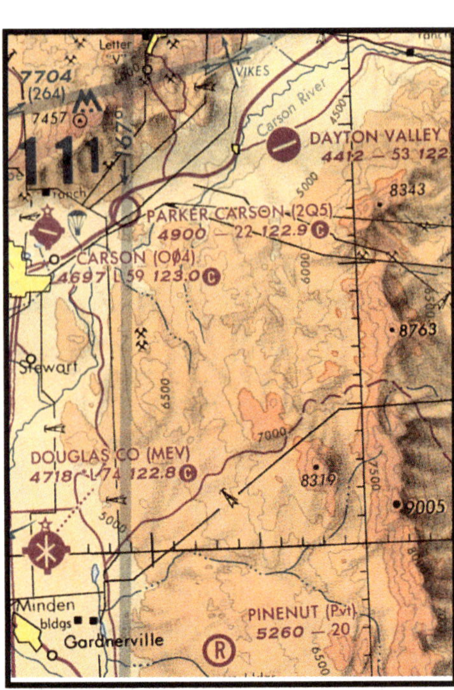

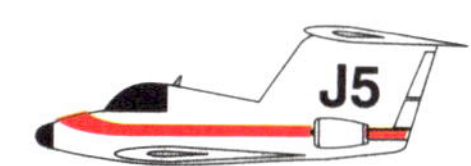

34. [J12/2/2]
Some airports are restricted in that they are private and not open to public use. These airports are identified by the airport symbol containing the letter ____.
A. P
B. UC
C. R

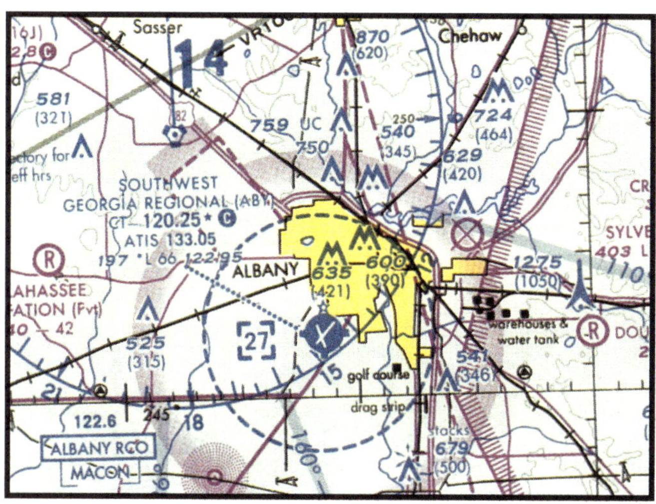

35. [J13/1/1]
Referring to the figure above, the airport data listed under Southwest Georgia Regional airport, what is the airport elevation?
A. 66 feet.
B. 133 feet.
C. 197 feet.

36. [J13/1/1]
Referring to the figure above, what is the length of the longest runway at Southwest Georgia Regional airport?
A. 660 feet.
B. 6,600 feet.
C. 19,700 feet.

37. [J13/1/2]
Referring to the figure above, what does the "*L" mean at Southwest Georgia Regional airport?
A. Runway lighting limitations exist.
B. Runway lighting is available only by prior arrangement.
C. Runway lighting is available if you flight a flight plan.

38. [J12/2/3 & J13/1/2&3]
Referring to the figure above, what are the ATIS and tower frequencies at Southwest Georgia Regional airport?
A. 120.25 MHz, 133.05 MHz.
B. 133.05 MHz, 122.95 MHz.
C. 133.05 MHz, 120.25 MHz.

Airways

39. [J13/2/1]
Referring to the figure at the right, what is the total airway distance between VORs for the airway named V66?
A. 66 nautical miles, on Victor airway 87.
B. 66 statute miles, on Victor airway 87.
C. 87 nautical miles, on Victor airway 66.

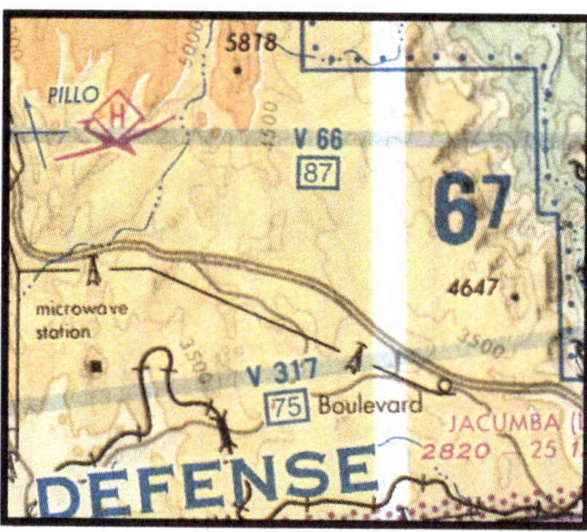

Visual Check Points

40. [J13/2/2]
Referring to the figure at the right, the flag symbol at Hooks Memorial airport (arrow A) represents a
A. compulsory reporting point for entering controlled airspace.
B. compulsory reporting point for Hooks Memorial airport.
C. visual checkpoint used to identify position for initial callup to an ATC facility.

Airborne Vehicle Symbols

41. [J13/3/2]
Referring to the figure at the right, the symbols identified by arrow A represent
A. airborne vehicles likely to be found in that airspace.
B. airborne vehicles in contact with the nearest ATC facility in that airspace.
C. airborne vehicles found only above 3,000 feet AGL in that airspace.

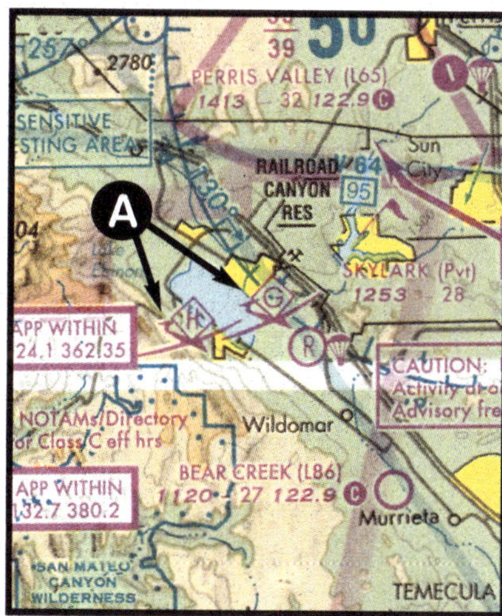

Park, Wildlife, Forest, Wilderness and Primitive Areas

42. [J14/1/1]
Pilots flying over a national wildlife refuge are requested to fly no lower than
A. 1,000 feet AGL.
B. 2,000 feet AGL.
C. 3,000 feet AGL.

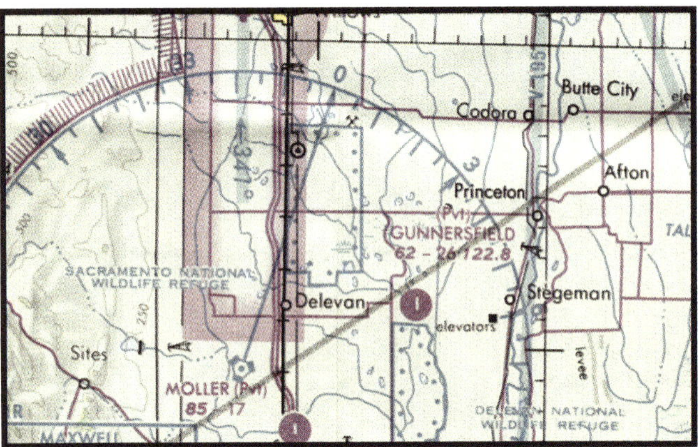

43. [J14/1/1]
What is the minimum altitude you should fly when heading northbound from Moller airport?
A. 2,000 feet AGL.
B. 1,000 feet AGL.
C. 2,000 feet MSL.

Postflight Briefing 10-2: Runway Patterns

44. [J14/Postflight Briefing #10-2]
Which runways at Long Beach airport have right hand patterns?
A. 7R, 16R, 25R, 34R.
B. 25L, 34L, 7L, 16L.
C. All traffic patterns are left hand in direction.

45. [J14/Postflight Briefing #10-2]
What do the letters represented by arrow "A" represent?
A. Secret Queen Mary code for overflights.
B. GPS identifier for this VFR waypoint.
C. VFR waypoint call letters which are given to ATC on initial call up.

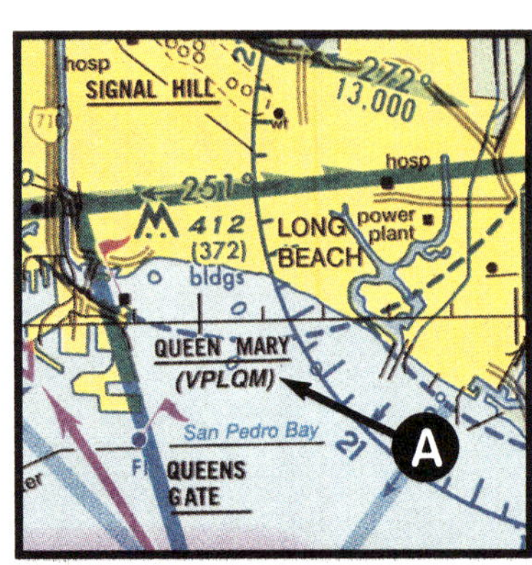

Chapter Ten Answers

1. B
2. A
3. standard
4. A
5. B
6. C
7. C
8. A
9. C
10. B
11. B
12. C
13. A
14. A
15. C
16. B
17. A
18. C
19. A
20. 5,350 feet
21. B
22. B
23. C
24. A
25. C
26. B
27. C
28. A
29. B
30. B
31. C
32. B
33. C
34. C
35. C
36. B
37. A
38. C
39. C
40. C
41. A
42. B
43. A
44. A
45. B

Note: To ensure that you have the most current answers to these questions, please check the *Book & Slide Updates* section at Rod Machado's web site: www.rodmachado.com

Radio Navigation:
The Frequency Flyer Program
Chapter Eleven

Pilotage

1. [K1/3/2]
Pilotage is navigation by
A. reference to flight instruments.
B. reference to landmarks.
C. reference to airborne satellites.

Electronic Elucidation

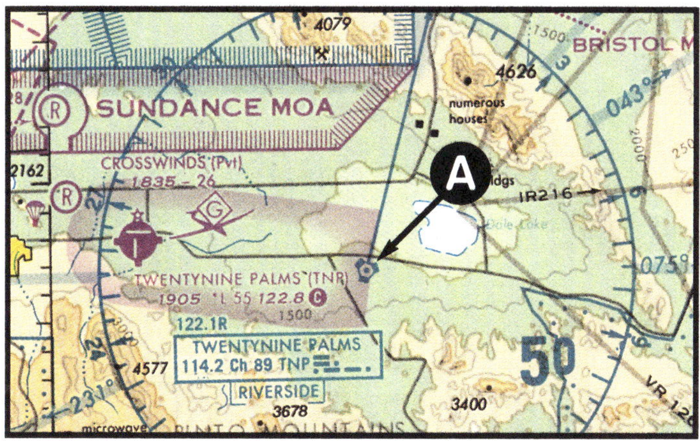

2. [K4/1/3]
The navigation station shown by arrow A in the middle of the figure above is
A. an NDB station.
B. a ground based GPS station.
C. a VOR station.

The Big Picture

3. [K4/2/1]
Each VOR station transmits its signals on
A. VHF between 134.95 MHz and 145.95 MHz.
B. VHF between 108.0 MHz and 117.95 MHz.
C. UHF between 118.0 MHz and 134.95 MHz.

4. [K4/Figure5]
Label the individual VOR radials as shown in the figure below:

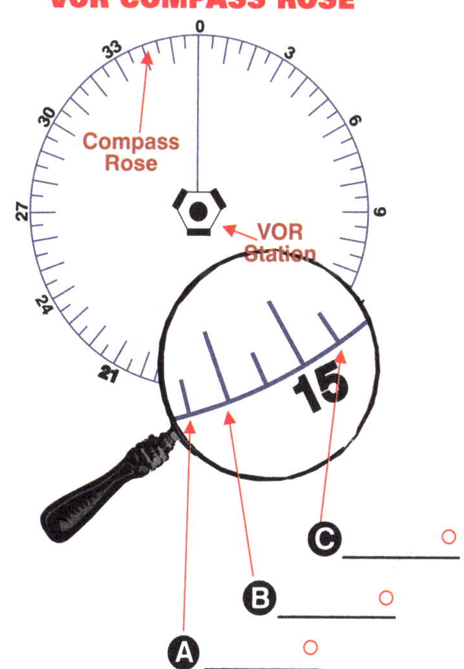

5. [K5/1/3]
VOR reception is
A. based on line of sight transmission of the signal.
B. unaffected by line of sight transmission of the signal.
C. LF and doesn't depend on altitude or line of sight.

Your VOR Equipment

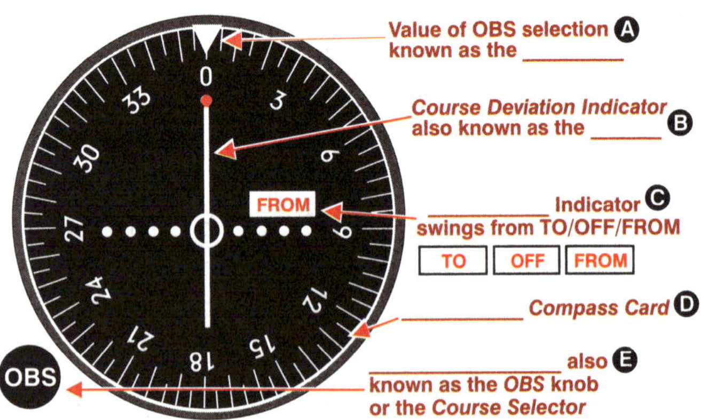

6. [K5/Figure 7]
Name the individual parts of the VOR display in the figure above.

VORs and Airborne Freeways

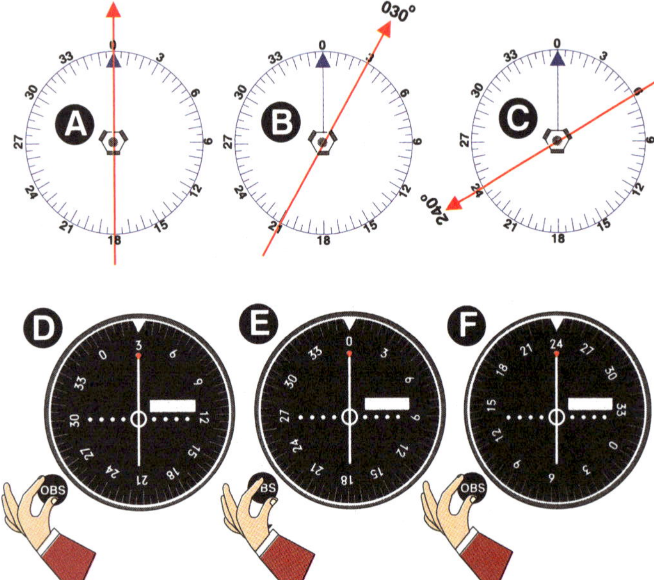

7. [K6/Figure9]
Referring to the figure above, match the individual OBS settings with their respective courses.
A. VORs: D, E, F go with Courses A, B, C respectively.
B. VORs: D, E, F go with Courses C, B, A respectively.
C. VORs: D, E, F go with Courses B, A, C respectively.

How to Navigate with VOR

8. [K7/All]
List, in order, the steps needed for navigating to a VOR station.
A. Tune and identify the VOR station, rotate the OBS until a "TO" flag appears, continue rotating the OBS to center the CDI then fly the heading shown under the index.
B. Tune and identify the VOR station, rotate the OBS and center the CDI, turn toward the station until a "TO" flag appears then fly the heading shown under the index.
C. Rotate the OBS until a "TO" flag appears, continue rotating the OBS to center the CDI then fly the heading shown under the index.

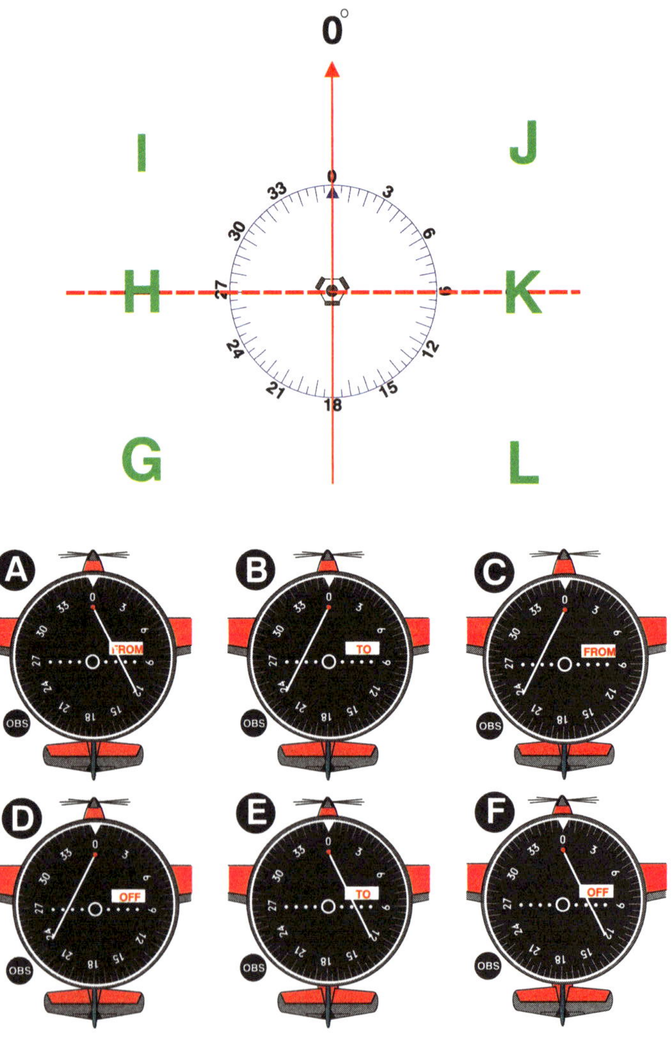

9. [K8/Figure 11]
Referring to the figure above, match the VOR indications with their respective positions in relation to the VOR station.
A. VORs A,B,C,D,E,F, go with positions G,J,L,K,H,I.
B. VORs A,B,C,D,E,F, go with positions I,L,J,K,G,H.
C. VORs A,B,C,D,E,F, go with positions H,G,K,J,L,I.

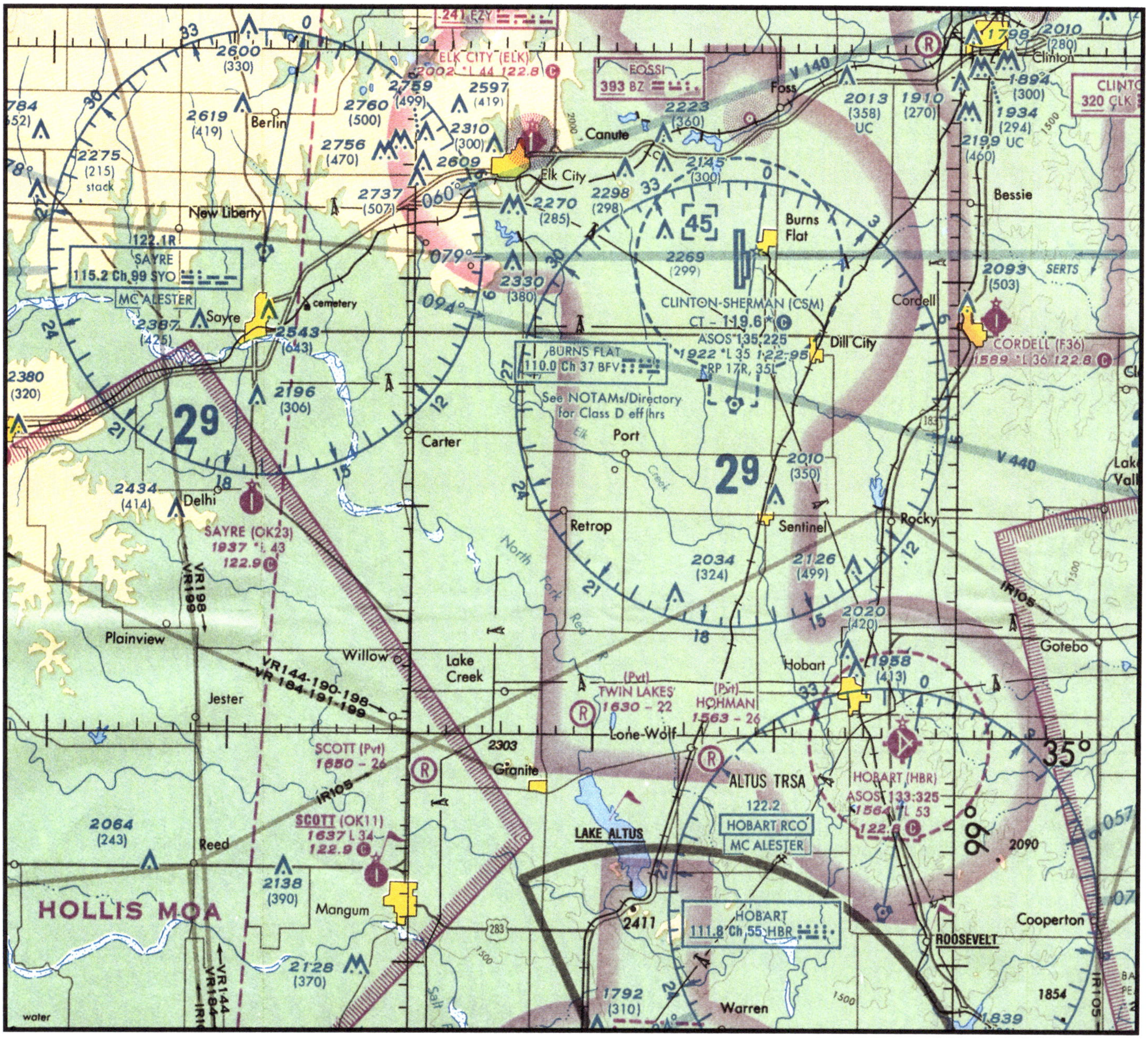

10. [K9/Figure 13]
Referring to the figure above, your VOR is tuned to the Burns Flat VORTAC. The omnibearing selector (OBS) is set on 045, with a TO indication, and a right course deviation indicator (CDI) deflection (as shown in the figure on the right). What is the aircraft's position from the VORTAC?

A. East-northeast.
B. North-northeast.
C. West-southwest.

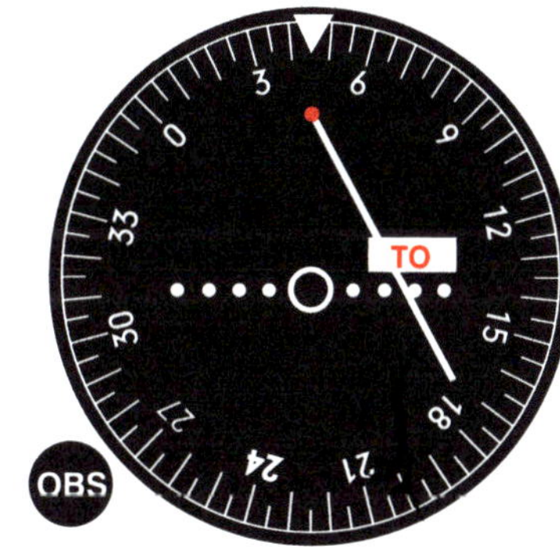

11. [K9/Figure 13]
Referring to the figure above, your VOR is tuned to the Sayre VORTAC. The omnibearing selector (OBS) is set on 145, with a FROM indication, and a right course deviation indicator (CDI) deflection. What is the aircraft's position from the VORTAC?

A. East-southeast.
B. North-northeast.
C. West-southwest.

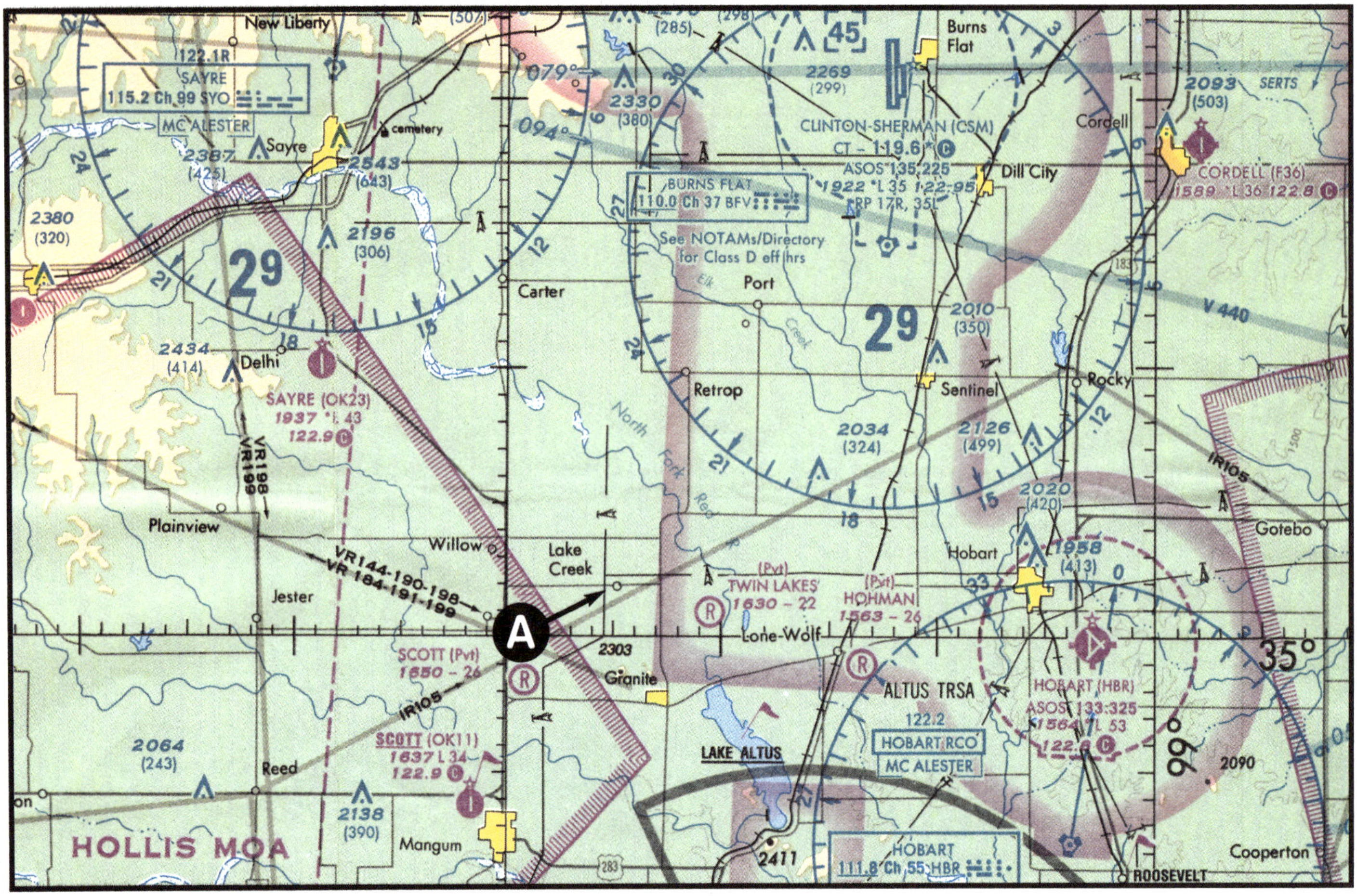

12. [K9/Figure 13]
Referring to the figure above, your VOR is tuned to the Burns Flat VORTAC. The omnibearing selector (OBS) is set on 240, with a FROM indication, and a left course deviation indicator (CDI) deflection as show to the right. What is the aircraft's position from the VORTAC?
A. East.
B. West.
C. South.

13. [K9/Figure 13]
Referring to the figure above, your #1 VOR is tuned to the Sayre VORTAC and the omnibearing selector (OBS) is set on 325, with a TO indication, and a right course deviation indicator (CDI) deflection. VOR #2 is tuned to the Burns Flat VORTAC and the omnibearing selector (OBS) is set on 225, with a FROM indication, and a right course deviation indicator (CDI) deflection (as shown below). What is the aircraft's position from the VORTAC? If you are over an airport with these VOR readings, which airport would it be?
A. Sayre airport.
B. Twin Lakes airport.
C. Scott airport.

14. [K9/Figure 13]
Refer to the figure above. Assume your airplane is positioned over the town of Lake Creek (arrow A). Which VOR indication below is correct?
A. 1.
B. 2.
C. 3.

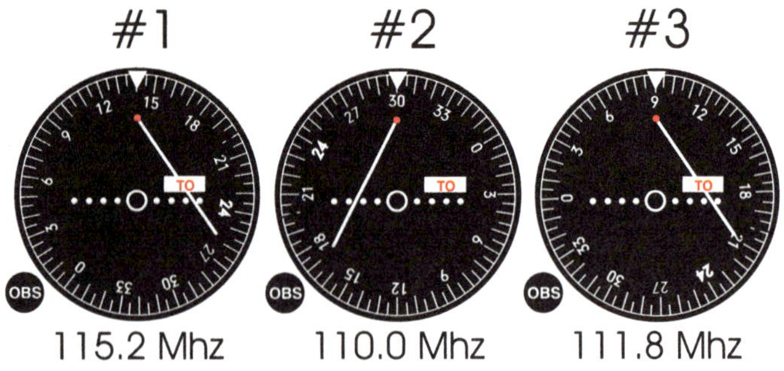

15. [K9/Figure 13]
Referring to the VOR indicator to the right, what radial is the aircraft crossing?
A. 220 degree radial.
B. 040 degree radial.
C. 22 degree radial.

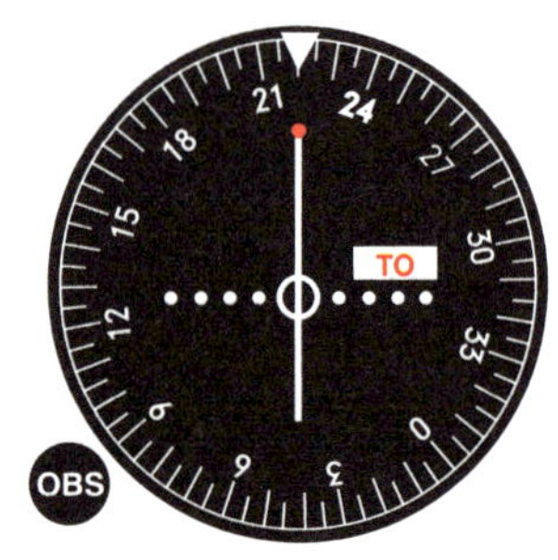

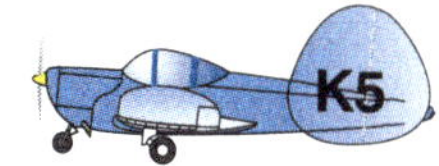

Intercepting a VOR Course

16. [K10/All & Figure 15]
Referring to VOR receiver #1 shown above, what heading should you fly to intercept and track outbound on the 330 degree radial at a 60 degree angle?
A. 270
B. 300
C. 030

17. [K10/All & Figure 15]
Referring to VOR receiver #2 shown above, what heading should you fly to intercept and track inbound on the 060 degree course at a 30 degree angle?
A. 030
B. 210
C. 090

18. [K10/All & Figure 15]
Referring to VOR receiver #3 shown above, what heading should you fly to intercept and track inbound on the 210 degree course at a 40 degree angle?
A. 170
B. 350
C. 250

19. [K10/All & Figure 15]
Referring to VOR receiver #4 shown above, what heading should you fly to intercept and track outbound on the 160 degree radial at a 25 degree angle?
A. 135
B. 185
C. 005

Flying from the VOR on a Selected Course

20. [K11/Figure 16]
Referring to the sectional chart excerpt at the top of the previous page, on what radial should the VOR receiver (OBS) be set to navigate direct from Burns Flat VORTAC to Cordell airport?
A. 064
B. 244
C. 144

21. [K11/Figure 16]
Referring to the sectional chart excerpt at the top of the previous page, to what should the VOR receiver's OBS be set so as to navigate direct from Scott airport to Hobart VORTAC?
A. 084.
B. 264.
C. 840.

Dual VORs for Position Fixing

22. [K12/1/3 & K12/Figure 18]
Referring to the sectional chart excerpt at the top of the previous page, what is your approximate position on low altitude airway Victor 440, east of Clinton-Sherman airport (arrow B), if the VOR receiver indicates you are on the 360 degree radial of Hobart VOR?
A. 22 nautical miles southeast of Hobart VORTAC.
B. 5 nautical miles south of Cordell airport.
C. 28 nautical miles southwest of Hobart VORTAC.

23. [K12/1/3 & K12/Figure 18]
Referring to the sectional chart at the top of the previous page, what is the approximate position of the aircraft if the VOR receivers indicate that you're on the 176 degree radial of Sayre VORTAC and the 221 degree radial of Burns Flat VOR?
A. Town of Reed.
B. Town of Lake Creek.
C. 5 miles east of Reed.

24. [K12/1/3 & K12/Figure 18]
Referring to the sectional chart excerpt at the top of the previous page, what is the approximate position of the aircraft if the VOR receivers indicate the 134 degree radial of Sayre VORTAC and the 292 degree radial of Hobart VORTAC?
A. Hohman Airport.
B. Town of Lake Creek.
C. Twin Lakes Airport.

Reverse Sensing

25. [K13/1/1]
Regarding VOR reverse sensing, don't try to navigate or orient yourself to a VOR station unless a _____ flag is showing. Similarly, don't try to navigate or orient yourself from a VOR station unless a _____ flag shows in the window.
A. FROM, TO
B. TO, FROM
C. OFF, TO

Tracking a Selected VOR Course

26. [K13/All]
What are the four recommended steps in tracking a VOR course?
A. Reintercept the course, identify the effect of wind, apply a wind correction, adjust the wind correction.
B. Identify the effect of wind, reintercept the course, apply a wind correction, adjust the wind correction.
C. Adjust the wind correction, identify the effect of wind, reintercept the course, apply a wind correction.

Chasing the Needle

27. [K14/2/2]
Rapid needle movements are not uncommon within _____ miles of a VOR transmitter.
A. 5
B. 50
C. 25

A Nifty Technique

28. [K14/2/3]
If you're having trouble keeping the VOR needle centered, try the following technique. As the needle is moving away from center, turn toward it to stop the needle's movement. As soon as the needle stops, note your heading. This value approximates the _____ heading required to stay on course.
A. intercept
B. wind correction
C. no wind

Proper Names

29. [K16/All, K16/Figure 22 & K9/Figure 13]
Referring to VOR #1 (upper right, next column), what radial is the aircraft crossing?
A. 030
B. 210
C. 330

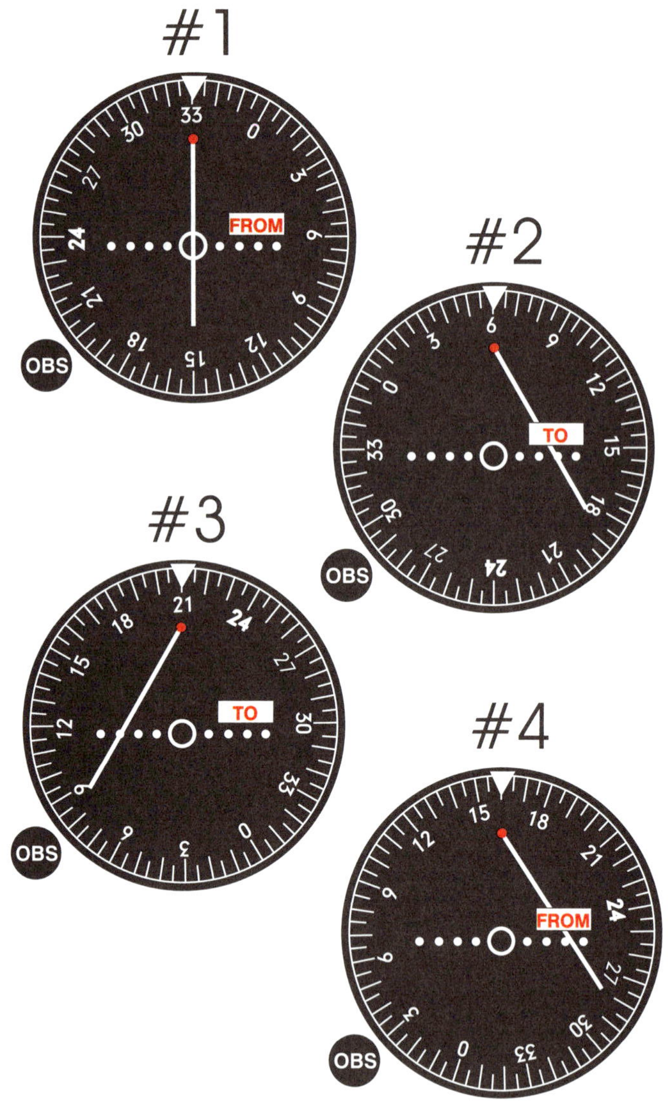

30. [K16/All, K16/Figure 22 & K9/Figure 13]
Referring to VOR #2 above, what is the aircraft's position relative to the station?
A. East.
B. West.
C. South.

31. [K16/All, K16/Figure 22 & K9/Figure 13]
Referring to VOR #3 above, what is the aircraft's position relative to the station?
A. North.
B. West.
C. South.

32. [K16/All, K16/Figure 22 & K9/Figure 13]
Referring to VOR #4 above, What is the aircraft's position relative to the station?
A. North.
B. West.
C. East.

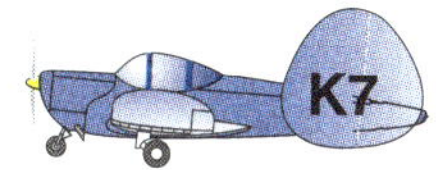

The Horizontal Situation Indicator

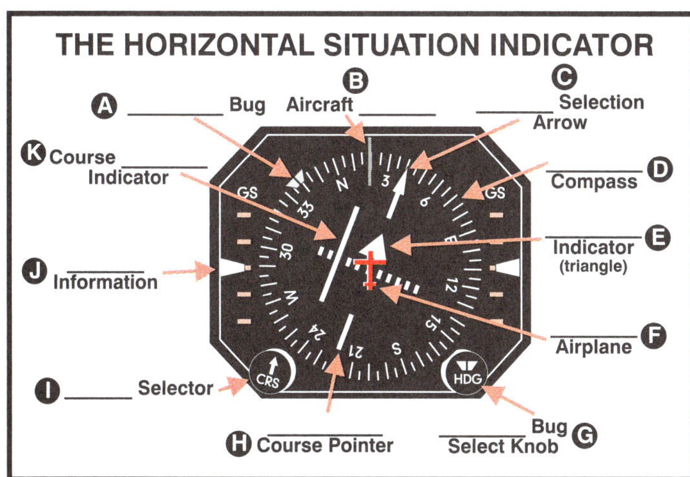

33. [K18/All] Fill in the blanks above:
Referring to the figure above, list all the components of the HSI (Horizontal Situation Indicator).

34. [K18/All]
Based on the two HSI indications immediately above, what is the airplane's position on the sectional chart shown in the top right hand column of this page?
A. The airplane is nowhere within the map borders.
B. Northeast of Tule VOR.
C. Directly over Delano airport.

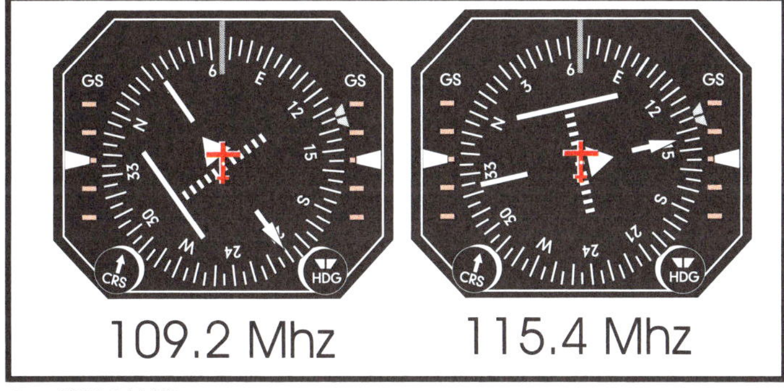

35. [K18/All]
Based on the HSI indications immediately above, what is the airplane's position on the sectional chart shown in the top right hand column of this page?
A. Near Cashen airport.
B. Directly over Delano airport.
C. Near the town of Shafter.

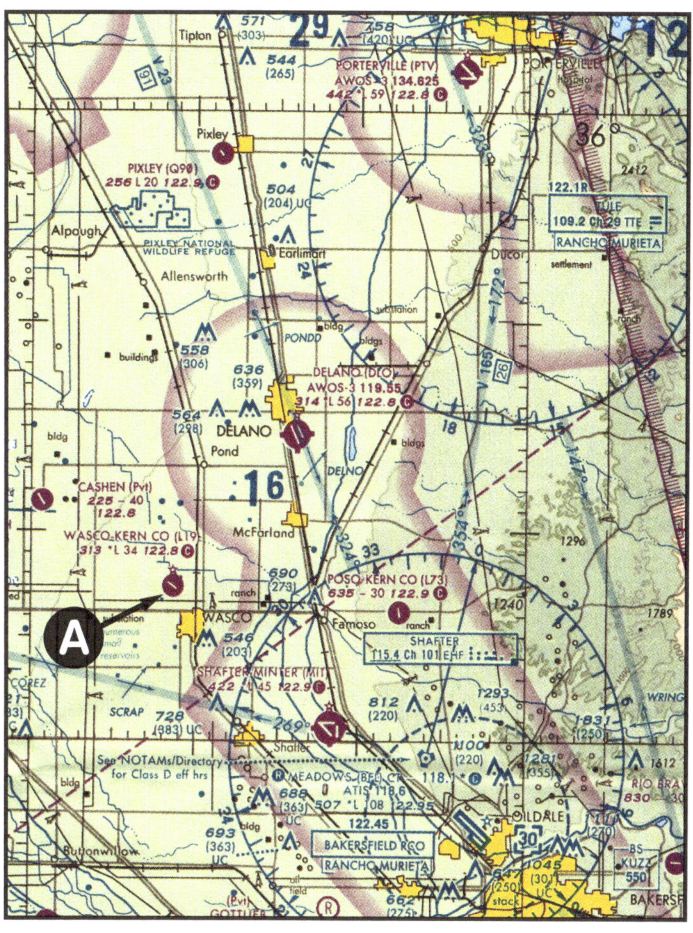

36. [K18/All]
Referring to the figure above, with the airplane positioned over Wasco-Kern airport (arrow A), which VOR indications below are correct?
A. D and A.
B. A and B.
C. B and C.

Distance Measuring Equipment (DME)

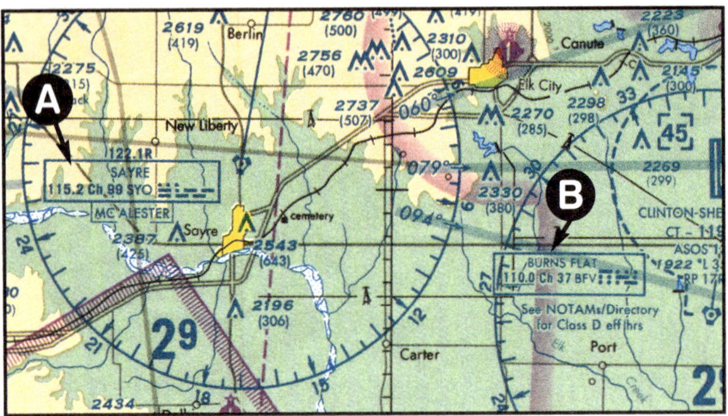

37. [K19/1/3]
In the operation of DME, a _____ sends a signal to an airborne DME receiver. This transmitter often coexists with a VOR. In fact, every specific VOR frequency has a(n) _____ DME frequency pair.
A. ground transmitter, assigned
B. ground transmitter, VHF
C. airborne transmitter, UHF

38. [K19/3/2]
Referring to the sectional chart excerpt above, which of the two VORs shown have DME capability?
A. Only Sayer VOR
B. Sayer and Burns Flat VOR
C. Only Burns Flat VOR

What DME Really Tells You

39. [K19/3/3]
DME gives you
A. actual distance in statute miles from the transmitter.
B. slant range distance from the transmitter.
C. actual horizontal distance from the transmitter.

The Global Positioning System – GPS

40. [K21/1/5 & R17/Column 3, Glossary]
A waypoint is a
A. location identified by VOR cross radials.
B. location identified by latitude/longitude/altitude coordinates.
C. location defined relative to a VORTAC station or in terms of latitude/longitude coordinates.

41. [K24/3/3]
GPS waypoints are defined by
A. latitude and longitude coordinates.
B. VOR radials and DME distances.
C. low frequency NDB stations.

42. [K24/Figure 34]
How many satellites are in the GPS constellation?
A. 25
B. 24
C. 22

43. [K24/Figure 34]
How many GPS satellites are capable of being received from almost any position on earth?
A. 4
B. 6
C. 5

44. [K24/Figure 34] Bonus Question
How many Global Positioning System (GPS) satellites are required to yield a three dimensional position (latitude, longitude, and altitude) and time solution?
A. 5
B. 4
C. 6

45. [K24/Figure 34]
GPS satellites are positioned at an orbit altitude of?
A. 22,500 miles above the earth.
B. 10,900 miles above the earth.
C. 10.9 miles above the earth.

Automatic Direction Finding (ADF) Navigation

46. [K25/3/2]
The ADF needle always points to a _____ beacon (NDB), which is the source of the ADF signal.
A. directional
B. nondirectional
C. omnidirectional

47. [K26/1/1]
When you turn the airplane to keep the ADF needle on the nose, you're performing what kind of navigation?
A. Dead reckoning.
B. Pilotage.
C. Homing.

Postflight Briefing 11-1: VOR Reverse Sensing

48. [K27/See Postflight Briefing #11-1]
VOR reverse sensing occurs when you try to go _____ a VOR with a_____ indication in the ambiguity indicator.
A. from, FROM
B. to, TO
C. to, FROM

49. [K27/See Postflight Briefing #11-1]
If you're experiencing VOR reverse sensing, you'll know it because as you turn toward the needle (the CDI) it will
A. not move at all.
B. move away from its center position.
C. move toward its center position.

Postflight Briefing 11-2: The Radio Magnetic Indicator

50. [K28/1/3]
The RMI is actually a _____ needle mounted over a _____ compass.
A. rotating, slaved
B. rotating, wet
C. stationary, fixed

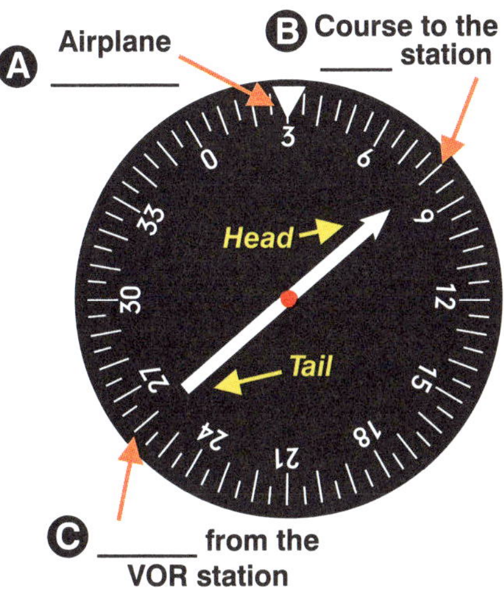

51. [K28/Figure 42]
Referring to the figure above, fill in the blanks.

52. [K28/2/1]
Based on the RMI above, what radial is the airplane on from the VOR station?
A. 330 degree radial.
B. 150 degree radial.
C. 030 degree radial.

53. [K28/2/1]
Based on the RMI above, what is the course to the VOR station:
A. 330 degrees.
B. 030 degrees.
C. 150 degrees.

Postflight Briefing 11-3: ADF: Bearing Down on Homing In

54. [K34/2/2 & K34/Figure 55]
Referring to ADF indicator #1 above, what is the relative bearing TO the station?
A. 030 degrees.
B. 330 degrees.
C. 150 degrees.

55. [K34/2/2 & K34/Figure 55]
Referring to ADF indicator #2 above, what is the relative bearing TO the station?
A. 100 degrees.
B. 280 degrees.
C. 360 degrees.

56. [K34/2/2 & K34/Figure 55]
Referring to ADF indicator #3 above, what is the relative bearing TO the station?
A. 200 degrees.
B. 220 degrees.
C. 040 degrees.

57. [K34/2/2 & K34/Figure 55]
Referring to ADF indicator #4 above, what is the relative bearing TO the station?
A. 140 degrees.
B. 320 degrees.
C. 360 degrees.

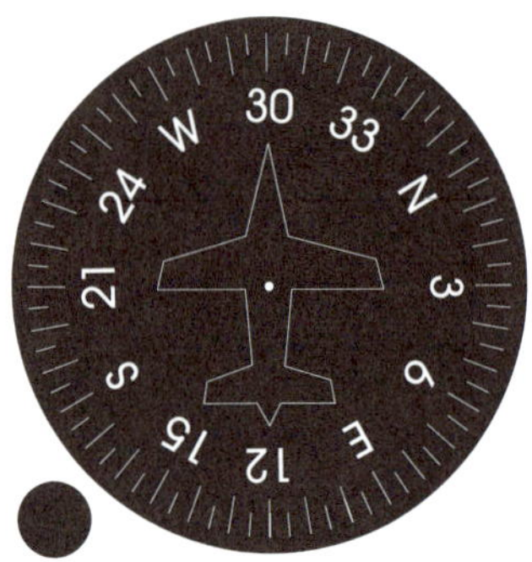

58. [K34/2/5 & Formula on K34]
Referring to the ADF and DG above, determine the magnetic bearing TO the station (MBTS).
A. 330 degrees MBTS.
B. 300 degrees MBTS.
C. 270 degrees MBTS.

59. [K34/2/5 & Formula on K34]
Referring to the ADF and DG above, determine the magnetic bearing TO the station.
A. 330 degrees MBTS.
B. 010 degrees MBTS.
C. 150 degrees MBTS.

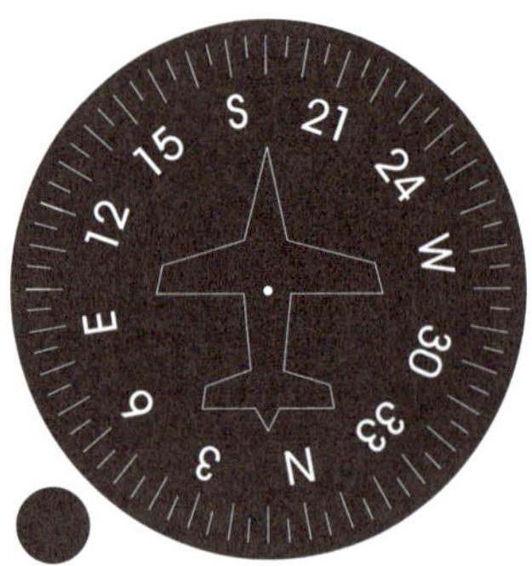

60. [K34/2/5 & Formula on K34]
Referring to the ADF and DG above, determine the magnetic bearing TO the station.
A. 330 degrees MBTS.
B. 070 degrees MBTS.
C. 190 degrees MBTS.

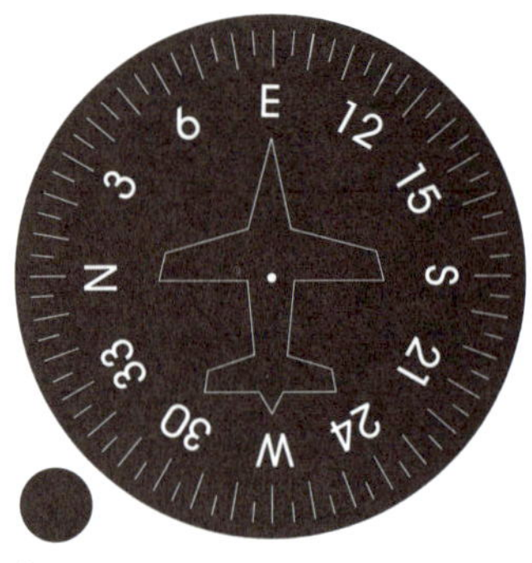

61. [K34/2/5 & Formula on K34]
Referring to the ADF and DG above, determine the magnetic bearing TO the station.
A. 360 degrees MBTS.
B. 270 degrees MBTS.
C. 090 degrees MBTS.

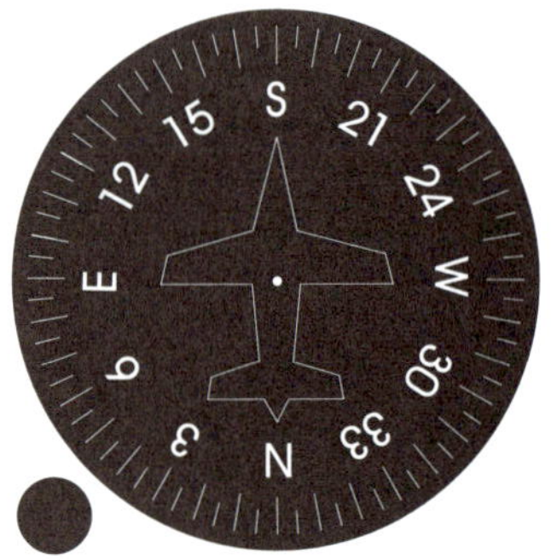

62. [K34/2/5 & Formula on K34]
Referring to the ADF and DG above, determine the magnetic bearing FROM the station.
A. 360 degrees MBFS.
B. 180 degrees MBFS.
C. 270 degrees MBFS.

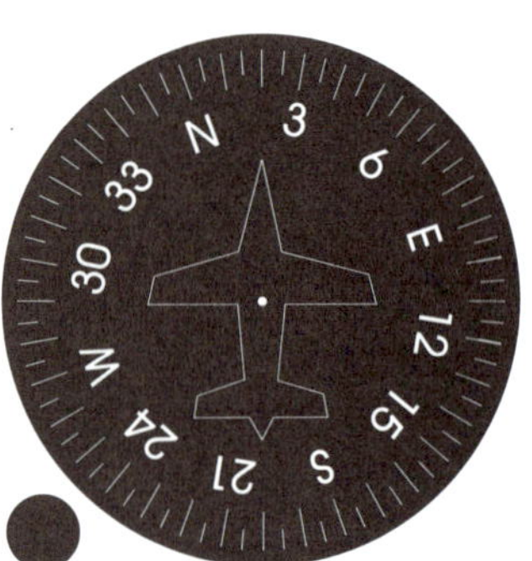

63. [K34/2/5 & Formula on K34]
Referring to the ADF and DG above, determine the magnetic bearing FROM the station.
A. 210 degrees MBFS.
B. 020 degrees MBFS.
C. 050 degrees MBFS.

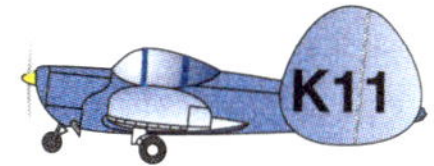

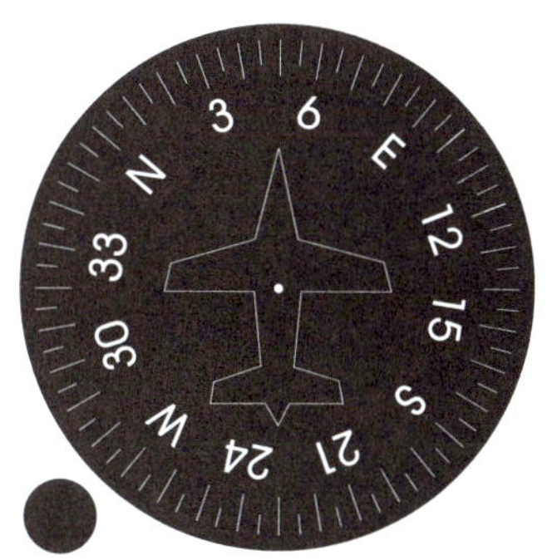

64. [K34/2/5 & Formula on K34]
Referring to the ADF and DG above, determine the magnetic bearing FROM the station.
A. 360 degrees MBFS.
B. 050 degrees MBFS.
C. 310 degrees MBFS.

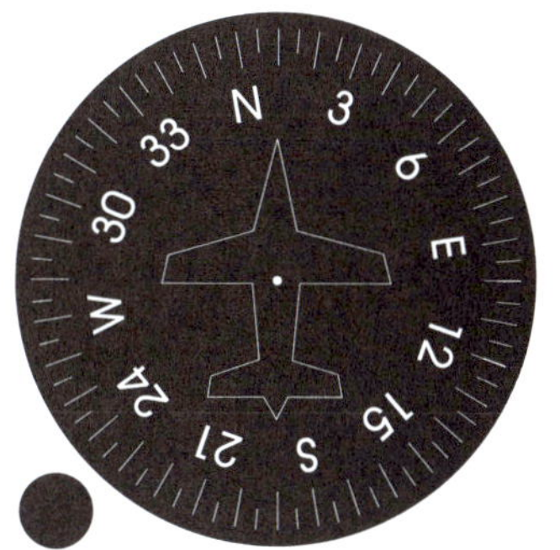

65. [K34/2/5 & Formula on K34]
Referring to the ADF and DG above, determine the magnetic bearing FROM the station.
A. 250 degrees MBFS.
B. 070 degrees MBFS.
C. 010 degrees MBFS.

VOR Test Signal

66. [K38/Postflight Briefing #11-5]
When the course deviation indicator (CDI) needle is centered during an omnireceiver check using a VOR test signal (VOT), the omnibearing selector (OBS) and the TO/FROM indicator should read
A. 180 degrees FROM only if the pilot is due north of the VOT.
B. 0 degrees TO or 180 degrees (FROM, regardless of the pilot's position from the VOT.
C. 0 degrees FROM or 180 degrees TO, regardless of the pilot's position from the VOT.

Chapter Eleven Answers

1. B
2. C
3. B
4. A/165, B/160, C/145
5. A
6. A/index, B/CDI, C/ambiguity, D/rotatable, E/omni bearing selector
7. C
8. A
9. B
10. C
11. A
12. B
13. C
14. C
15. B
16. A
17. C
18. A
19. A
20. A
21. A
22. B
23. A
24. C
25. B
26. B
27. A
28. B
29. C
30. B
31. A
32. C
33. A/heading, B/heading, C/course, D/slaved, E/To-From, F/symbolic, G/heading, H/reciprocal, I/course, J/glideslope, K/deviation
34. C
35. C
36. A
37. A
38. B
39. B
40. C
41. A
42. B
43. C
44. B
45. B
46. B
47. C
48. C
49. B
50. A
51. A/heading, B/VOR C/radial
52. A
53. C
54. B
55. A
56. B
57. A
58. C
59. B
60. B
61. A
62. C
63. C
64. A
65. A
66. C

Note: To ensure that you have the most current answers to these questions, please check the *Book & Slide Updates* section at Rod Machado's web site: www.rodmachado.com

Chapter Twelve

Understanding Weather: Looking for Friendly Skies

Introduction

1. [L1/3/2]
Weather accounts for _____ of the aviation accidents and almost _____ of all aviation fatalities.
A. 40%, 95%
B. 25%, 40%
C. 80%, 10%

Atmospheric Circulation

2. [L2/1/2]
Every physical process of weather is accompanied by, or is the result of, a
A. movement of air.
B. pressure differential.
C. heat exchange.

3. [L2/1/2 & L25/3/2]
What causes variations in altimeter settings between weather reporting points?
A. unequal heating of the Earth's surface.
B. variation of terrain elevation.
C. Coriolis force.

4. [L2/1/5]
Up to a certain altitude in the atmosphere (called the tropopause), temperature generally _____ with altitude.
A. remains constant
B. decreases
C. increases

The Coriolis Force

5. [L2/2/1]
Air is forced to curve _____ by something known as the Coriolis force.
A. to the left
B. upward
C. to the right

6. [L4/Figure5] Fill in the blanks below:

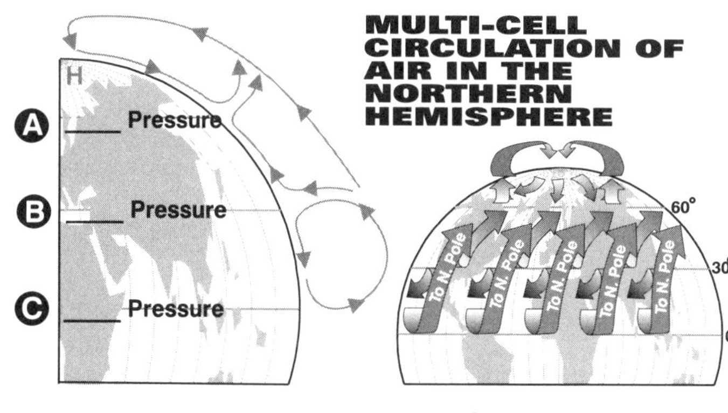

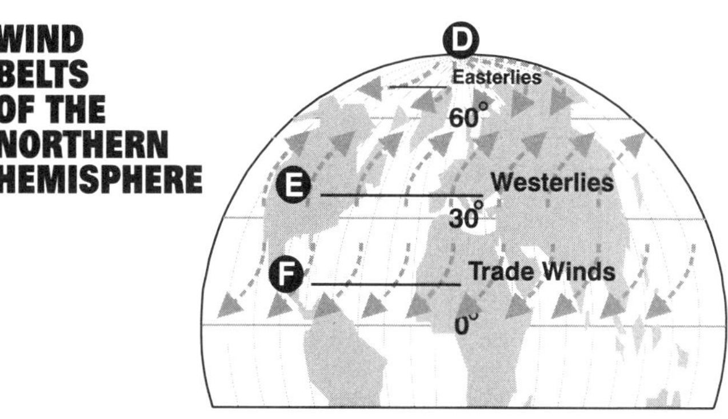

Air Pressure and Vertical Air Movement

7. [L5/1/3]
Air has _____.
A. no weight
B. no molecules
C. weight

8. [L5/1/4]
Along the equator there exist areas of _____ air.
A. colder (rising)
B. warmer (descending)
C. warmer (rising)

9. [L5/2/1]
Rising air certainly _____ create as much pressure on the surface as descending air.
A. does
B. does not
C. cannot

10. [L5/2/2]
Permanent high pressure areas exist at the _____. Colder (more dense) air descends, creating _____ pressure on the earth's surface.
A. poles, low
B. equator, high
C. poles, high

Getting Water in the Air

11. [L6/1/1]
What are the processes by which moisture is added to unsaturated air?
A. Evaporation and sublimation.
B. Heating and condensation.
C. Supersaturation and evaporation.

12. [L6/1/4]
The air is said to be _____ when additional water vapor is unable to enter the air.
A. evaporated
B. saturated
C. humidified

13. [L6/1/4]
The amount of water vapor which air can hold depends on the
A. dewpoint.
B. air temperature.
C. stability of the air.

The Water Content of Warm and Cold Air

14, [L6/3/2]
Clouds, fog, or dew will always form when
A. water vapor condenses.
B. water vapor is present.
C. relative humidity reaches 100 percent.

15. [L6/3/2]
The process of water vapor becoming visible is called _____. When water vapor condenses it means that the air is _____ and can't hold any more water in vapor form.
A. condensation, saturated
B. condensation, moisturization
C. dewpoint, saturated

Two Ways to Cool Air

16. [L7/1/3]
There are two ways to cool air and condense water vapor. You can put the air in contact with a _____ surface or environment. Second, you can _____ the air and let it cool by expansion.
A. warmer, lower
B. cooler, lower
C. cooler, lift

17. [L7/2/2]
Fog formation is often based on air coming into contact with a _____ surface.
A. warmer
B. cooler
C. wetter

18. [L7/3/1]
Air that rises always _____ and _____.
A. expands, heats up
B. expands, cools
C. compresses, heats up

Relative Humidity

19. [L8/1/4]
Relative humidity is a number that tells you how much _____ the air is holding in relationship to how much it could theoretically hold at its current _____.
A. water vapor, temperature
B. water vapor, humidity
C. pressure, volume

20. [L8/Figure 15]
Cooling the air _____ its relative humidity.
A. increases
B. decreases
C. doesn't affect

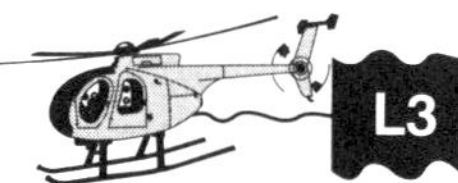

21. [L8/Figure 15]
Heating the air _____ its relative humidity.
A. increases
B. decreases
C. doesn't affect

22. [L9/1/5]
While relative humidity informs us about the air's capacity to hold _____, it doesn't allow us to predict when _____ will form.
A. air molecules, clouds
B. water vapor, clouds
C. particulate matter, rain

The Dewpoint

23. [L9/2/2]
What is meant by the term "dewpoint"?
A. The temperature at which condensation and evaporation are equal.
B. The temperature at which dew will always form.
C. The temperature to which air must be cooled to become saturated.

24. [L9/2/3]
Meteorologists refer to the actual temperature difference between the temperature and the dewpoint as the
A. subtraction.
B. temperature-dewpoint spread.
C. temperature-humidity spread.

25. [L9/3/2]
The dewpoint is a great indicator of the atmosphere's _____. High dewpoint temperatures indicate a lot of water in the air; _____ dewpoint temperatures indicate little water in the air.
A. pressure, low
B. water content, low
C. water content, high

26. [L10/1/1]
If the temperature/dewpoint spread is small and decreasing, and the temperature is 62 degrees F, what type weather is most likely to develop?
A. Freezing precipitation.
B. Thunderstorms.
C. Fog or low clouds.

27. [L10/1/2]
Be aware that fog or low cloud formation can even occur with temperature-dewpoint spreads of as much as _____ degrees F. Also, water vapor condenses more easily when the air contains an abundance of _____.
A. 10, condensation nuclei
B. 6, water
C. 6, condensation nuclei

Condensation and Cloud Formation

28. [L10/1/6]
For significant precipitation to occur, clouds must generally be around _____ feet thick.
A. 4,000
B. 10,000
C. 2,000.

Lapse Rates and Temperature Inversions

29. [L10/2/2]
The sun heats the _____. Very little of the sun's energy directly heats the _____.
A. earth, air
B. air, earth
C. water, air

30. [L10/2/2]
It's the _____ that heats or cools the air directly.
A. sun
B. moisture in the air
C. surface of the earth

31. [L11/1/4]
The _____ is essentially a vertical temperature profile of the atmosphere.
A. environmental temperature lapse rate
B. wet pseudo adiabatic lapse rate
C. vertical temperature scale

32. [L11/1/6]
The development of thermals depends upon
A. a counterclockwise circulation of air.
B. temperature inversions.
C. solar heating (of the earth).

33. [L11/Figure 20]
Referring to the figure below, which temperature profile lines represent a temperature drop with an increase in altitude?
A. line C only.
B. lines C and A.
C. line B only.

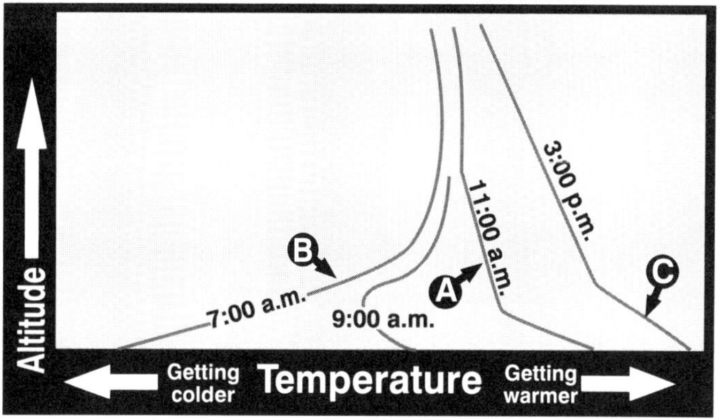

Temperature Inversions

Surface Inversions

34. [L12/1/2]
A temperature inversion occurs when the normal decreasing temperature lapse rate is turned upside down, or inverted. There are two types of temperature inversions: a _____ and an _____.
A. warm on top inversion, inverted inversion
B. surface inversion, inversion aloft
C. top cooling, over the water inversion

35. [L12/1/3]
The most frequent type of ground- or surface-based temperature inversion is produced by
A. terrestrial radiation on a clear, relatively still night.
B. warm air being lifted rapidly aloft in the vicinity of mountainous terrain.
C. the movement of colder air under warm air, or the movement of warm air over cold air.

36. [L12/Figure 22]
A temperature inversion would most likely result in which weather condition?
A. Clouds with extensive vertical development above an inversion aloft.
B. Good visibility in the lower levels of the atmosphere and poor visibility above an inversion aloft.
C. An increase in temperature as altitude is increased.

37. [L12/Figure 21]
Fill in the blanks in the figure below:

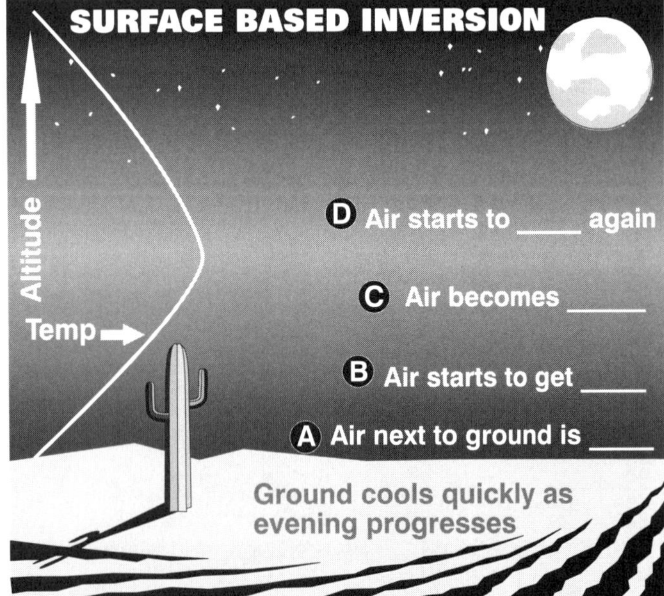

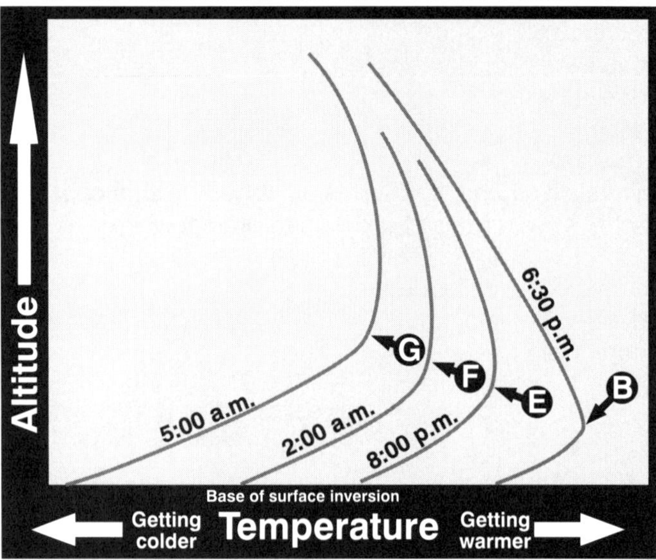

38. [L12/Figure 22]
Referring to the figure above, positions B, E, F and G represent the
A. bottom of the surface inversion at different times of the day.
B. top of the surface inversion at different times of the day.
C. base of the inversion aloft.

39. [L13/1/3]
Which conditions result in the formation of frost?
A. The temperature of the collecting surface is at or below freezing when small droplets of moisture fall on the surface.
B. The temperature of the collecting surface is at or below the dewpoint of the adjacent air and the dewpoint is below freezing.
C. The temperature of the surrounding air is at or below freezing when small drops of moisture fall on the collecting surface.

Inversion Aloft

40. [L13/3/1]
Inversions aloft are caused by _____ air moving horizontally over _____ air or by air aloft that is descending and _____ by compression.
A. colder, warmer, heated
B. warmer, colder, heated
C. warmer, colder, cooled

41. [L14/1/2]
When warm, _____ air subsides (lowers) onto the cool, moist inflow of air from the ocean, _____ forms.
A. high pressure, an inversion aloft
B. low pressure, a surface-based inversion always
C. high pressure, a mid-layer inversion

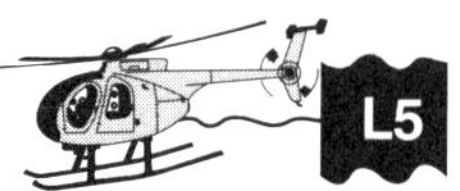

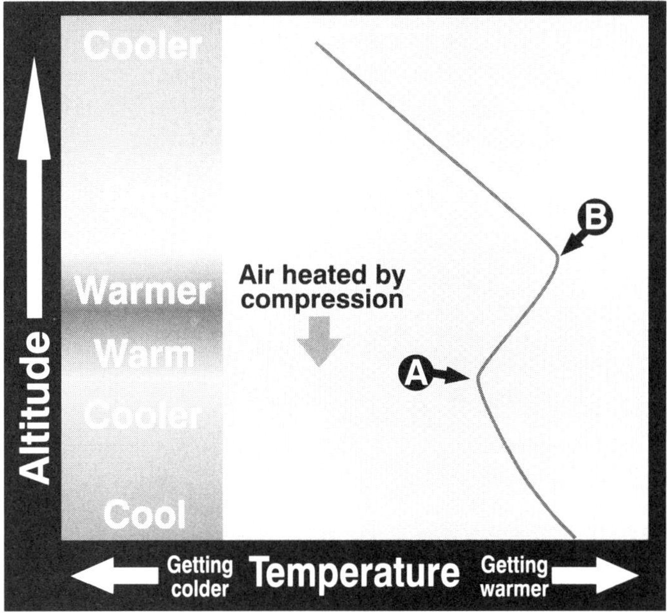

42. [L14/Figure 24]
Referring to the figure above, what do positions A and B represent?
A. Position A is the top of the surface-based inversion and position B is the base of the inversion aloft.
B. Position A is the base of the inversion aloft and position B is the top of the inversion aloft.
C. Position A is where the air begins to cool and position B is where the air begins heating up.

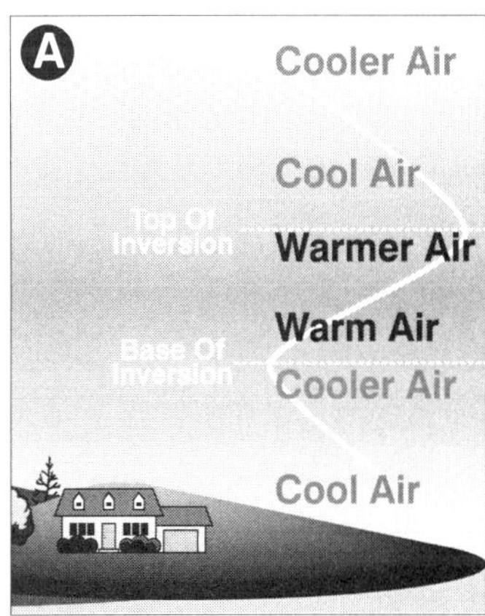

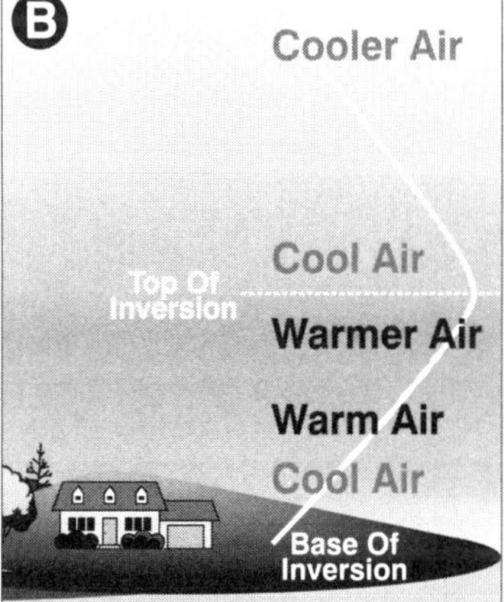

43. [L14/Figure 25]
Referring to the figure above, what type of temperature inversion does the profile in each box represent?
A. Box A represents an inversion aloft while box B represents a surface-based inversion.
B. Box A represents a surface-based inversion while box B represents an inversion aloft.
C. Box A represents a surface-based inversion and box B also represents a surface-based inversion.

Effects of Temperature Inversions

44. [L14/1/6 & L15/1/1&2]
Which weather conditions should be expected beneath a low-level temperature inversion layer when the relative humidity is high?
A. Smooth air, poor visibility, fog, haze, or low clouds.
B. Light wind shear, poor visibility, haze, and light rain.
C. Turbulent air, poor visibility, fog, low stratus type clouds, and showery precipitation.

What to Expect in an Inversion

45. [L16/1/1]
The visible signs of a temperature inversion should cue pilots to expect
A. the chance of heavy rain.
B. the possibility of cumulus clouds.
C. increased likelihood of cloud formation.

46. [L16/1/2]
The visible signs of a temperature inversion should cue pilots to expect
A. increased visibility beneath the inversion.
B. decreased visibility beneath the inversion.
C. increased chance of virga.

47. [L16/1/5]
In the presence of an inversion, if surface winds are calm and winds at a few hundred to a few thousand feet AGL are in the 25 knot range, be prepared for possible _____.
A. thunderstorms
B. wind shear
C. icing

48. [L16/2/1]
Where can wind shear occur?
A. Only at higher altitudes.
B. Only at lower altitudes.
C. At all altitudes, in all directions.

49. [L16/2/2]
Temperature inversions indicate the presence of
A. stable air.
B. unstable air.
C. neutrally stable air.

Atmospheric Stability: Warm Over Cold, and Cold Over Warm

50. [L17/2/2]
The atmosphere is stable when it _____ the upward movement of air. It is unstable when it _____ or promotes this upward movement.
A. amplifies, limits
B. permits, resists
C. resists, permits

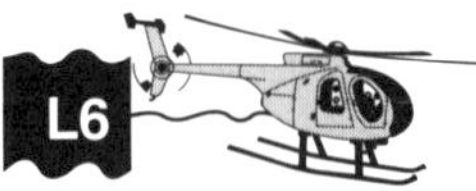

51. [L17/3/1]
Warm air resting on top of a cold layer of air would be considered
A. a stable condition.
B. an unstable condition.
C. a neutrally stable condition.

52. [L17/3/2]
What feature is associated with a temperature inversion?
A. A stable layer of air.
B. An unstable layer of air.
C. Chinook winds on mountain slopes.

53. [L17/3/2]
What would increase the stability of an air mass?
A. Warming from below.
B. Cooling from below.
C. An increase in water vapor.

54. [L17/3/3]
Cold air resting on top of a warmer layer of air would be considered
A. a stable condition.
B. an unstable condition.
C. a neutrally stable condition.

55. [L17/3/3]
What would decrease the stability of an air mass?
A. Warming from below.
B. Cooling from below.
C. Decrease in water vapor.

STABLE AND UNSTABLE AIR

56. [L17/Figure 31]
Referring to the figure above, which box represents stable air?
A. Box A.
B. Box B.
C. Both boxes.

The Environmental Lapse Rate

57. [L18/1/3]
The environmental temperature lapse rate is the actual rate at which _____ temperature changes with altitude.
A. atmospheric
B. a moving air parcel's
C. Both of the above.

58. [L18/2/2&3]
How do meteorologists determine the environmental lapse rate?
A. They estimate it at 3.5 degrees F per thousand feet.
B. One way is to use a rawinsonde
C. The environmental lapse rate never changes.

59. [L18/3/3]
If the temperature decreases with altitude, then the _____ is above you and the _____ is below. Some _____ is sure to be present.
A. warm air, colder air, instability
B. colder air, warmer air, instability
C. neutral air, cooler air, stability

Rising Parcels of Air

60. [L 20/1/4]
What measurement can be used to determine the stability of the atmosphere?
A. Atmospheric pressure.
B. Actual lapse rate.
C. Surface temperature.

61. [L19/1/2]
A rising parcel of air will
A. compress.
B. expand.
C. remain unchanged.

62. [L19/1/4]
As a parcel of air rises it
A. cools.
B. heats up.
C. experiences no temperature change.

63. [L19/1/4]
A parcel of air, moving vertically, _____ and _____ with predictability.
A. expands, cools
B. expands, heats up
C. compresses, cools

64. [L19/1/6]
A rising parcel of air expands and cools at a constant rate of _____ for every thousand feet of ascent as long as that parcel remains unsaturated (doesn't form a cloud).
A. 3.5 degrees C
B. 5.4 degrees F
C. 3.5 degrees F

65. [L19/2/1]
While a rising parcel of unsaturated air expands and cools at a constant rate, the rate at which the atmosphere changes temperature _____.
A. is consistent
B. is inconsistent
C. is incapable of being measured

EXPANSION & COOLING OF RISING AIR

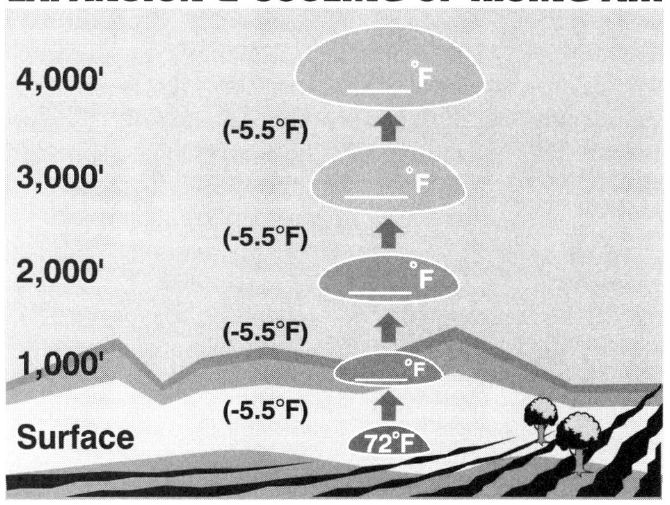

66. [L19/Figure 32]
Based on the figure above, assume a parcel of unsaturated air starts at the surface at a temperature of 72 degrees. After the parcel rises to 4,000 feet, what will its temperature be? (Assume a lapse rate of 5.5 degrees F per thousand feet for convenience of calculating.)
A. 58 degrees F.
B. 94 degrees F.
C. 50 degrees F.

67. [L19/Figure 33]
Referring to the figure below, assume that the environmental lapse rate is 6.5 degrees F/1000 feet. If a bubble of unsaturated air starts at the surface at 68 degrees F, will it continue to rise and reach point C or will it stop rising before point C?
A. It will continue to rise to point C.
B. It will stop rising before point C.
C. You are unable to tell with the given information.

ENVIRONMENTAL VS. RISING AIR PARCEL LAPSE RATES

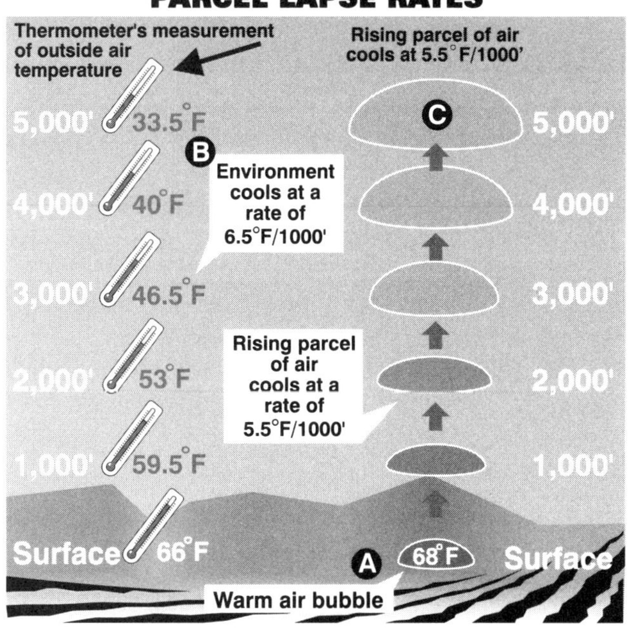

68. [L19/2/4 & L20/1/5]
If the environmental lapse rate is 6 degrees F per thousand feet and a parcel of unsaturated air begins rising, it will most likely become
A. stable.
B. unstable.
C. neutrally stable.

69. [L19/Figure 32 & L20/1/5]
During a climb you notice that the airplane's thermometer indicates an environmental temperature decrease of 6 degrees F per thousand feet. Therefore, the atmosphere is potentially
A. stable.
B. unstable.
C. neutrally stable.

70. [L20/1/3 & 6]
During a climb you notice that the airplane's thermometer indicates an environmental temperature decrease of 4 degrees F per thousand feet. Therefore, the atmosphere is potentially
A. stable.
B. unstable.
C. neutrally stable.

71. [L20/1/7] Fill in the blank:
During a climb you notice that the airplane's thermometer indicates an environmental temperature decrease of 5.5 degrees F per thousand feet. Therefore, the atmosphere is likely to be ____________ stable.

Saturated Parcels of Rising Air

72. [L21/1/2]
Once rising parcels cool to within a few degrees of their dewpoint, _____ occurs and _____ form.
A. condensation, clouds
B. wind, wind shear
C. rain, thunderstorms

73. [L21/1/3]
A rising air parcel that's forming clouds doesn't _____ as quickly as a rising unsaturated air parcel that isn't forming clouds.
A. heat up
B. cool
C. sublimate

74. [L21/1/3]
_____ parcels of air cool at rates between 2 degrees F and 5 degrees F per thousand feet, depending on how much water vapor (and thus how much trapped heat) was in the air to begin with.
A. Unsaturated
B. Dry
C. Saturated

75. [L21/See *Water Vapor & Instability*]
If the atmosphere contains a great deal of water vapor, then the temperature of a saturated, rising parcel of air within the atmosphere might decrease at only 2 degrees F per thousand feet. This means the air parcel is being kept warm by the release of its _____ as water vapor condenses.
A. latent cooling
B. latent heat
C. latent humidity

76. [L21/See *Water Vapor & Instability*]
A moist atmosphere is potentially more _____ than a dry one. Instability, however, is predicated on the condition that a rising air parcel within the atmosphere is _____ enough to reach saturation.
A. unstable, lowered
B. stable, compressed
C. unstable, lifted

77. [L21/See *Blowing in the Wind*]
Think of evaporation as a process opposite that of condensation. While evaporation puts water into the atmosphere, condensation gives the water back. If _____ takes heat away, condensation must give heat back. The heat given back is known as the _____.
A. evaporation, latent heat of condensation
B. latent heat, evaporation factor
C. water, latent heat of fusion

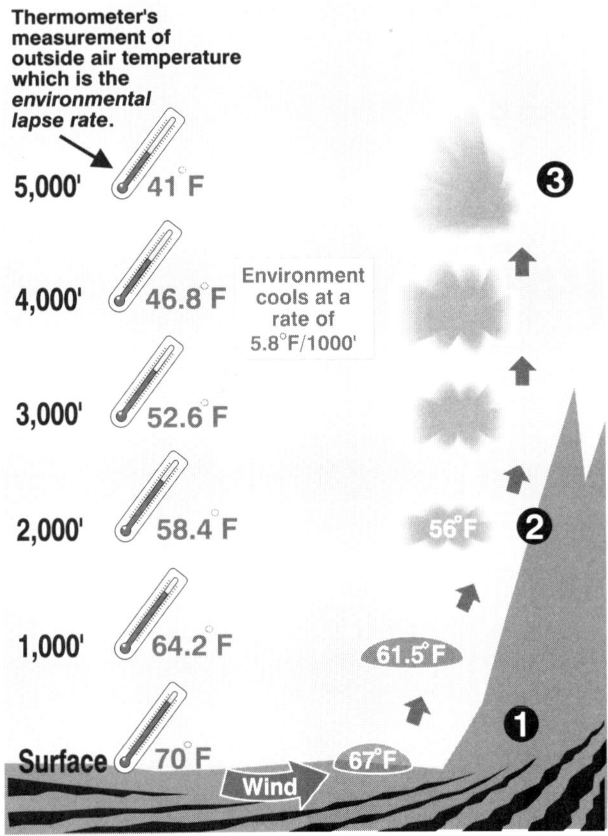

78. [L22/Figure 36]
Referring to the figure in the bottom left hand corner of this page, assume that the environmental lapse rate is 5.5 degrees F per thousand feet. Also assume that the saturated parcel in this example cools at a rate of 2.8 degrees F per thousand feet. As the parcel of air shown in position 1 is blown up the side of the mountain and begins to condense at position 2, what will happen to the parcel?
A. It will eventually become unstable and cumulus clouds may form.
B. It will remain stable and stratus clouds may form.
C. The parcel will begin sinking by the time it reaches position 3.

Clouds and Atmospheric Stability

79. [L23/1/1]
The conditions necessary for the formation of cumulonimbus clouds are a lifting action and
A. unstable air containing an excess of condensation nuclei.
B. unstable, moist air.
C. either stable or unstable air.

80. [L23/1/2]
What cloud types would indicate convective turbulence?
A. Cirrus clouds.
B. Nimbostratus clouds.
C. Towering cumulus clouds.

81. L23/1/1 & Figure 37]
If an unstable air mass is forced upward, what type clouds can be expected?
A. Stratus clouds with little vertical development.
B. Stratus clouds with considerable associated turbulence.
C. Clouds with considerable vertical development and associated turbulence.

82. [L23/Figure 37]
What are characteristics of unstable air?
A. Turbulence and good surface visibility.
B. Turbulence and poor surface visibility.
C. Nimbostratus clouds and good surface visibility.

83. [L23/Figure 37]
What are characteristics of a moist, unstable air mass?
A. Cumuliform clouds and showery precipitation.
B. Poor visibility and smooth air.
C. Stratiform clouds and showery precipitation.

84. [L23/Figure 38]
What is a characteristic of stable air?
A. Stratiform clouds.
B. Unlimited visibility.
C. Cumulus clouds.

85. [L23/Figure 38]
Moist, stable air flowing upslope can be expected to
A. produce stratus type clouds.
B. cause showers and thunderstorms.
C. develop convective turbulence.

86. [L23/Figure 38]
A stable air mass is most likely to have which characteristic?
A. Showery precipitation.
B. Turbulent air.
C. Smooth air.

87. [L23/Figure 38]
Steady precipitation preceding a front is an indication of
A. stratiform clouds with moderate turbulence.
B. cumuliform clouds with little or no turbulence.
C. stratiform clouds with little or no turbulence.

88. [L23/See *Nearer My Cloud to Thee*]
At approximately what altitude above the surface would the pilot expect the base of cumuliform clouds if the surface air temperature is 82 degrees F and the dewpoint is 38 degrees F?
A. 9,000 feet AGL.
B. 10,000 feet AGL.
C. 11,000 feet AGL.

89. [L23/See *Nearer My Cloud to Thee*]
What is the approximate base of the cumulus clouds if the surface air temperature at 1,000 feet MSL is 70 degrees F and the dewpoint is 48 degrees F?
A. 4,000 feet MSL.
B. 5,000 feet MSL.
C. 6,000 feet MSL.

90. [L24/1/1]
Stratus clouds are the result of water vapor condensing in air parcels that have little vertical movement. For instance, if the atmosphere becomes _____ with height (such as in an inversion), parcels of warm air are prevented from rising.
A. colder
B. wetter
C. warmer

91. [L24/1/2]
Stratus clouds often signal the presence of a _____. Limited vertical air movement in stable air means _____.
A. thunderstorms, great visibility
B. temperature inversion, poor visibility
C. convective turbulence, reduced visibility

92. [L24/Figure 39]
One square inch of air standing from the earth's surface to the top of the atmosphere weighs
A. 147 pounds.
B. 14.7 pounds.
C. .147 pounds.

High and Low Pressure Areas

93. [L25/3/2]
Variations in temperature, over water and land, cause changes in _____.
A. the Coriolis force
B. gravity
C. atmospheric pressure

94. [L25/3/2]
The basic reason for the existence of high and low pressure areas is
A. the Coriolis force.
B. atmospheric moisture differences.
C. temperature differences.

Sea and Land Breeze Circulation

95. [L26/1/2]
It's called a *sea breeze* because the air blows from the _____ to the _____.
A. sea, land
B. land, sea
C. sea, sea

96. [L26/1/2]
Convective circulation patterns associated with sea breezes are caused by
A. warm, dense air moving inland from over the water.
B. water absorbing and radiating heat faster than the land.
C. cool, dense air moving inland from over the water.

97. [L26/Figure 43]
Referring to the figure below, describe circulations A and B.
A. Circulations at position A are high pressure while circulation at position B is low pressure.
B. Circulations at position A are low pressure while circulation at position B is high pressure.
C. Circulation A and B are at equal pressure.

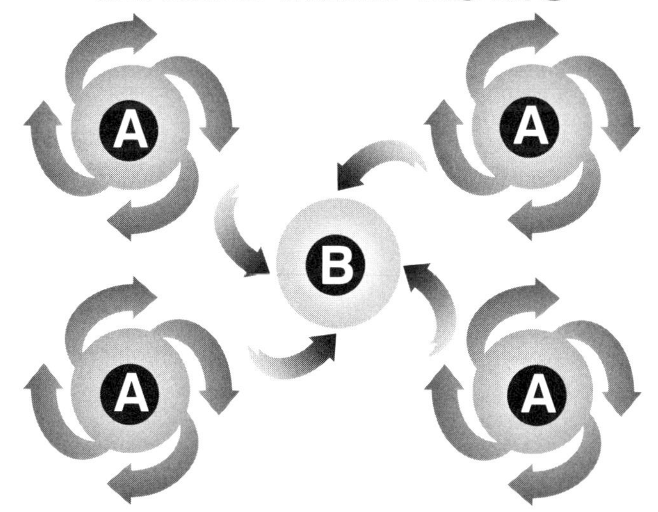

Keeping It in Perspective

98. [L27/1/2]
It's important to think about high and low pressure centers as large oceans of air. These oceans of air are the _____ within which smaller parcels of air move.
A. environments
B. parcels
C. same size parcels

Highs and Lows on Weather Maps

99. [L27/1/3]
High pressure centers are cooler and _____ masses of air. Moving _____ and outward, they rotate in a _____ direction.
A. more dense, upward, clockwise
B. more dense, downward, clockwise
C. thinner, downward, counterclockwise.

100. [L27/1/3]
Low pressure centers are typically _____, less dense masses of air. Air moves _____ and upward and rotates _____ in a low pressure system.
A. warmer, inward, counterclockwise
B. cooler, inward, clockwise
C. warmer, inward, clockwise

101. [L27/1/3 & Figure 44]
On weather maps, high pressure and low pressure systems are represented by a series of contour lines called
A. pressure altitudes.
B. isocreams.
C. isobars.

102. [L27/2/1]
Isobars connect areas of
A. equal barometric pressure.
B. equal pressure altitudes.
C. level altitude levels.

Circulation in Highs and Lows: Going With the Flow

103. [L28/1/1]
Air should flow directly from the high to the low because of something known as _____.
A. pressure gradient force
B. the Jedi force
C. pressure altitude force

104. [L28/1/1]
As the descending high pressure air settles and spreads outward, the _____ force adds a right curve to its motion. This explains the _____ circulation of the air.
A. wind's, counterclockwise
B. Coriolis, clockwise
C. pressure, clockwise

105. [L28/1/1]
Air _____ in a low pressure system. Eventually, the air is forced into a _____ circulation around the low.
A. diverges away from the center, clockwise
B. diverges away from the center, counterclockwise
C. converges toward the center, counterclockwise

106. [L28/1/2]
Because of these two forces—pressure gradient and the Coriolis force—air circulation around a high or low flows _____ to the isobars instead of _____ them.
A. parallel, across
B. perpendicular, across
C. across, perpendicular to

107. [L28/1/3&4]
Surface friction means the right curve added to the air diminishes slightly. This gives the wind the appearance of turning slightly _____ of the isobars within 2,000 feet of the surface.
A. perpendicular
B. right
C. left

108. [L28/1/3&4]
The wind at 5,000 feet AGL is southwesterly while the surface wind is southerly. This difference in direction is primarily due to
A. a stronger pressure gradient at higher altitudes.
B. friction between the wind and the surface.
C. the Coriolis force, which is stronger at the surface.

The Answer is Flowin' in the Wind

109. [L30/1/3&4]
As high pressure air descends and makes contact with the surface it _____ and as low pressure air rises, the air around it _____ toward the center of the low.
A. diverges, moves perpendicular
B. converges, diverges
C. diverges, converges

110. [L30/1/5]
Closely spaced isobars indicate a _____ pressure change which means faster winds. Isobar comparison around the high and low provides a good indication of _____ in that system.
A. rapid, frontal movement
B. slow, temperature change
C. rapid, wind speed

111. [L30/1/6]
Weak pressure gradients on a weather map are shown by
A. solid lines.
B. dashed lines.
C. hooked lines.

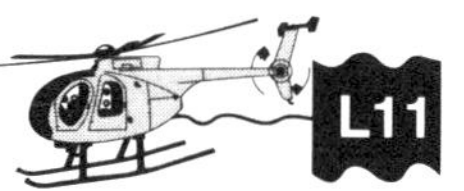

Weather Associated With Highs and Lows

112. [L30/1/7]
As high pressure air descends, it tends to _____ slightly, decreasing the relative humidity.
A. cool down
B. warm up
C. expand

113. [L30/1/7]
Clouds are _____ likely to form with decreasing relative humidity, and clouds that are present are likely to dissipate. Highs are generally associated with _____ skies.
A. more, clear
B. more, dark
C. less, clear

114. [L31/1/1]
Areas of lows pressure have _____ air that expands and cools. If the air within the low approaches its dewpoint, condensation occurs. If this happens, clouds appear and the weather usually _____.
A. lowering, gets worse
B. lowering, gets better
C. rising, gets worse

115. [L31/1/1]
Air that condenses in a low pressure system releases its latent heat, which _____ the heat content of the low pressure system. The low pressure area is now slightly warmer, causing it to rise _____. This further increases the chance that the weather in a low pressure area will continue deteriorating.
A. adds to, slower
B. subtracts from, slower
C. adds to, faster

Ridges and Troughs

116. [L31/1/6 & L32/1/1] Fill in the blanks:
Elongated low pressure area A is known as a _____ and elongated high pressure area B is known as a _____.

TROUGHS AND RIDGES

117. [L31/1/6]
An elongated area of low pressure is known as a _____.
A. ridge
B. trough
C. pressure pothole

118. [L32/1/2]
An elongated area of high pressure is known as a _____.
A. ridge
B. trough
C. pressure pothole

Frontal Systems

119. [L32/1/5]
The science of how fronts form is known as
A. frontogenesis.
B. frontolysis.
C. frontostartysis.

120. [L32/2/1]
The boundary between two different air masses is referred to as _____.
A. frontolysis
B. frontogenesis
C. a front

121. [L32/2/2]
Frontal systems start with large air masses. These masses of air take on the characteristics of the _____ over which they form and originate.
A. source region
B. altitudes
C. pressure centers

122. [L32/Figure 54] Fill in the blanks:
Referring to the figure below, name the four most common source regions for air masses
A. ____________________ C. ____________________
B. ____________________ D. ____________________

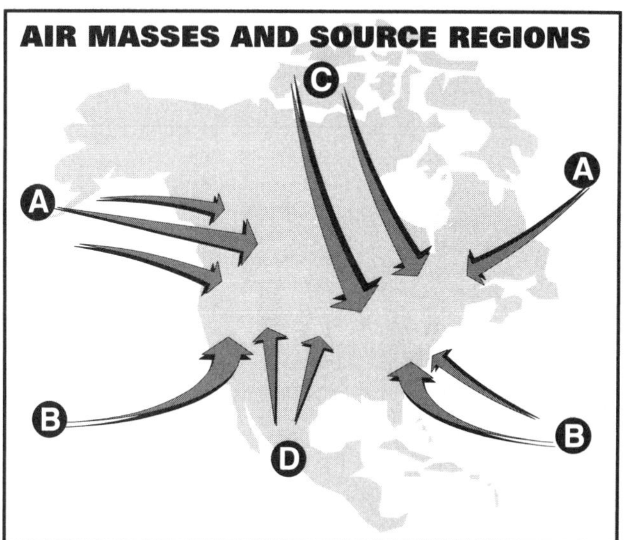

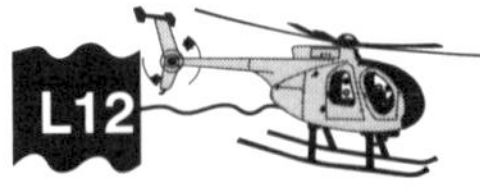

The Polar Front

123. [L33/1/1]
The _____ is the zone between the cold polar easterlies and the warmer prevailing westerlies.
A. colar front
B. equatorial front
C. polar front

124. [L33/1/2]
Several protrusions or waves of cold air occur along the polar frontal zone. Typically, there can be _____ to _____ long waves existing globally at any one time.
A. one, fifty
B. seven, thirty-five
C. three, seven

THE POLAR FRONT

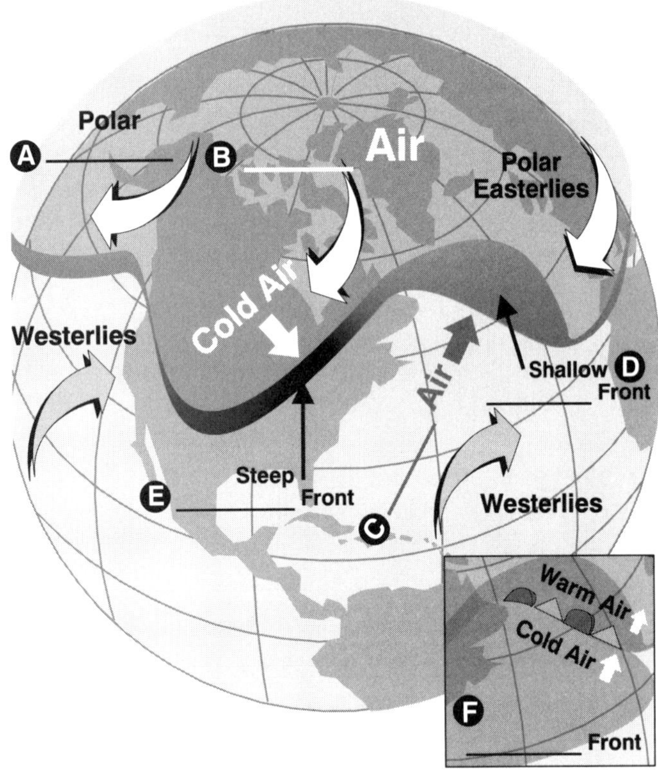

125. [L33/Figure 55]
Fill in the blanks in the figure above.

Different Types of Fronts

126. [L33/2/3]
As a plunging long wave of cold (polar) air moves southward, it overtakes warmer, moister air. This is called a _____ front.
A. cold
B. stationary
C. warm

127. [L33/2/3]
When warm tropical air fills in the receding side of the long cold wave, this is called a _____ front.
A. cold
B. stationary
C. warm

128. [L33/2/3]
Sometimes the warm and cold air butt up against one another and neither moves. This is called a _____ front.
A. cold
B. stationary
C. warm

129. [L33/2/4 & L34/1/1]
As a front approaches, the pressure _____ and as it passes the pressure _____ .
A. rises, falls
B. falls, rises
C. falls, falls some more

130. [L34/1/3]
The trough is a line where the pressure is _____ than on either side of the line, and where the _____ form a counter-clockwise curvature but closed circulation doesn't occur.
A. lower, isobars
B. higher, isobars
C. constant, isotherms

131. [L33/Figure 56]
Referring to the figure below, name the four different types of fronts.
A. ____________________ C. ____________________

B. ____________________ D. ____________________

A.

B.

C.

D.

132. [L34/1/4]
Polar cold fronts tend to move toward the _____.
A. low pressure trough
B. high pressure ridge
C. area of continuity

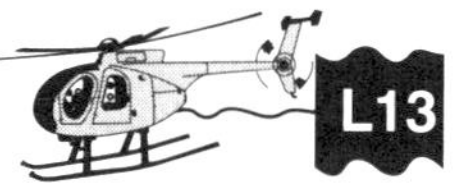

133. [L35/1/1]
You can think of the front as being drawn along with the _____ much like a horse is drawn to a carrot.
A. ridge
B. trough
C. clouds

Discontinuities Across a Front

134. [L35/2/1]
One of the most easily recognized discontinuities across a front is
A. a change in temperature.
B. an increase in cloud coverage.
C. an increase in relative humidity.

135. [L35/2/3]
One weather phenomenon which will always occur when flying across a front is a change in the
A. wind direction.
B. type of precipitation.
C. stability of the air mass.

136. [L35/2/5]
Aside from temperature, _____, and _____, _____ barometer is another good indication of frontal passage.
A. relative humidity, wind shift, a falling then rising
B. dewpoint, thunderstorms, a rising then falling
C. dewpoint, wind shift, a falling then rising

Cold Front Characteristics

137. [L36/1/1]
A cold air mass overtaking a warm air mass is called a(n) _____ front.
A. warm
B. cold
C. occluded

138. [L36/1/1]
Because of the consistency of a cold air mass along with surface friction, the leading edge of the cold air tends to stick to the surface, forming a _____ sloped frontal edge.
A. steeply
B. shallow
C. barely

139. [L36/1/1]
Faster moving cold fronts have _____ frontal slopes than slower ones.
A. steeper
B. shallow
C. straight

140. [L36/1/2]
Cold frontal slopes range on a meteorological scale of _____ (steep) to _____ (not too steep) and average about _____.
A. 1/10, 1/1550, 1/800
B. 1/50, 1/150, 1/80
C. 1/2, 1/7, 1/15

141. [L36/2/1]
In the northern hemisphere, strong cold fronts are usually oriented in a _____ to southwest direction and move toward the east and _____.
A. northeast, southeast
B. northeast, north
C. northwest, southeast

Two Types of Cold Fronts

142. [L36/3/4]
Most of the cloudiness and precipitation associated with a cold front are located _____ the area where warm and cold air meet.
A. 300 miles behind
B. along and ahead of
C. way ahead of

143. [L37/1/2]
Fast-moving cold fronts can generate squall lines _____ to _____ miles in advance of and parallel to the front.
A. 20, 1000
B. 40, 800
C. 30, 180

144. [L37/1/2]
Thunderstorms that generally produce the most intense hazard to aircraft are
A. squall line thunderstorms.
B. steady-state thunderstorms.
C. warm front thunderstorms.

145. [L37/1/2]
A nonfrontal, narrow band of active thunderstorms that often develops ahead of a cold front is a known as a
A. prefrontal system.
B. squall line.
C. dry line.

146. [L37/See *Cloud Families*]
Clouds are divided into four families according to their
A. outward shape.
B. height range.
C. composition.

147. [L37/See *Cloud Families*]
The suffix "nimbus," used in naming clouds, means
A. a cloud with extensive vertical development.
B. a rain cloud.
C. a middle cloud containing ice pellets.

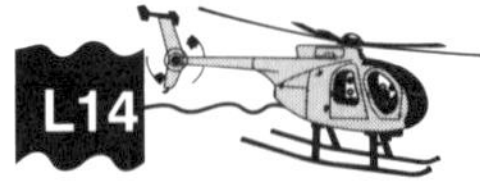

Warm Fronts

148. [L38/2/2]
Warm fronts are typically associated with the small wave patterns moving along the _____ front.
A. polar
B. stationary
C. occluded

149. [L38/2/2]
Retreating cool air in the upper part of a small frontal wave is replaced by warmer, moister air from the _____.
A. north
B. south
C. east

150. [L38/3/1]
Warm front orientation is more _____, with the frontal position moving in a _____ direction.
A. east-west, southeasterly
B. north-south, westerly
C. north-south, northeasterly

151. [L39/1/1]
Warm fronts typically move _____ than cold fronts.
A. faster
B. slower
C. much faster

152. [L39/1/1]
Warm front slopes typically range between _____ and _____ with an average of _____.
A. 1/2, 1/50, 1/76
B. 1/50, 1/200, 1/100
C. 1/50, 1/150, 1/80

153. [L39/1/1]
A shallower slope means that warm frontal weather is distributed over a _____ area than that of a cold front.
A. larger
B. smaller
C. more horizontal

154. [L39/1/2]
The sequence of cloud formation encountered in advance of the warm front is: _____, _____, _____ and _____.
A. nimbostratus, cirrus, cumulus, altostratus
B. cirrus, cirrostratus, altostratus, nimbostratus
C. cirrus, altostratus, cirrostratus, nimbostratus

155. [L39/2/3]
The presence of ice pellets at the surface is evidence that there
A. are thunderstorms in the area.
B. has been cold frontal passage.
C. is a temperature inversion with freezing rain at a higher altitude.

156. [L39/2/3]
In which environment is aircraft structural ice most likely to have the highest accumulation rate?
A. Cumulus clouds with below freezing temperatures.
B. Freezing drizzle.
C. Freezing rain.

Stationary Fronts

157. [L39/2/4]
Sometimes the opposing forces exerted by air masses of different densities are of similar strength and little or no movement occurs between them. We call this a(n) _____ front
A. warm
B. occluded
C. stationary

158. [L39/2/4]
Wind on either side of a stationary front blows _____ to the front rather than _____ it.
A. parallel, across
B. perpendicular, parallel to
C. across, parallel to

The Jet Stream

159. [L40/1/1]
The jet stream consists of one or more tubes of very fast moving air flowing _____ to _____ across the United States.
A. west, east
B. east, west
C. north, south

Thunderstorms

160. [L40/2/4]
What conditions are necessary for the formation of thunderstorms?
A. High humidity, lifting force, and unstable conditions.
B. High humidity, high temperature, and cumulus clouds.
C. Lifting force, moist air, and extensive cloud cover.

161. [L40/2/5]
Moisture is a necessary ingredient for thunderstorms because it contains trapped energy in the form of _____.
A. lift
B. frozen heat
C. latent heat

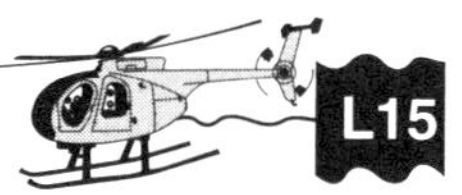

The (Not So Secret) Life of a Thunderstorm

162. [L40/3/3]
Thunderstorms have life cycles averaging from _____ minutes.
A. precisely 30 to 35
B. 20 to 90
C. 20 to no more than 60

163. [L40/3/3]
Thunderstorm cells have three stages:
A. cumulus, mature, dissipating.
B. nimbus, mature, dissipating.
C. updraft, downdraft, backdraft.

164. [L41/1/2]
What feature is normally associated with the cumulus stage of a thunderstorm?
A. Roll cloud.
B. Continuous updraft.
C. Frequent lightning.

165. [L41/1/4]
Thunderstorms reach their greatest intensity during the
A. mature stage.
B. downdraft stage.
C. cumulus stage.

166. [L41/2/2]
Which weather phenomenon signals the beginning of the mature stage of a thunderstorm?
A. The appearance of an anvil top.
B. Precipitation begins to fall.
C. Maximum growth rate of the clouds.

167. [L42/1/3]
During the life cycle of a thunderstorm, which stage is characterized predominately by downdrafts?
A. Cumulus.
B. Dissipating.
C. Mature.

Thunderstorm Types

168. [L42/2/2]
The lifting action required for thunderstorm formation can be furnished by any of four sources (name them):
1. _____
2. _____
3. _____
4. _____

169. [L42/2/3]
Air mass thunderstorms generally form within a warm, moist_____ and are not associated with _____.
A. fog layer, drops in pressure
B. air mass, fronts
C. front, air masses

170. [L42/2/3]
Air mass thunderstorms are usually classified as _____ (heated from below) or _____ (mountain induced) thunderstorms.
A. convective, orographic
B. orographic, convective
C. warm front, latent heat

171. [L42/3/3]
Orographic thunderstorms occur when moist, unstable air is forced _____.
A. across mountain slopes
B. down mountain slopes
C. up mountain slopes

172. [L43/1/1]
Faster moving fronts usually produce the _____ thunderstorms.
A. weakest
B. nastiest
C. most benign type of

173. [L43/1/2]
Thunderstorms associated with _____ fronts are normally the worst, except for those found in _____.
A. warm, squall lines
B. occluded, cold fronts
C. cold, squall lines

174. [L43/1/4]
Thunderstorms can also be found in _____ fronts.
A. warm
B. fog
C. nocturnal

175. [L43/1/4]
The gentleness of warm frontal lifting produces _____ type clouds. These clouds can hide thunderstorms known as _____ thunderstorms.
A. stratiform, embedded
B. cumulus, elevated
C. cirrus, thinnis

176. [L43/1/4]
Embedded thunderstorms are _____ to pilots. Fortunately, because of the shallow lifting, thunderstorms associated with _____ fronts are usually the least severe of all frontal-type thunderstorms.
A. always visible, fog
B. easily visible, cold
C. not easily visible, warm

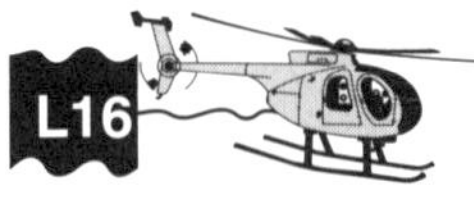

Thunderstorm Turbulence

177. [L44/2/2 & Figure 75]
If there is thunderstorm activity in the vicinity of an airport at which you plan to land, which hazardous atmospheric phenomenon might be expected on the landing approach?
A. Precipitation static.
B. Wind-shear turbulence.
C. Steady rain.

178. [L44/2/2]
One of the significant hazards produced by thunderstorms is something known as the _____, which forms at the beginning of the _____ stage of thunderstorm development.
A. first gust, mature
B. fog puff, cumulus
C. stratiform gust, mature

179. [L44/3/3]
Seasoned pilots watch for several signs of the first gust when thunderstorms are present. They keep an eye out for the first signs of _____ beneath the cell. This usually indicates rapid changes in _____.
A. rain, wind direction or velocity
B. clouds, air temperature
C. reduced visibility, humidity

Virga

180. [L45/1/3]
As rain falls into drier and warmer air beneath a cloud, the rain _____.
A. evaporates
B. condenses
C. humidifies

181. [L45/1/3]
It takes heat to evaporate water. Where does this heat come from? It comes from the _____ the rain falls through, which makes the air cold.
A. wind
B. cloud
C. air

182. [L45/1/3]
The cold air falls, _____ as it descends. You can expect _____ turbulence and high-velocity downdrafts beneath virga.
A. picking up speed, moderate
B. reducing its speed, no
C. maintaining its speed, very light

Thunderstorm Avoidance

183. [L46/1/2]
Avoid thunderstorm cells by at least _____ miles.
A. 40
B. 10
C. 20

184. [L46/1/3]
Flying between two cells is recommended only if enough distance separates them. Most pilots use a minimum of _____ miles separation between big cells for this minimum distance.
A. 80
B. 20
C. 40

Lightning

185. [L46/1/4]
Which weather phenomenon is always associated with a thunderstorm?
A. Lightning.
B. Heavy rain.
C. Hail.

186. [L46/1/4]
A reduction in _____ vision occurs when your pupils close as a reaction to intense lightning. Turning the white cockpit lights _____ as possible is usually the best remedy for acclimating yourself to future flashes.
A. day, up as bright
B. night, down as low
C. night, up as bright

Turbulence and Wind Shear

187. [L46/3/2]
Wind shear occurs when wind makes a rapid change in _____ or _____ (or both).
A. direction, pressure
B. direction, velocity
C. velocity, temperature

188. [L47/1/3]
The most common form of turbulence you're likely to encounter in flight comes from _____ currents.
A. convective
B. stratiform
C. latent heat

189. [L47/1/4]
The uneven heating of land causes variable concentrations of _____ near the surface.
A. "alto" type clouds
B. heated air
C. variable air

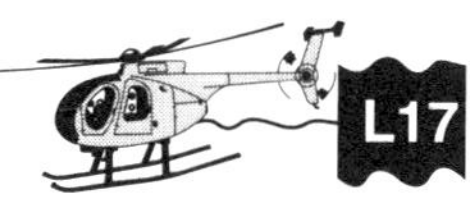

190. [L47/2/2]
Surfaces that are _____, such as plowed fields and paved roads, warm up quickly, producing rising currents of air. Conversely, green fields and small bodies of water _____, creating downward currents of air.
A. darker, remain cooler
B. lighter, remain warmer
C. mid tones, absorb heat

191. [L47/2/3]
Fair weather _____ clouds are generally visible indications of rising and condensing air.
A. cirrus
B. stratus
C. cumulus

Mountain Waves

192. [L48/1/3]
Very stable air, moving above the level of surface friction, usually flows in a _____ or _____ layered pattern.
A. vertical, chaotic
B. laminar, smooth
C. turbulent, vertical

193. [L48/1/3]
When stable air encounters a large enough obstacle (a mountain range, for example), a _____ or _____ wave pattern is established in the mass of moving air.
A. standing, mountain
B. moving, vertical
C. thunderstorm, cumulus

194. [L48/1/5]
Directly over the mountain top, a _____ or _____ forms above the rough mountain surface
A. pileus, cap cloud
B. cumulus, lens-type cloud
C. stratus, lens-type cloud

195. [L48/Figure 82]
Fill in the blanks in the figure below.

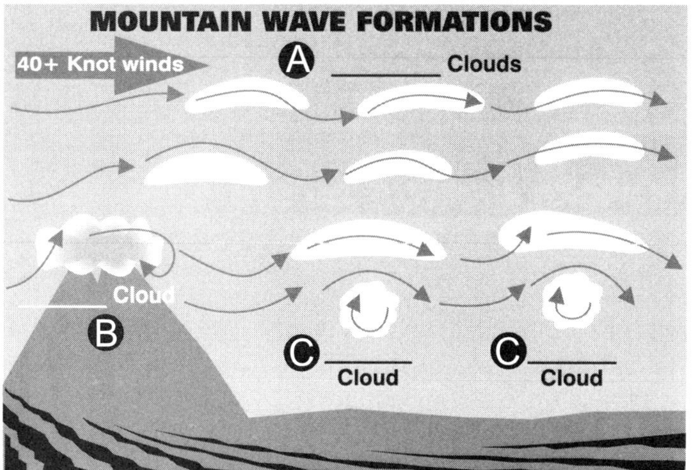

196. [L48/1/5]
An almond or lens-shaped cloud which appears stationary, but which may contain winds of 50 knots or more, is referred to as
A. an inactive frontal cloud.
B. a funnel cloud.
C. a lenticular cloud.

197. [L48/1/5 & L48/2/1]
Crests of standing mountain waves may be marked by stationary, lens-shaped clouds known as
A. mammatocumulus clouds.
B. standing lenticular (wave) clouds.
C. roll clouds.

198. [L48/Figure 82]
Possible mountain wave turbulence could be anticipated when winds of 40 knots or greater blow
A. across a mountain ridge, and the air is stable.
B. down a mountain valley, and the air is unstable.
C. parallel to a mountain peak, and the air is stable.

199. [L48/3/2]
Directly underneath the lenticular clouds, a small _____ cloud can frequently be found below mountain peak altitudes.
A. stratus
B. cumulus
C. rotor

200. [L49/1/3]
Approaching a mountain on the windward side (the upwind side), you might find yourself _____ at several thousand feet per minute no matter how you set the power.
A. climbing
B. descending
C. remaining stationary

201. [L49/2/2]
Approaches to mountains with strong winds present should be made with caution, because the strong _____ can easily exceed the ability of a light airplane to _____.
A. updrafts, climb
B. downdrafts, descend
C. downdrafts, climb

202. [L49/3/1]
One way to minimize the risk if you are uncertain of conditions is to approach a ridge at a _____ degree angle.
A. 45
B. 90
C. 135

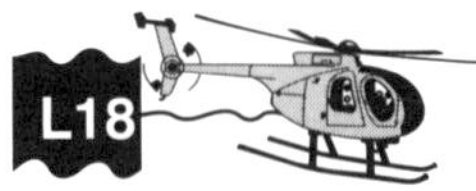

Temperature Inversions and Wind Shear

203. [L50/1/1]
Temperature inversions are common on the surface during clear, calm _____ with little or no _____ winds.
A. days, surface
B. nights, high altitude
C. nights, surface

204. [L50/1/1]
If a temperature inversion exists at the surface, strong winds (if they exist) can be expected
A. just below the top of the inversion.
B. just above the inversion.
C. right at the inversion layer.

205. [L50/1/2]
When may hazardous wind shear be expected?
A. When stable air crosses a mountain barrier, where it tends to flow in layers and form lenticular clouds.
B. In areas of low-level temperature inversion, frontal zones, and clear air turbulence.
C. Following frontal passage when stratocumulus clouds form, indicating mechanical mixing.

206. [L50/1/2&3]
A pilot can expect a wind shear zone in a temperature inversion whenever the windspeed at 2,000 to 4,000 feet above the surface is at least
A. 10 knots.
B. 15 knots.
C. 25 knots.

207. [L50/1/3]
If the inversion occurs within a few hundred feet above the ground, _____ can be a significant hazard.
A. wind shear
B. thunderstorms
C. cumulus clouds

208. [L50/2/2]
A pilot descending through a temperature inversion into a tailwind might expect the airspeed to momentarily
A. increase.
B. decrease.
C. remain unaffected.

209. [L50/3/1]
A pilot descending through a temperature inversion into a headwind might expect the airspeed to momentarily
A. increase.
B. decrease.
C. remain unaffected.

210. [L50/3/2]
Given the proper conditions, you can anticipate wind shear within the first few hours _____.
A. after sunset
B. after midnight
C. after noontime

Fog

211. [L51/1/1]
Fog is a cloud that
A. touches the ground.
B. remains at least 200 feet above the ground.
C. doesn't touch the ground.

212. [L51/1/3]
Small _____ spreads are conducive to fog formation.
A. temperature-relative humidity
B. temperature-pressure
C. temperature-dewpoint

Radiation Fog

213. [L51/2/2]
What situation is most conducive to the formation of radiation fog?
A. Warm, moist air over low, flatland areas on clear, calm nights.
B. Moist, tropical air moving over cold, offshore water.
C. The movement of cold air over much warmer water.

214. [L51/3/2]
Areas of _____ are especially conducive to the formation of radiation fog.
A. moderate humidity (ground with no vegetation)
B. high humidity (rain soaked ground and vegetation)
C. low humidity (dry, parched ground)

215. [L51/3/4]
Winds up to about _____ knots tend to mix and deepen the layer of radiation fog, while _____ winds tend to disperse radiation fog. Solar heating of the earth after sunrise tends to _____ radiation fog.
A. 25, lower, increase
B. 15, higher, dissipate
C. 5, higher, dissipate

Advection Fog

216. [L52/1/2]
Convection means to move something _____; advection means moving it _____.
A. vertically, sideways
B. sideways, vertically
C. upslope, vertically

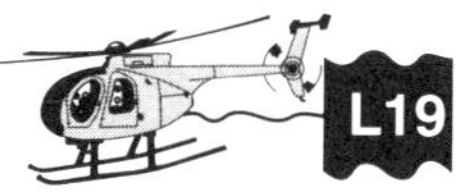

217. [L51/1/2]
Advection fog is sometimes called
A. bog fog.
B. sea fog.
C. mountain fog.

218. [L52/1/2]
In which situation is advection fog most likely to form?
A. A warm, moist air mass on the windward side of mountains.
B. An air mass moving inland from the coast in winter.
C. A light breeze blowing colder air out to sea.

219. [L52/1/2]
Advection fog deepens when winds increase to about _____ knots.
A. 35
B. 25
C. 15

Upslope Fog

220. [L52/2/3]
What types of fog depend upon wind in order to exist?
A. Radiation fog and ice fog.
B. Steam fog and ground fog.
C. Advection fog and upslope fog.

Precipitation-Induced Fog

221. [L52/3/2]
Warm rain, falling through cooler air, can bring the air to the point of _____, forming precipitation-induced fog. Commonly associated with _____, it can occur in slow-moving cold fronts and stationary fronts.
A. low humidity, warm fronts
B. saturation, squall lines
C. saturation, warm fronts

Ice Fog

222. [L52/3/3]
Ice fog forms under conditions similar to those that cause _____ fog, except the air temperature is well below freezing. Instead of fog, water _____ directly into the air as ice crystals.
A. radiation, sublimates
B. advection, transfers
C. precipitation-induced, condenses

Steam Fog

223. [L53/1/2]
As dry, cold air passes over a body of warm water, moisture _____ rapidly from the surface. Condensation takes place as the cold air is quickly saturated. Water droplets often _____, falling back into the water as ice crystals. Steam fog is quite conducive to low level turbulence and _____.
A. sublimes, evaporate, icing
B. condenses, freeze, rain
C. evaporates, freeze, icing

224. [L53/1/2]
Low-level turbulence can occur and icing can become hazardous in which type of fog?
A. Rain-induced fog.
B. Upslope fog.
C. Steam fog.

Postflight Briefing 12-1: Wave Cyclones (Frontal Waves)

225. [L54/1/3]
Sometimes a front forms as a series of small waves along one of the three to seven larger waves of the _____ front.
A. equatorial
B. polar
C. westerly

226. [L54/1/4]
The word cyclone indicates low atmospheric pressure having a _____ rotation.
A. clockwise
B. counterclockwise
C. normal

227. [L54/1/4]
Small wave cyclones are also simply called _____ because they bring inclement weather.
A. precipitation
B. fog
C. storms

228. [L54/1/5]
Wave cyclones usually form in slow moving _____ or _____ fronts.
A. cold, stationary
B. warm, stationary
C. squall lines, fast moving cold

229. [L54/1/6]
Any small disturbance in a stationary frontal pattern, caused by _____ heating, irregular _____ or high altitude _____, can start a wave-like bend in the front.
A. uneven, terrain, winds (the jet stream)
B. frontal, winds, fog
C. irregular, air, winds (the jet stream)

230. [L54/1/6]
If a wave-like bend in a stationary front is energetic enough, _____ air rises over the retreating _____ air. This often leads to _____, and the release of latent heat into the atmosphere which intensifies the low pressure system.
A. cold, warmer, evaporation
B. cooler, warmer, sublimation
C. warm, colder, condensation

231. [L54/1/8]
In a wave cyclone, advancing cold air (cold front) usually moves _____ than retreating cool air.
A. much slower
B. slower
C. faster

232. [L54/2/2]
The faster moving cold front in a wave cyclone eventually catches up with and overtakes the slower warm front. Meteorologists call this an _____.
A. occlusion
B. evolution
C. inversion

233. [L54/2/5]
Sometimes multiple storm systems (several wave cyclones) form along the polar front. These move _____, with the flow of the westerlies and in the same direction as the high altitude _____ .
A. westward, jet stream
B. eastward, jet stream
C. eastward, Coriolis force

Wave Cyclone Weather Patterns Cold Occlusions

234. [L56/1/2]
What clouds have the greatest turbulence?
A. towering cumulus.
B. cumulonimbus.
C. nimbostratus.

235. [L56/1/3]
In wave cyclones, _____ fronts usually catch up to and overtake slower moving _____ fronts. This overtaking produces what is known as a cold-type occluded front.
A. cold, warm
B. warm, cold
C. warm, stationary

236. [L56/1/6]
As the cold occlusion develops, _____ air is lifted higher and higher.
A. cold
B. warm
C. stationary

237. [L56/2/1]
Cold occlusions form predominantly over continents or along the _____ and are more common than warm occlusions.
A. northern territories
B. east coast
C. west coast

Warm Occlusions

238. [L56/2/2]
Warm-type occluded fronts (or warm occlusions) begin with _____ masses from the Pacific overtaking a retreating colder air mass from the Arctic while _____ air is caught in between these moving air masses.
A. cold air, warmer
B. stationary, colder
C. warm, cold

239. [L57/1/1]
Weather associated with a warm front occlusion has characteristics of both _____ and _____ fronts.
A. stationary, cold
B. warm, cold
C. warm, dewpoint

How the Jet Stream Forms

240. [L57/2/2]
The tropopause is the boundary between _____ and the stratosphere.
A. the troposphere
B. space
C. 29,000 feet

241. [L57/2/2]
Temperatures typically decrease until reaching the top of the _____. Above the tropopause, however, temperatures remain steady and then start to _____.
A. troposphere, increase
B. stratosphere, decrease
C. troposphere, become random

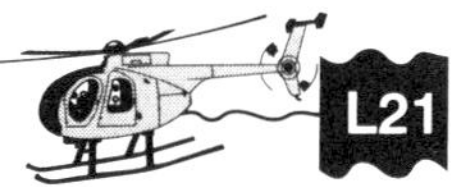

242. [L57/3/2]
The tropopause tends to _____ at the middle of the earth.
A. bulge
B. shrink
C. flatten

243. [L57/3/2]
Normally, the tropopause is located around _____ feet MSL in the northern latitudes and bulges at the equator to altitudes of _____ feet MSL.
A. 15,000, 65,000
B. 30,000, 65,000
C. 65,000, 115,000

244. [L57/3/2]
As we move northward the air cools and shrinks slightly, thus _____ the height of the tropopause.
A. lowering
B. raising
C. flattening

245 [L57/3/3]
The jet stream can be thought of as a _____ in the tropopause.
A. break
B. seal or patch
C. fish habitat for jet pilots

246. [L57/3/3]
A strong vertical _____ gradient can cause large differences in pressure, resulting in a bulge or tear in the tropopause.
A. temperature
B. tropopause
C. humidity

247. [L57/3/4]
Pressure differences along the break in the tropopause occur because warmer air _____ and colder air _____.
A. shrinks, expands
B. expands, shrinks
C. expands, spreads horizontally

248. [L58/1/2]
Warm air circulates upward and northward toward the cold polar air. Then it circulates downward as it's cooled. As it circulates upward and downward, it's curved to the _____ by the Coriolis force.
A. outside
B. left
C. right

Chapter Twelve Answers

1. B
2. C
3. A
4. B
5. C
6. A/low, B/high
   C/low, D/polar
   E/prevailing
   F/northeast
7. C
8. C
9. B
10. C
11. A
12. B
13. B
14, A
15. A
16. C
17. B
18. B
19. A
20. A
21. B
22. B
23. C
24. B
25. B
26. C
27. C
28. A
29. A
30. C
31. A
32. C
33. B
34. B
35. A
36. C
37. A/cool, B/warm
    C/warmer, D/cool
38. B
39. B
40. B
41. A
42. B
43. A
44. A
45. C
46. B
47. B
48. C
49. A
50. C
51. A
52. A
53. B
54. B
55. A
56. A
57. A
58. B
59. B
60. B
61. B
62. A
63. A
64. B
65. B
66. C
67. A
68. B
69. B
70. A
71. neutrally
72. A
73. B
74. C
75. B
76. C
77. A
78. A
79. B
80. C
81. C
82. A
83. A
84. A
85. A
86. C
87. C
88. B
89. C
90. C
91. B
92. B
93. C
94. C
95. A
96. C
97. A
98. A
99. B
100. A
101. C
102. A
103. A
104. B
105. C
106. A
107. C
108. B
109. C
110. C
111. B
112. B
113. C
114. C
115. C
116. A/trough, B/ridge
117. B
118. A
119. A
120. C
121. A
122.
   A/maritime polar
   B/maritime tropical
   C/continental polar
   D/continental tropical
123. C
124. C
125. A/easterlies,
     B/cold, C/warm
     D/warm, E/cold
     F/occluded
126. A
127. C
128. B
129. B
130. A
131. A/cold front
     B/warm front
     C/stationary front
     D/occluded front
132. A
133. B
134. A
135. A
136. C
137. B
138. A
139. A
140. B
141. A
142. B
143. C
144. A
145. B
146. B
147. B
148. A
149. B
150. C
151. B
152. B
153. A
154. B
155. C
156. C
157. C
158. A
159. A
160. A
161. C
162. B
163. A
164. B
165. A
166. B
167. B
168.
   1/lifting by a front
   2/heating from below
   3/movement up a
     mountain
   4/convergence of air
169. B
170. A
171. C
172. B
173. C
174. A
175. A
176. C
177. B
178. A
179. A
180. A
181. C
182. A
183. C
184. C
185. A
186. C
187. B
188. A
189. B
190. A
191. C
192. B
193. A
194. A
195. A/lenticular
     B/cap, C/rotor
196. C
197. B
198. A
199. C
200. A
201. C
202. A
203. C
204. B
205. B
206. C
207. A
208. B
209. A
210. A
211. A
212. C
213. A
214. B
215. C
216. A
217. B
218. B
219. C
220. C
221. C
222. A
223. C
224. C
225. B
226. B
227. C
228. A
229. A
230. C
231. C
232. A
233. B
234. B
235. A
236. B
237. B
238. A
239. B
240. A
241. A
242. A
243. B
244. A
245. A
246. A
247. B
248. C

Note: To ensure that you have the most current answers to these questions, please check the *Book & Slide Updates* section at Rod Machado's web site: www.rodmachado.com

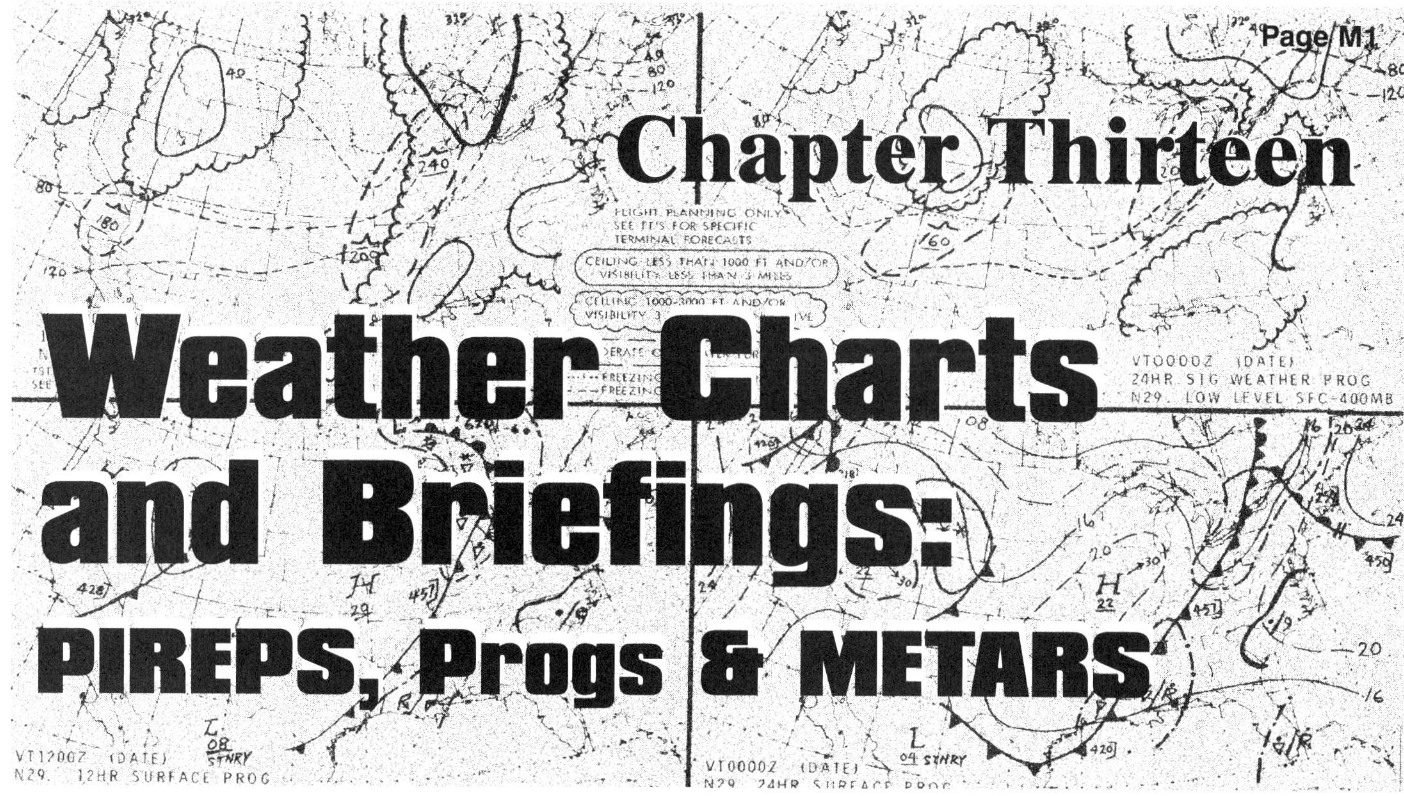

Chapter Thirteen

Weather Charts and Briefings: PIREPS, Progs & METARS

The Telephone Briefing

1. [M4/1/9]
When telephoning a weather briefing facility for preflight weather information, pilots should
A. identify themselves as pilots.
B. tell the number of hours they have flown within the preceding 90 days.
C. state the number of occupants on board and the color of the aircraft.

2. [M5/1/2]
To get a complete weather briefing for the planned flight, the pilot should request
A. a general briefing.
B. an abbreviated briefing.
C. a standard briefing.

3. [M5/1/3]
Which type weather briefing should a pilot request when departing within the hour, if no preliminary weather information has been received?
A. Outlook briefing.
B. Abbreviated briefing.
C. Standard briefing.

4. [M5/1/3]
Which type of weather briefing should a pilot request to supplement mass-disseminated data?
A. An outlook briefing.
B. A supplemental briefing.
C. An abbreviated briefing.

5. [M5/1/3]
To update a previous weather briefing, a pilot should request
A. an abbreviated briefing.
B. a standard briefing.
C. an outlook briefing.

6. [M5/1/4]
When requesting weather information for the following morning, a pilot should request
A. an outlook briefing.
B. a standard briefing.
C. an abbreviated briefing.

7. [M5/1/4]
A weather briefing that is provided when the information requested is 6 or more hours in advance of the proposed departure time is
A. an outlook briefing.
B. a forecast briefing.
C. a prognostic briefing.

8. [M5/2/2]
When telephoning a weather briefing facility for preflight weather information, pilots should state
A. the full name and address of the formation commander.
B. that they possess a current pilot certificate.
C. whether they intend to fly VFR only.

9. [M5/3/1]
When telephoning a weather briefing facility for preflight weather information, pilots should state
A. the full name and address of the pilot in command.
B. the intended route, destination, and type of aircraft.
C. the radio frequencies to be used.

10. [M5/3/1]
When telephoning a weather briefing facility for preflight weather information, pilots should state
A. the aircraft identification or the pilot's name.
B. true airspeed.
C. fuel on board.

Other Sources of Weather Information

11. [M7/1/2]
PATWAS is
A. a regularly scheduled weather broadcast on a VOR frequency.
B. a continuous recording of weather and aeronautical information for pilots.
C. VHF radio receiver tuned to an Automatic Terminal Information Service (ATIS) frequency.

Telephone Information Briefing Service (TIBS)

12. [M7/1/3]
TIBS is
A. a regularly scheduled weather broadcast on a VOR frequency.
B. a continuous telephone briefing service consisting of prerecorded weather information.
C. VHF radio receiver tuned to an Automatic Terminal Information Service (ATIS) frequency.

Transcribed Weather Broadcast (TWEB)

13. [M7/2/2]
To obtain a continuous transcribed weather briefing, including winds aloft and route forecasts for a cross country flight, a pilot should monitor a
A. Transcribed Weather Broadcast (TWEB) on an ADF radio receiver.
B. VHF radio receiver tuned to an Automatic Terminal Information Service (ATIS) frequency.
C. regularly scheduled weather broadcast on a VOR frequency.

14. [M7/2/2]
Transcribed Weather Broadcasts (TWEBs) may be monitored by tuning the appropriate radio receiver to certain
A. airport advisory frequencies.
B. VOR and NDB frequencies.
C. ATIS frequencies.

15. [M7/2/2]
Individual forecasts for specific routes of flight can be obtained from which weather source?
A. Transcribed Weather Broadcasts (TWEB's).
B. Terminal Forecasts.
C. Area Forecasts.

Hazardous In-flight Weather Advisory Service

16. [M7/2/4]
HIWAS is a
A. VHF radio receiver tuned to an Automatic Terminal Information Service (ATIS) frequency.
B. Transcribed Weather Broadcast (TWEB) on an ADF radio receiver.
C. continuous broadcast of in-flight weather advisories.

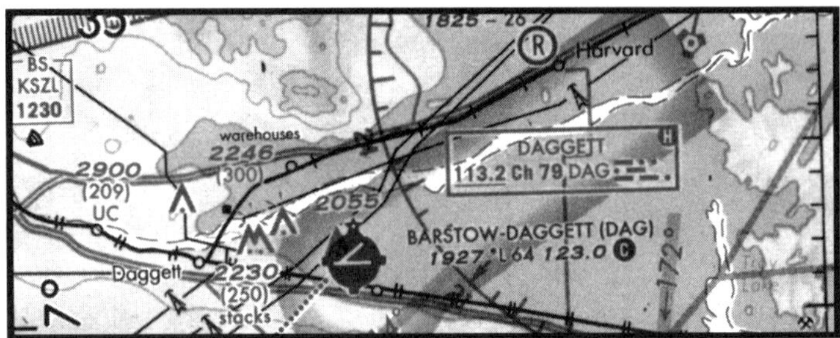

17. [M7/2/4]
On what frequency can a pilot receive Hazardous Inflight Weather Advisory Service (HIWAS) in the vicinity of Daggett VOR?
A. 113.2 MHz.
B. 121.5 MHz.
C. 122.0 MHz.

Enroute Flight Advisory Service (EFAS)

18. [M8/1/1]
How should contact be established with an Enroute Flight Advisory Service (EFAS) station, and what service would be expected?
A. Call EFAS on 122.2 for routine weather, current reports on hazardous weather, and altimeter settings.
B. Call flight assistance on 122.5 for advisory service pertaining to severe weather.
C. Call Flight Watch on 122.0 for information regarding actual weather and thunderstorm activity along proposed route.

19. [M8/1/1]
What service should a pilot normally expect from an Enroute Flight Advisory Service (EFAS) station?
A. Actual weather information and thunderstorm activity along the route.
B. Preferential routing and radar vectoring to circumnavigate severe weather.
C. Severe weather information, changes to flight plans, and receipt of routine position reports.

20. [M8/1/1]
Below FL180, enroute weather advisories should be obtained from an FSS on
A. 122.0 MHz.
B. 122.1 MHz.
C. 123.6 MHz.

Weather Reports

21. [M10/1/2]
METAR observations are taken
A. weekly.
B. hourly
C. daily

22. [M10/1/2]
A METAR report preceded by the letters *SPECI* means this is
A. always a late report.
B. a normal METAR report.
C. an unscheduled report as a result of a significant weather change.

23. [M10/2/3]
One way to remember the information sequence for the METAR is to use the acronym STWVRWCTDA which stands for:

S ____________________

T ____________________

W ____________________

V ____________________

R ____________________

W ____________________

C ____________________

T ____________________

D ____________________

A ____________________

24. [M10/2/3]
Referring to Figure 1 below, fill in the names of each segment of the METAR.

Station

25. [M10/3/1]
Referring to Figure 2 below, the first four letters (KINK) at the beginning of the METAR represent
A. the reporting station.
B. the weather at the front of the airport.
C. the visibility.

26. [M10/3/1]
Referring to Figure 2 below, the letter K at the beginning of each four letter METAR identifier is an international designator that precedes all
A. domestic U.S. location identifiers.
B. airports located in Kansas.
C. domestic locations in the U.S. and Canada.

Time

27. [M10/3/2]
Referring to Figure 2 below, the second sequence of terms in the METAR such as 081955 indicates
A. that the report was filed on the eighth month, 19th day and 55 minutes past 0000 Zulu.
B. that this METAR observation was made on the 8th day of the month, at 1955 Zulu.
C. that this station had winds from 080 degrees at 19 knots, gusting to 55 knots.

Wind

28. [M10/3/3]
The wind direction and velocity at Boise, Idaho (KBOI) is from
A. 230° true at 8 knots.
B. 023° magnetic at 8 knots.
C. 230° magnetic at 8 knots.

METAR SURFACE AVIATION WEATHER OBSERVATION FORMAT

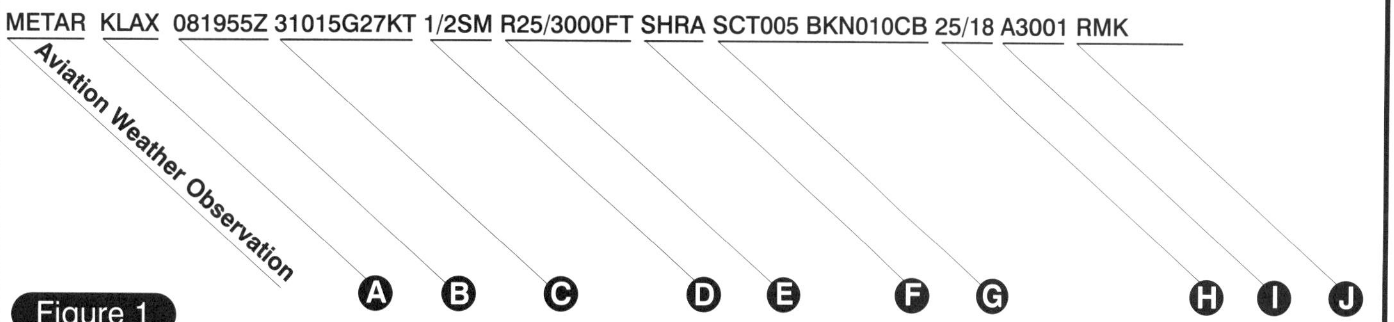

METAR KLAX 081955Z 31015G27KT 1/2SM R25/3000FT SHRA SCT005 BKN010CB 25/18 A3001 RMK

METAR: Aviation Weather Observation

KLAX: A
081955Z: B
31015G27KT: C
1/2SM: D
R25/3000FT: E
SHRA: F
SCT005 BKN010CB: G
25/18: H
A3001: I
RMK: J

Figure 1

Figure 2

METAR WEATHER REPORTING FORMAT

METAR KINK 081955Z 32014G20KT 1/2SM R30R/2400FT DZ FG OVC006 13/12 A3004
SPECI KMKC 081936Z 20014G24KT 1/2SM R34/2600FT -SHRA OVC008 04/03 A2898
METAR KBOI 081953Z 23008KT 5SM SCT015 19/13 A2994 RMK SLP156 T01930128
METAR KLAX 081955Z 01013G20KT 3SM HZ SKC 18/11 A2995

METAR WEATHER REPORTING FORMAT

```
METAR  KINK  081955Z  32014G20KT  1/2SM  R30R/2400FT  DZ FG  OVC006  13/12  A3004
SPECI  KMKC  081936Z  20014G24KT  1/2SM  R34/2600FT  -SHRA  OVC008  04/03  A2898
METAR  KBOI  081953Z  23008KT  5SM  SCT015  19/13  A2994 RMK SLP156 T01930128
METAR  KLAX  081955Z  01013G20KT  3SM  HZ  SKC 18/11 A2995
```

29. [M10/3/3]
Referring to the figure above, what are the wind conditions at Los Angeles (KLAX)?
A. Calm.
B. 110° at 1.3 knots, gusts 2 knots.
C. 010° at 13 knots, gusts 20 knots.

Visibility

30. [M11/1/2]
The sequence "1/2SM" in the METAR for KMKC indicates
A. the surface visibility is 1 to 2 miles in smoke and mist.
B. the surface visibility is 1/2 statute miles.
C. the surface winds are blowing 10 to 20 statute miles per hour.

31. [M11/1/2]
The sequence "3SM" in the METAR for KLAX indicates
A. the surface visibility is 3 miles in smoke and mist.
B. the surface visibility is 3 statute miles.
C. the surface winds are blowing 3 statute miles per hour.

Runway Visual Range

32. [M11/1/3]
Referring to the figure above, the sequence "R34/2600FT" in the METAR for KMKC indicates
A. Runway 34 is 2,600 feet long.
B. Runway 34 prevailing visibility is 2,600 feet.
C. Runway 34 RVR is 2,600 feet.

Weather

33. [M11/2/2]
Referring to the figure above, the letters "-SHRA" found in the METAR for KMKC indicate
A. that light snow showers and rain existed at the time this observation was made.
B. that showers of rain will exist at some time in the future at this station.
C. that light rain showers existed at the time this observation was made.

Clouds

34. [M12/2/2]
Referring to the figure above, what are the current conditions for Wink, Texas (KINK)?
A. sky 600 feet overcast, visibility 1/2SM, drizzle, fog.
B. sky 6000 feet overcast, visibility 1 1/2SM, drizzle, fog.
C. sky 60 feet overcast, visibility 1 1/2SM, drizzle, fog.

35. [M12/2/2]
Referring to the figure above, which of the reporting stations have VFR weather?
A. All.
B. KINK, KBOI, and KLAX.
C. KBOI and KLAX.

36. [M12/4/3]
For aviation purposes, ceiling is defined as the height above the earth's surface of the
A. lowest reported obscuration and the highest layer of clouds reported as overcast.
B. lowest broken or overcast layer or vertical visibility into an obscuration.
C. lowest layer of clouds reported as scattered, broken, or thin.

Temperature/Dewpoint

37. [M13/1/1]
Referring to the figure above, the temperature and dewpoint for Boise, Idaho (KBOI) are
A. 19 degrees C and 13 degrees C, respectively.
B. 13 degrees F and 19 degrees F, respectively.
C. 01 degrees C and 93 degrees F, respectively.

Altimeter

38. [M13/1/2]
Referring to the figure above, which station listed in the METAR below has the lowest altimeter setting?
A. KMKC.
B. KBOI.
C. KLAX.

39. [M11/Figure 8]
Referring to the figure below, fill in the blanks for the weather codes used in METAR & other reports.

WEATHER CODES

Qualifiers	
Intensity or proximity 1	Descriptor 2
- ________	MI ________
________ (no qualifier)	BC ________
+ ________	PR ________
VC means: in the ________ (METAR: between 5 & 10 sm of observation point(s) TAF: between 5 to 10 sm from center of runway complex)	DR ________
	BL ________
	SH ________
	TS ________
	FZ ________

40. [M11/Figure 8]
Referring to the figure below, fill in the blanks for the weather codes used in METAR and other reports.

WEATHER CODES

Weather Phenomena		
Precipitation 3	Obscuration 4	Other 5
DZ ________	BR ________	PO ________
RA ________	FG ________	
SN ________	FU ________	SQ ________
SG ________	VA ________	FC ________
IC ________	DU ________	
PL ________		+FC ________
GR ________	SA ________	SS ________
GS ________	HZ ________	
UP ________	PY ________	DS ________

41. [M12/Figure 9]
Referring to the figure below, fill in the blanks for the remarks used to append METARS:

Remarks appended to METARS

Remarks	*Definition*
Sky and Ceiling	
FEW CU	Few ____________ clouds.
BINOVC	____________in overcast.
LWR CLDS NE	Lower clouds ____________
CIG 14V19	Ceiling _______between 1,400 feet and 1,900 feet.
Obscuring Phenomena	
FG7	_____ obscuring 7/10 of the sky.
BLSA3	_______ sand obscuring 3/10 of the sky.
THN FG NE	_________ fog northeast from reporting station.
Visibility	
VSBY S1W1/4	_________ south is 1 mile, west is 1/4 mile.
SFC VSBY 1/2	_______ visibility is 1/2 mile.

42. [M12/Figure 9]
Referring to the figure below, fill in the blanks for the remarks used to append METARS:

Remarks appended to METARS

Remarks	*Definition*
Weather and Obstruction to Vision	
RAB30	Rain ______ 30 minutes after the hour.
RAE30	Rain ______30 minutes after the hour.
OCNL DST LTG NW	_________distant lightning ________of reporting station.
T OVHD MOVG NE	Thunderstorm _____, moving northeast.
Wind	
WND 20V26	Wind variable between ___degrees and ___degrees.
PK WND 15027/35 ("PK WND" is used whenever the peak winds exceed 25 knots)	____wind within the past hour from ___ degrees at ___ knots occurred ___ minutes past the hour.
Pressure	
PRESSR	Pressure ______rapidly.
PRESFR	Pressure ______rapidly.

METAR WEATHER REPORTING FORMAT

METAR KINK 100655Z 35016G22KT 1/2SM R30R/2000FT -RA BKN010 09/06 A3000
SPECI KMKC 100655Z 25014G28KT 3/4SM R34/2600FT -SHRA OVC003 03/02 A2998
METAR KBOI 100655Z 30010KT 4SM SCT025 18/11 A3004 RMK SLP158 T01820114
METAR KLAX 100655Z 01011G18KT 3SM FUHZ SKC 18/11 A2995

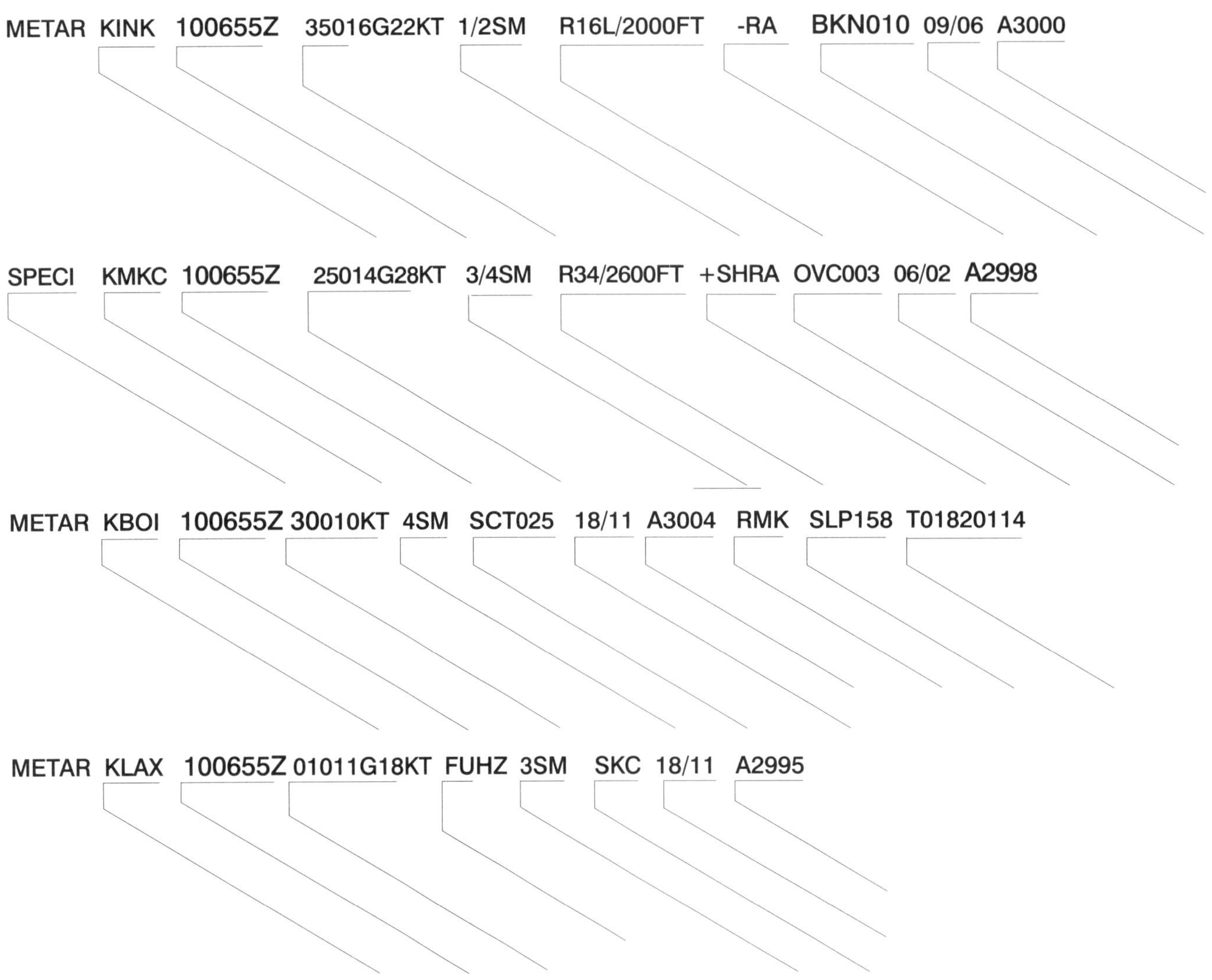

METAR KINK 100655Z 35016G22KT 1/2SM R16L/2000FT -RA BKN010 09/06 A3000

SPECI KMKC 100655Z 25014G28KT 3/4SM R34/2600FT +SHRA OVC003 06/02 A2998

METAR KBOI 100655Z 30010KT 4SM SCT025 18/11 A3004 RMK SLP158 T01820114

METAR KLAX 100655Z 01011G18KT FUHZ 3SM SKC 18/11 A2995

43. [M13/Figure 10]
Referring to the figure above, fill in the diagonal blank lines of the accompanying METAR.

Automatic Weather Observing Programs

44. [M14/1/3]
ASOS transmissions are usually receivable within _____ nautical miles of the airport and at below 10,000 feet AGL.
A. 50
B. 25
C. 100

45. [M14/1/5]
Automated stations cannot report clouds above _____ feet.
A. 12,000
B. 5,000
C. 2,000

46. [M14/1/5]
The ASOS will report _____ (meaning no clouds are reported below 12,000 feet) instead of the SKC value shown in the METAR
A. NCL
B. CLR
C. OVC

47. [M14/1/6]
If you see or hear something missing from a ASOS report, it doesn't mean it's not there. It just means that the ASOS unit _____.
A. must be activated by the pilot
B. is inoperative
C. can't or didn't detect it

48. [M14/2/1]
You can tell that an observation in a METAR was derived by an automated source because the word _____ appears directly after the location of the date/time signifier.
A. AUTO
B. MANU
C. AUTOSRCE

49. [M14/3/3]
Which AWOS level provides cloud-ceiling data?
A. AWOS-2.
B. AWOS-3.
C. AWOS-4.

Whither the Weather?

Terminal Aerodrome Weather Forecasts (TAF)

50. [M16/1/1]
Referring to the numbers at the beginning of the TAF (i.e., 101740Z), the first two numbers represent the _____ and the last four numbers represent the _____ the report was issued.
A. time, date
B. date, time
C. local time, Zulu time

51. [M16/1/4]
Referring to KLAX in the figure below, what does "SHRA" stand for?
A. Rain showers.
B. A shift in wind direction is expected.
C. A significant change in precipitation is possible.

52. [M16/Figure 16]
Referring to the figure below, fill in the diagonal blank lines of the accompanying TAF:

DECODED TERMINAL AERODROME WEATHER FORECAST (TAF)

TAF
KLAX 201740Z 201818 24018KT 3SM RA BKN012
FM2030 32018G25KT 3SM +SHRA OVC015 PROB40 2223 1/2SM TSRA OVC006CB
FM2400 28011KT 5SM SHRA BKN015 OVC030 TEMPO 0306 00000KT 1/2SM -RA FG
FM0800 VRB05KT 5SM -SHRA OVC035 BECMG 1315 25015KT P6SM NSW SKC

KLAX 201740Z 201818 24018KT 3SM RA BKN012

FM2030 32018G25KT 3SM +SHRA OVC015 PROB40 2223 1/2SM TSRA OVC006CB

FM2400 28011KT 5SM SHRA BKN015 OVC030 TEMPO 0306 00000KT 1/2SM -RA FG

FM0800 VRB05KT 5SM -SHRA OVC035 BECMG 1315 25015KT P6SM NSW SKC

TERMINAL AERODROME WEATHER FORECAST (TAF)

TAF
KMEM 121720Z 121818 20012KT 5SM HZ BKN030 PROB40 2022 1SM TRSA OVC008CB
FM2200 33015G20KT P6SM BKN015 OVC025 PROB40 2202 3SM SHRA
FM0200 35012KT OVC008 PROB40 0205 2SM -RASN BECMG 0608 02008KT BKN012
BECMG 1012 00000KT 3SM BR SKC TEMPO 1214 1/2SM FG
FM1600 VRB04KT P6SM SKC

KOKC 051130Z 051212 14008KT 5SM BR BKN030 TEMPO 1316 1 1/2SM BR
FM1600 18010KT P6SM SKC BECMG 2224 20013G20KT 4SM SHRA OVC020
PROB40 0006 2SM TRSA OVC008CB BECMG 0608 21015KT P6SM SCT040

53. [M16/Figure 16]
Referring to the figure above, during the time period from 0600Z to 0800Z, what significant weather is forecast for KOKC?
A. Wind - 210° at 15 knots.
B. Visibility - possibly 6 statute miles with scattered clouds at 4,000 feet.
C. No significant weather is forecast for this time period.

54. [M16/1/1]
Referring to the figure above, what is the valid period for the TAF for KMEM?
A. 1200Z to 1200Z.
B. 1200Z to 1800Z.
C. 1800Z to 1800Z.

55. [M16/1/3]
Referring to the figure above, between 1000Z and 1200Z, the visibility at KMEM is forecast to be?
A. 1/2 statute mile.
B. 3 statute miles.
C. 6 statute miles.

56. [M16/2/1]
Referring to the figure above, in the TAF from KOKC, the clear sky becomes
A. overcast at 2,000 feet during the forecast period between 2200Z and 2400Z.
B. overcast at 200 feet with a 40% probability of becoming overcast at 600 feet during the forecast period between 2200Z and 2400Z.
C. overcast at 200 feet with the probability of becoming overcast at 400 feet during the forecast period between 2200Z and 2400Z.

57. [M17/1/4]
Referring to the figure above, what is the forecast wind for KMEM from 1600Z until the end of the forecast?
A. no significant wind.
B. 020° at 8 knots.
C. variable in direction at 4 knots.

58. [M17/1/4]
Referring to the figure above, in the TAF from KOKC, the "FM (FROM) Group" is
A. forecast for the hours from 1600Z to 2200Z with the wind from 180° at 10 knots.
B. forecast for the hours from 1600Z to 2200Z with the wind from 180° at 10 knots, becoming 220° at 13 knots with gusts to 20 knots.
C. forecast for the hours from 1600Z to 2200Z with the wind from 180° at 10 knots, becoming 210° at 15 knots.

59. [M16/Figure 16]
The only cloud type forecast in TAF reports is
A. nimbostratus.
B. cumulonimbus.
C. scattered cumulus.

60. [M17/1/3] Fill in the blank:
Regarding the TAF, when gradual changes in the prevailing conditions are expected, the abbreviation ______________ is used.

61. [M17/1/3] Fill in the blanks:
Any weather condition forecast prior to the term "BECMG" which is not revised following the change time period, is expected to ______________ the ______________.

62. [M17/1/4] Fill in the blank:
FM is followed by a four digit number which indicates the _________ time of a self contained portion of the forecast.

63. [M17/2/2] Fill in the blank:
______________ indicates that fluctuations (temporary and usually lasting less than one hour) from the predominant weather conditions are expected.

64. [M17/2/3] Fill in the blanks:
PROB30 is used when the likely occurrence of any weather phenomena falls in the _____% to _____% range of expectations. PROB40 is used when the likely occurrence of any weather phenomena falls in the _____% to _____% range of expectations. If the probability of the condition is _____% or higher, the terms BECMG, TEMPO and FM are used.

Area Forecasts (FA)

65. [M18/1/1]
To best determine general forecast weather conditions over several states, the pilot should refer to
A. aviation Area Forecasts.
B. Weather Depiction Charts.
C. satellite maps.

66. [M18/1/1]
From which primary source should information be obtained regarding expected weather at the estimated time of arrival if your destination has no terminal forecast?
A. Low-level prognostic chart.
B. Weather depiction chart.
C. Area forecast.

67. [M18/1/1] Fill in the blank:
The Area Forecast is a general forecast covering a region that includes several ____________.

68. [M18/1/1] Fill in the blank:
To find the forecast weather between reporting stations you should refer to the ____________ Forecast.

69. [M18/1/2] Fill in the blanks:
Area Forecasts are issued ____________ times daily for ____________ specific areas in the lower 48 states.

70. [M18/1/2] Fill in the blanks:
Area Forecasts are ____________ hour forecasts with an additional ____________ hour outlook.

71. [M18/1/2] Fill in the blank:
All heights mentioned in the Area Forecast are ________.

72. [M18/1/3] Fill in the blank:
The Area Forecast consists of ____________ sections.

73. [M18/2/1] Fill in the blank:
As with other weather reports, the times listed in the Area Forecast are in ____________.

74. [M18/1/3] Fill in the blanks:
The sections of the Area Forecast are:
1. ____________
2. ____________
3. ____________
4. ____________

75. [M19/2/2]
The section of the Area Forecast entitled "SIG CLDS AND WX" contains a summary of
A. cloudiness and weather significant to flight operations broken down by states or other geographical areas.
B. forecast sky cover, cloud tops, visibility, and obstructions to vision along specific routes.
C. weather advisories still in effect at the time of issue.

THE AREA FORECAST

BOSC FA 241845
SYNOPSIS AND VFR CLDS/WX
SYNOPSIS VALID UNTIL 251300
CLDS/WX VALID UNTIL 250700...OTLK VALID 250700-251300
ME NH VT MA RI CT NY LO NJ PA OH LE WV MD DC DE VA AND CSTL WTRS
.
SEE AIRMET SIERRA FOR IFR CONDS AND MTN OBSCN.
TS IMPLY SEV OR GTR TURB SEV ICE LLWS AND IFR CONDS.
NON MSL HGTS DENOTED BY AGL OR CIG.
.
SYNOPSIS...19Z CDFNT ALG A 160NE ACK-ENE LN...CONTG AS A QSTNRY FNT ALG AN END-50SW MSS LN. BY 13Z...CDFNT ALG A 140ESE ACK-HTO LN...CONTG AS A QSTNRY FNT ALG A HTO-SYR-YYZ LN. TROF ACRS CNTRL PA INTO NRN VA.
...REYNOLDS...
.
OH LE
NRN HLF OH LE...SCT-BKN025 OVC045. CLDS LYRD 150. SCT SHRA. WDLY SCT TSRA. CB TOPS FL350. 23-01Z OVC020-030. VIS 3SM BR. OCNL -RA. OTLK...IFR CIG BR FG.
SWRN QTR OH...BKN050-060 TOPS 100. OTLK...MVFR BR.
SERN QTR OH...SCT-BKN040 BKN070 TOPS 120. WDLY SCT -TSRA. 00Z SCT-BKN030 OVC050. WDLY SCT -TSRA. CB TOPS FL350. OTLK...VFR SHRA.

76. [M20/Figure 21]
Referring to the figure above, what is the forecast ceiling for the northern half of Ohio and Lake Erie from 2300Z to 0100Z?
A. Ceilings 2,000 to 3,000 feet overcast.
B. Ceilings 200 to 300 feet overcast.
C. No ceiling is forecast for this area during this time period.

77. [M20/Figure 21]
Referring to the figure above, what precipitation (if any) is expected in the southeastern quarter of Ohio from 0700Z to 1300Z?
A. No precipitation is expected.
B. Moderate rain showers.
C. Thunderstorms, but no rain is expected.

78. [M20/Figure 21]
What are the tops of the cumulonimbus clouds expected during the forecast period in southeastern Ohio?
A. 20,000 feet.
B. 3,500 feet.
C. 35,000 feet.

79. [M19/Figure 18B & M11/Figure 8]
In the southwestern quarter of Ohio, the outlook calls for
A. marginal VFR with mist.
B. mostly VFR with blowing rain.
C. marginal VFR with blowing rain.

80. [M20/1/4] Fill in the blanks:
Only surface visibilities of ____________ or less miles and sustained winds of ____________ knots or greater are shown in the Area Forecast.

THE AREA FORECAST

BOSC FA 241845
SYNOPSIS AND VFR CLDS/WX
SYNOPSIS VALID UNTIL 251300
CLDS/WX VALID UNTIL 250700...OTLK VALID 250700-251300
ME NH VT MA RI CT NY LO NJ PA OH LE WV MD DC DE VA AND CSTL WTRS
.
SEE AIRMET SIERRA FOR IFR CONDS AND MTN OBSCN.
TS IMPLY SEV OR GTR TURB SEV ICE LLWS AND IFR CONDS.
NON MSL HGTS DENOTED BY AGL OR CIG.
.
SYNOPSIS...19Z CDFNT ALG A 160NE ACK-ENE LN...CONTG AS A QSTNRY FNT ALG AN END-50SW MSS LN. BY 13Z...CDFNT ALG A 140ESE ACK-HTO LN...CONTG AS A QSTNRY FNT ALG A HTO-SYR-YYZ LN. TROF ACRS CNTRL PA INTO NRN VA.
...REYNOLDS...
.
OH LE
NRN HLF OH LE...SCT-BKN025 OVC045. CLDS LYRD 150. SCT SHRA. WDLY SCT TSRA. CB TOPS FL350. 23-01Z OVC020-030. VIS 3SM BR. OCNL -RA. OTLK...IFR CIG BR FG.
SWRN QTR OH...BKN050-060 TOPS 100. OTLK...MVFR BR.
SERN QTR OH...SCT-BKN040 BKN070 TOPS 120. WDLY SCT -TSRA. 00Z SCT-BKN030 OVC050. WDLY SCT -TSRA. CB TOPS FL350. OTLK...VFR SHRA.

81. [M20/Figure 21]
Referring to the figure above, what type of pressure system exists across central Pennsylvania into northern Virginia?
A. A ridge
B. A trough
C. A Reynolds system.

82. [M20/Figure 21]
Referring to the synopsis section in the figure above, where is the cold front located?
A. At 16,000 feet over Nebraska.
B. Along a line that begins at a point 160 nautical miles northeast of ACK and extends to ENE.
C. At 160 nautical miles from NE that extends along a line from ACK to ENE.

83. [M19/Figure 18B & M11/Figure 8]
In the southeastern quarter of Ohio, the outlook calls for
A. VFR with moderate rain showers.
B. mostly VFR with snow and rain.
C. VFR with blowing rain.

84. [M20/Figure 21]
Where does the scattered to broken layer begin in the northern half of Ohio and Lake Erie?
A. 2,500 feet AGL
B. 2,500 feet MSL
C. 250 feet MSL

Winds Aloft Forecasts (FD)

85. [M21/1/2] Fill in the blanks:
Winds aloft forecasts are issued _____ daily and are valid for either _____, _____ or _____ hours as indicated.

FD WBC 151745
BASED ON 151200Z DATA
VALID 1600Z FOR USE 1800-0300Z. TEMPS NEG ABV 24000

FT	3000	6000	9000	12000	18000	24000	30000	34000
ALS			2420	2635–08	2535–18	2444–30	245945	246755
AMA		2714	2725+00	2625–04	2531–15	2542–27	265842	256352
DEN			2321–04	2532–08	2434–19	2441–31	235347	236056
HLC		1707–01	2113–03	2219–07	2330–17	2435–30	244145	244854
MKC	0507	2006+03	2215–01	2322–06	2338–17	2348–29	236143	841652
STL	2113	2325+07	2332+02	2339–04	2356–16	2373–27	239440	730649

86. [M21/2/1]
Winds aloft forecasts are provided in _____ (knots? mph?), their direction is always referenced to _____ north and their temperatures are always in degrees _____.

87. [M21/2/1]
Referring to the figure above, what wind is forecast for STL at 18,000 feet?
A. 230 degrees true at 56 knots.
B. 235 degrees true at 06 knots.
C. 235 degrees magnetic at 06, peak gusts to 16 knots.

88. [M21/2/2]
Referring to the figure above, determine the wind and temperature aloft forecast for 3,000 feet at MKC.
A. 050 degrees true at 7 knots, temperature missing.
B. 360 degrees magnetic at 5 knots, temperature -7 degrees C.
C. 360 degrees true at 50 knots, temperature +7 degrees C.

89. [M21/2/2]
Referring to the figure above, what wind is forecast for STL at 6,000 feet?
A. 210 degrees magnetic at 13 knots.
B. 230 degrees true at 25 knots.
C. 232 degrees true at 5 knots.

90. [M21/2/2]
Referring to the figure above, determine the wind and temperature aloft forecast for DEN at 30,000 feet.
A. 023 degrees magnetic at 53 knots, temperature 47 degrees C.
B. 230 degrees true at 53 knots, temperature -47 degrees C.
C. 235 degrees true at 34 knots, temperature -7 degrees C.

91. [M21/Figure 22]
Referring to the figure above, what wind is forecast for STL at 34,000 feet?
A. 007 degrees magnetic at 30 knots.
B. 073 degrees true at 6 knots.
C. 230 degrees true at 106 knots.

92. [M21/Figure 22]
Referring to the winds aloft forecast in the figure above, what are the winds expected to be over MKC at 34,000 feet?
A. 84 degrees at 16 knots.
B. 840 degrees at 116 knots.
C. 340 degrees at 116 knots.

93. [M21/3/1]
If winds are expected to be light and variable, the identifier _____ is used; if they're expected to be calm, the identifier _____ is used.

94. [M21/3/1]
When the term "light and variable" is used in reference to a winds aloft forecast, the coded group and windspeed are
A. 0000 and less than 7 knots.
B. 9900 and less than 5 knots.
C. 9999 and less than 10 knots.

Weather Charts: Getting the Big Picture

Weather Depiction Chart

95. [M23/1/1]
Referring to the figure below, of what value is the Weather Depiction Chart to the pilot?
A. For determining general weather conditions on which to base flight planning.
B. As a forecast of cloud coverage, visibilities, and frontal activity.
C. For determining frontal trends and air mass characteristics.

96. [M23/1/1]
The Weather Depiction Chart is issued every _____ starting at _____ Z.
A. six hours, 1200
B. three hours, 0100
C. four hours, 0600

97. [M23/1/3]
On the Weather Depiction Chart, areas of marginal VFR conditions are shown by contoured areas _____ shading.
A. without
B. with
C. with and without

98. [M23/1/3]
On the Weather Depiction Chart, areas of IFR conditions are shown by contoured _____ areas.
A. non shaded
B. shaded
C. shaded and non shaded

99. [M23/1/3]
An MVFR area has ceilings of _____ to _____ feet and visibilities of _____ to _____ miles.
A. 3,000, 5000, 3, 5
B. 500, 1000, 1, 3
C. 1,000, 3000, 3, 5

The Weather Depiction Chart

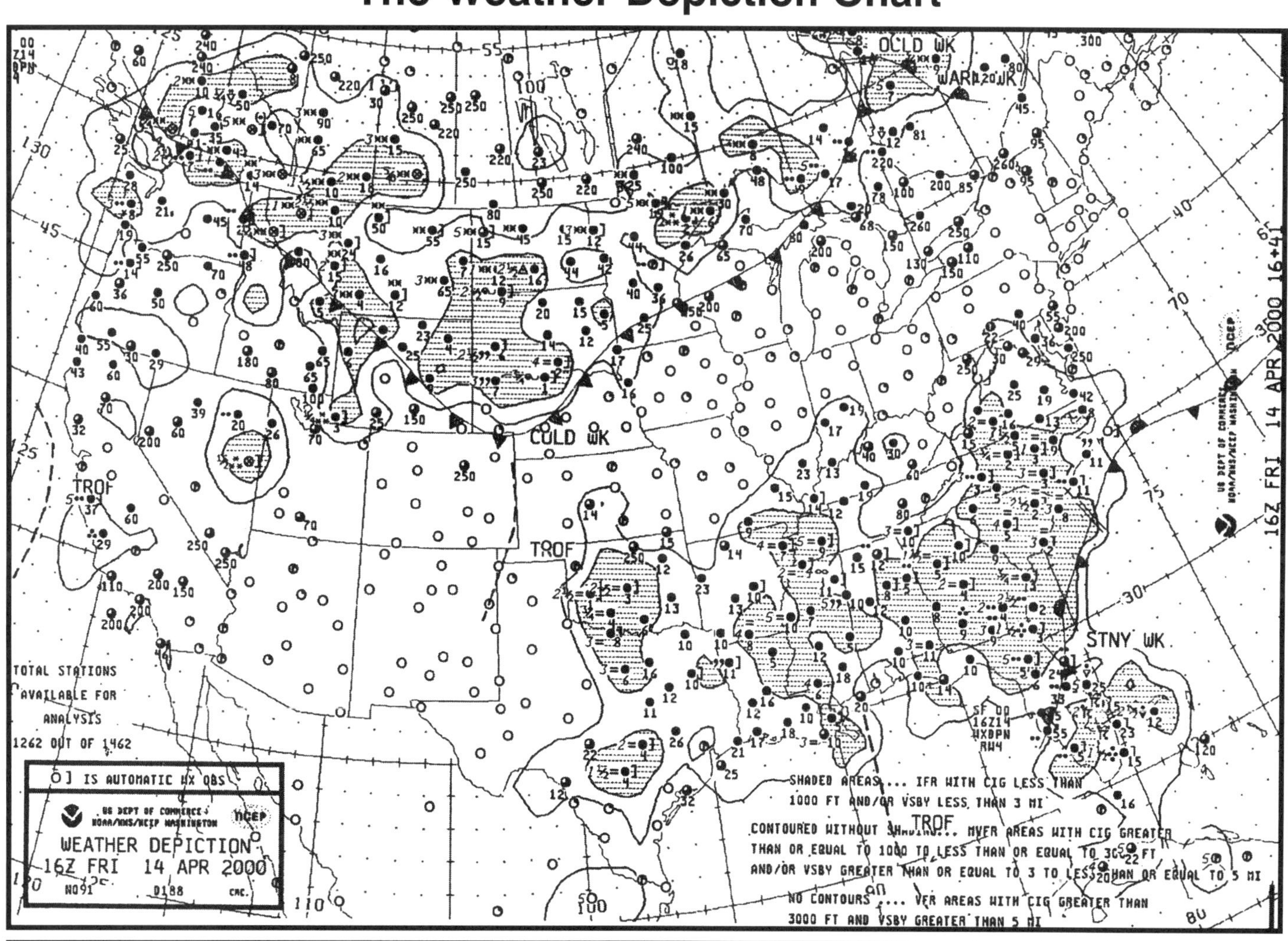

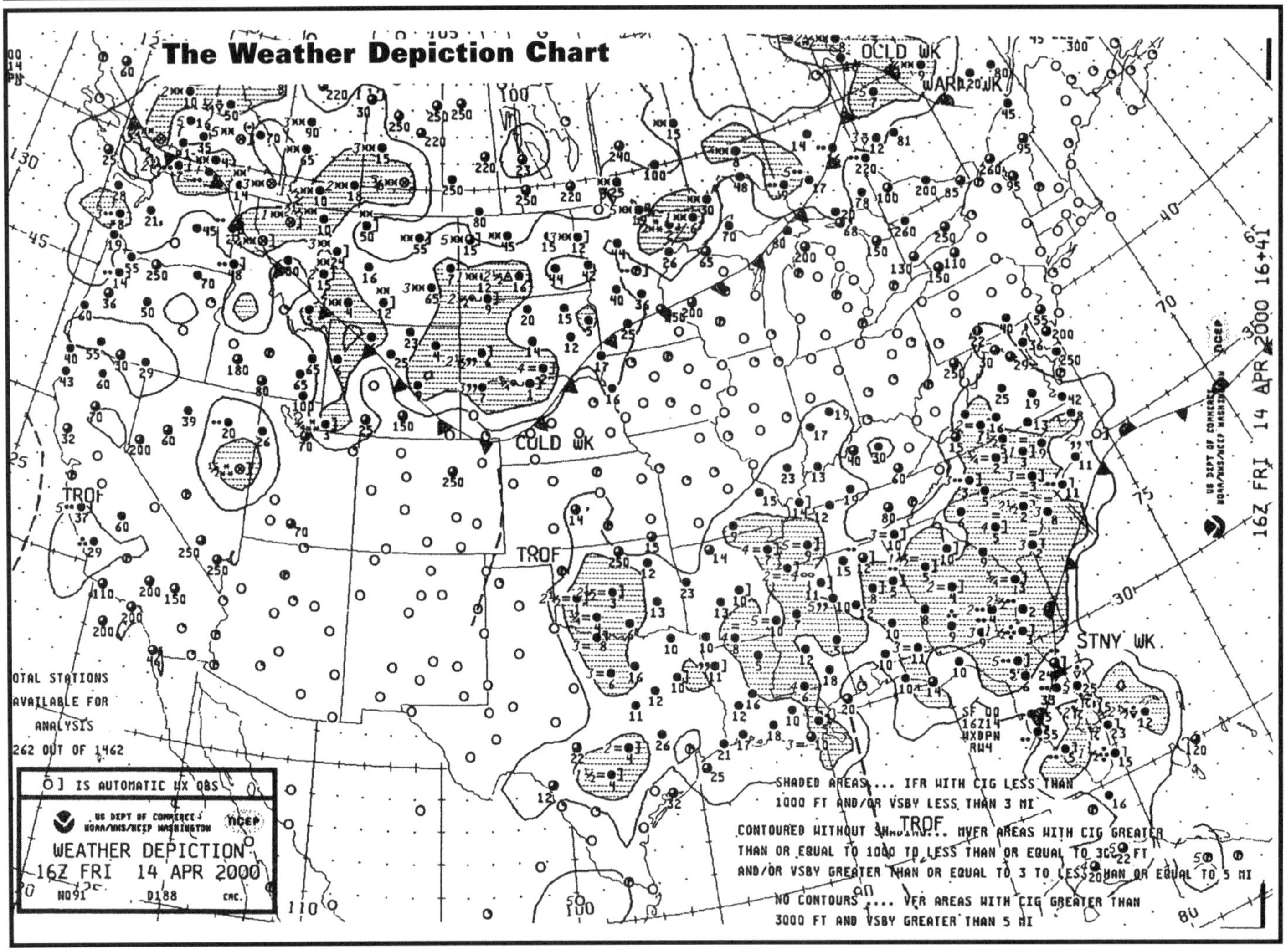

The Weather Depiction Chart

SHADED AREAS.... IFR WITH CIG LESS THAN 1000 FT AND/OR VSBY LESS THAN 3 MI

CONTOURED WITHOUT SHADING.... MVFR AREAS WITH CIG GREATER THAN OR EQUAL TO 1000 TO LESS THAN OR EQUAL TO 3000 FT AND/OR VSBY GREATER THAN OR EQUAL TO 3 TO LESS THAN OR EQUAL TO 5 MI

NO CONTOURS.... VFR AREAS WITH CIG GREATER THAN 3000 FT AND VSBY GREATER THAN 5 MI

WEATHER DEPICTION 16Z FRI 14 APR 2000

100. [M23/1/2 & M23/Figures 24 & 25]
Referring to the figure above, according to the Weather Depiction Chart the weather for a flight from southern Michigan to north Indiana has ceilings

A. 1,000 to 3,000 feet and/or visibility 3 to 5 miles.
B. less than 1,000 feet and/or visibility less than 3 miles.
C. greater than 3,000 feet and visibility greater than 5 miles.

101. [M23/1/3 & M23/Figures 24 & 25]
Referring to the figure above, the IFR weather in central Texas is due to

A. intermittent rain.
B. fog.
C. dust devils.

102. [M23/1/3 & M23/Figures 24 & 25]
Referring to the figure above, what weather phenomenon is causing IFR conditions in central Oklahoma?

A. Low visibility only.
B. Low ceilings and visibility.
C. Heavy rain showers.

103. [M23/1/3 & M23/Figures 24 & 25]
Referring to the figure above, the marginal weather in central Kentucky is due to low

A. visibility.
B. ceilings and visibility.
C. ceilings.

104. [M23/2/4]
Referring to the figure above, what is the status of the front that is off the coast of Virginia, the Carolinas and Georgia?

A. Stationary.
B. Occluded.
C. Retreating.

105. [M23/2/4]
Referring to the figure above, the front that cuts across Wisconsin is a

A. stationary front.
B. cold front.
C. warm front.

106. [M23/Figure 24]
Referring to the figure below, fill in the blanks with the appropriate fraction in eights represented by the symbol:

SYMBOLS FOR SKY COVER

SYMBOL	TOTAL SKY COVER
○	Sky ____________
◔	FEW - (more than ___ to ___)
◔	SCT - Scattered (____ to ____)
◕	BKN - Broken (_____ to _____)
●	OVC - Overcast (_____)

107. [M23/Figure 25]
Referring to the figure below, fill in the blanks with the appropriate word which represents the symbol:

Radar Summary Chart

108. [M24/1/1]
The radar summary chart is issued _____ times during a 24 hour period.
A. 24
B. 16
C. 8

109. [M24/1/1]
The radar summary chart presents information on the location of radar echoes resulting from _____ suspended in _____.
A. precipitation, clouds
B. clouds, the air
C. moisture, the stratosphere

110. [M24/1/1]
Radar energy goes right through _____, but reflects off_____.
A. hail, water
B. rain, hail
C. a cloud, water or hail

111. [M24/1/1&2]
What information is provided by the radar summary chart that is not shown on other weather charts?
A. Lines and cells of hazardous thunderstorms.
B. Ceilings and precipitation between reporting stations.
C. Types of clouds between reporting stations.

WX DEPICTION PLOTTING SYMBOLS

PLOTTING SYMBOL	MEANING	PLOTTING SYMBOL	MEANING
◑ 8	_____ clouds, base 800 feet, visibility more than _____ miles	2= ○	Sky _______visibility 2, ground fog or fog
◕ 12	_________ sky cover, ceiling ______ feet, rain shower, visibility more than _______ miles	1/2 ⊗	Sky partially _________, visibility ______, blowing snow, no cloud layers observed
5 ∞ ◑	Thin _________ with breaks, visibility ____ in ______	2= ⊗ 200	Sky _________obscured, visibility 2, fog, cloud layer at ________ feet.
▲ ◔ 30	Scattered at _______ feet, clouds topping _______, visibility more than 6 miles	1/4 ★ ⊗ 5	Sky ___________, ceiling ______ feet, visibility ¼, snow
		1 ● 12	________, ceiling ______ feet, ________________, rain showers, visibility 1

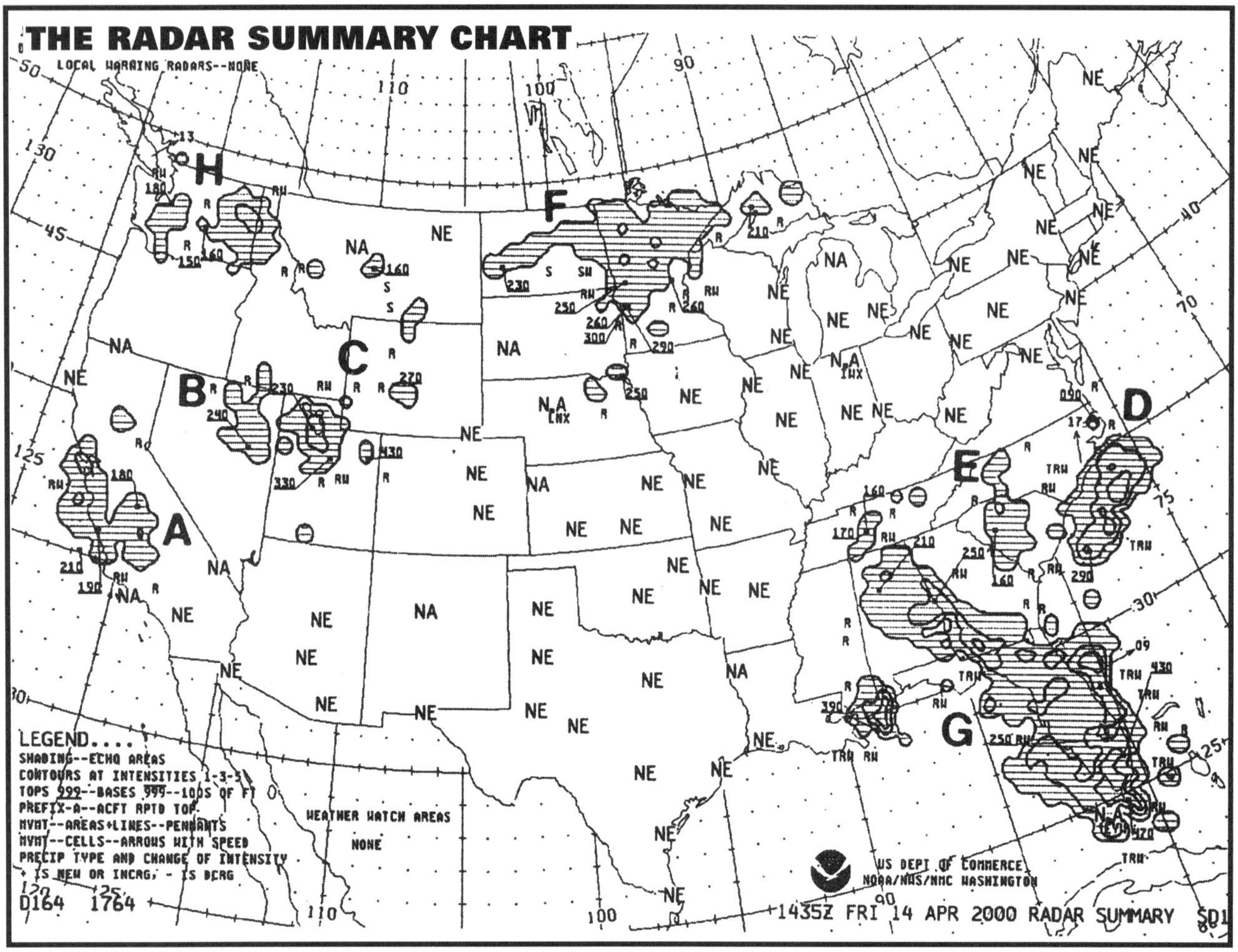

THE RADAR SUMMARY CHART

LOCAL WARNING RADARS--NONE

LEGEND....
SHADING--ECHO AREAS
CONTOURS AT INTENSITIES 1-3-5
TOPS 999--BASES 999--100S OF FT
PREFIX-A--ACFT RPTD TOP
MVMT--AREAS+LINES--PENNANTS
MVMT--CELLS--ARROWS WITH SPEED
PRECIP TYPE AND CHANGE OF INTENSITY
+ IS NEW OR INCRG; - IS DCRG

WEATHER WATCH AREAS
NONE

US DEPT OF COMMERCE
NOAA/NWS/NMC WASHINGTON

1435Z FRI 14 APR 2000 RADAR SUMMARY

112. [M24/1/1&2, M24/Figure 27]
Radar weather reports are of special interest to pilots because they indicate
A. large areas of low ceilings and fog.
B. location of precipitation along with type, intensity, and trend.
C. location of broken to overcast clouds.

113. [M25/1/1, M24/Figure 27]
Referring to the figure above, what type of weather is occurring in the radar return at position B?
A. Rain showers increasing in intensity.
B. Light to moderate rain showers.
C. Continuous rain.

114. [M25/1/2, M24/Figure 27]
Referring to the figure above, what is the direction and speed of movement of the cell at position D?
A. South at 17 knots.
B. North at 17 knots.
C. North at 17 MPH.

115. [M25/1/2, M24/Figure 27]
Referring to the figure above, the top of the precipitation of the cell at position E is
A. 16,000 feet AGL.
B. 25,000 feet MSL.
C. 16,000 feet MSL.

116. [M25/1/3, M24/Figure 27]
Referring to the figure above, what is the top for precipitation of the radar return at position B?
A. 24,000 feet AGL.
B. 24,000 feet MSL.
C. 2,400 feet MSL.

117. [M25/1/4, M24/Figure 27]
What does the heavy dashed line that forms a large rectangular box on a radar summary chart refer to?
A. Areas of heavy rain.
B. Severe weather watch area.
C. Areas of hail 1/4 inch in diameter.

Low Level Significant Weather Prognostic Chart

118. [M26/1/3]
The Significant Weather Prognostic Chart provides a picture forecast for
A. the eastern or western half of the United States.
B. one of six areas in the United States.
C. the entire United States.

119. [M26 & M27/All]
Referring to the figure below, how are Significant Weather Prognostic Charts best used by a pilot?
A. For overall planning at all altitudes.
B. For determining areas to avoid (freezing levels and turbulence).
C. For analyzing current frontal activity and cloud coverage.

120. [M27/1/1]
Referring to the figure below, what weather is forecast for the Florida area just ahead of the stationary front during the first 12 hours?
A. Ceiling 1,000 to 3,000 feet and/or visibility 3 to 5 miles with continuous precipitation.
B. Ceiling 1,000 to 3,000 feet and/or visibility 3 to 5 miles with intermittent precipitation.
C. Ceiling less than 1,000 feet and/or visibility less than 3 miles with continuous precipitation.

121. [M27/1/2]
Referring to the figure below, at what altitude is the freezing level over the middle of Florida on the 12-hour Significant Weather Prognostic Chart?
A. 4,000 feet.
B. 12,000 feet.
C. 8,000 feet.

122. [M27/1/2]
Referring to the figure below, at what altitude is the freezing level in the northeastern tip of Montana on the 24-hour Significant Weather Prognostic Chart?
A. At the surface.
B. 8,000 feet.
C. 12,000 feet.

123. [M27/2/1]
Referring to the figure below, interpret the weather symbol depicted in Utah on the 12-hour significant weather prognostic chart.
A. Moderate turbulence, surface to 18,000 feet.
B. Thunderstorm tops at 18,000 feet.
C. Base of clear air turbulence, 18,000 feet.

124. [M27/2/2]
Referring to the figure below, the enclosed shaded area associated with the low pressure system over northern Utah is forecast to have
A. continuous snow.
B. intermittent snow.
C. continuous snow showers.

THE LOW LEVEL SIGNIFICANT WEATHER PROGNOSTIC CHART

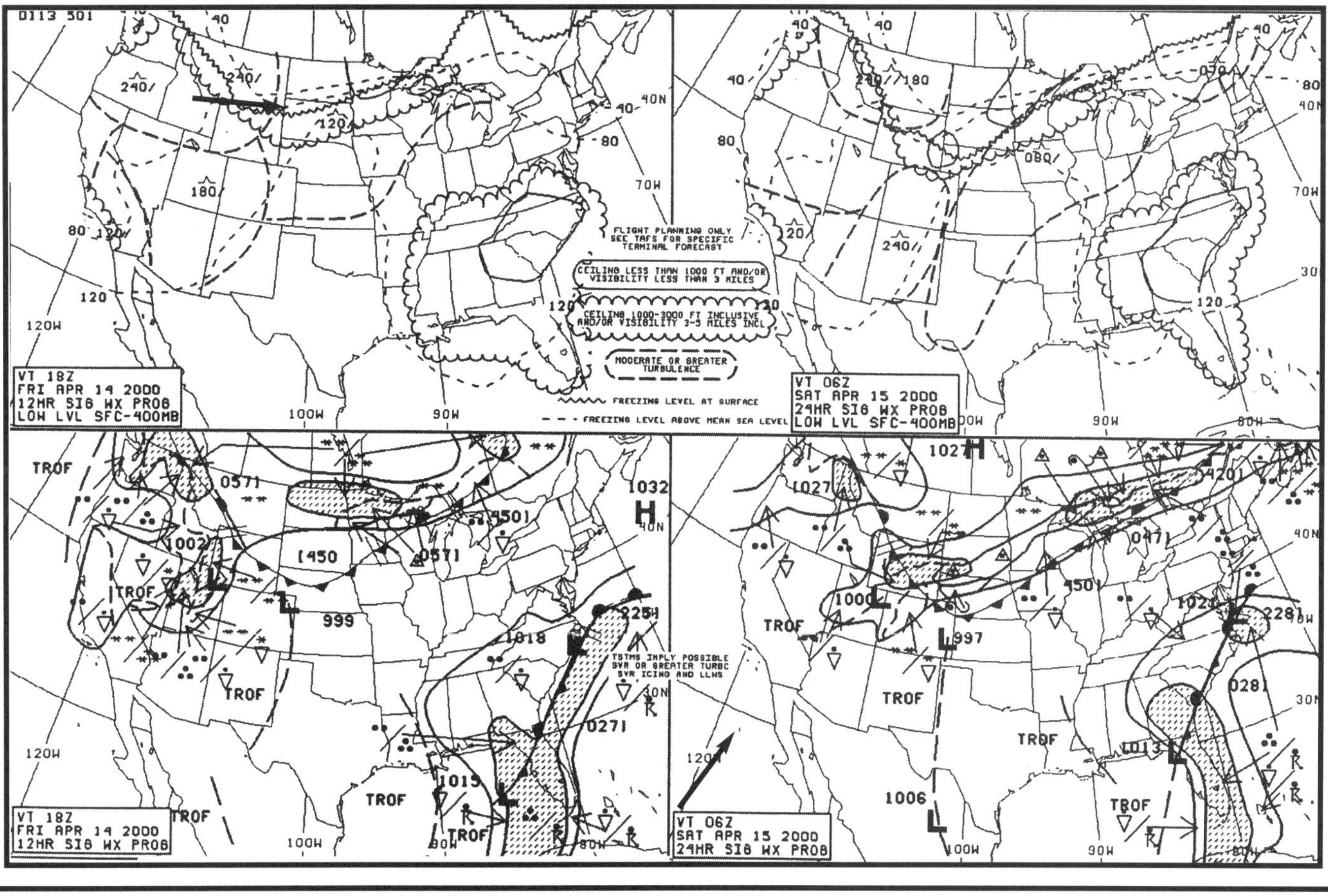

Surface Analysis

125. [M28/1/1]
The surface analysis chart shows the location of _____ and _____.
A. pressure patterns, fronts
B. freezing levels, radar echoes.
C. Airmets, Sigmets

126. [M28/1/1]
This chart is based on reported weather data and is issued every _____ hours.
A. twelve
B. three
C. six

127. [M28/1/2]
You can think of the surface analysis as a very large visual portrayal of hundreds of _____.
A. METARs
B. Area Forecasts
C. TAFs

128. [M28/1/2]
The surface analysis chart shows the temperature/dew point spread over a large area. From this you can get a good feel for the likelihood of _____ formation before and after sunset.
A. thunderstorm
B. wind shear
C. fog or low cloud

In-flight Aviation Weather Advisories

SIGMET (WS)

129. [M28/2/2]
In-flight aviation weather advisories are forecasts which advise pilots of potentially _____ weather.
A. hazardous
B. helpful
C. rainy

130. [M28/2/2]
In-flight advisories may also be obtained from _____ and _____ broadcasts over specific VORs and NDBs.
A. EFAS, TCA
B. TWEB, HIWAS
C. MOA, UB12

131. [M29/1/2]
SIGMETs are issued as a warning of weather conditions hazardous to which aircraft?
A. Small aircraft only.
B. Large aircraft only.
C. All aircraft.

132. [M29/1/2]
When are SIGMETs issued?
A. Every six hours, on the hour.
B. When, in the opinion of a forecaster, the weather will be marginal.
C. Whenever any of the specified hazardous conditions occur or are expected to occur.

133. [M29/1/2]
Which of the following would not be covered by a SIGMET:
1) severe icing not associated with thunderstorms
2) severe or extreme turbulence or clear air turbulence (CAT) not associated with thunderstorms
3) dust storms
4) sandstorms
5) volcanic ash that lowers surface visibilities to less than 3 miles
6) any volcanic eruption

A. 5 and 6
B. 3 and 6
C. All of the above are covered

134. [M29/1/2]
Which in-flight advisory would contain information on severe icing that's not associated with a thunderstorm?
A. Convective SIGMET.
B. SIGMET.
C. AIRMET.

135. [M29/2/2]
SIGMETs are named by the phonetic alphabet using designators from _____ through _____, excluding Sierra, Tango and Zulu.
A. Oscar, November
B. Alpha, Yankee
C. November, Yankee

AIRMET (WA)

136. [M29/2/5]
AIRMETs are advisories of significant weather phenomena but of lower intensities than SIGMETs and are intended for dissemination to
A. only IFR pilots.
B. only VFR pilots.
C. all pilots.

137. [M29/2/6]
AIRMETs are issued once every _____ hours on a scheduled basis with unscheduled amendments issued as required.
A. 2
B. 12
C. 6

138. [M29/2/6]
AIRMETs are valid for a period of _____ hours.
A. 6
B. 12
C. 18

139. [M29/2/6]
Which of the following would not be covered by an AIRMET:
1) severe icing
2) moderate turbulence
3) sustained winds of 30 knots or more at the surface
4) ceilings less than 1,000 feet and/or visibilities less than 3 miles affecting over 50% of the area at one time
5) extensive mountain obscuration

A. 5 only
B. 2 and 5
C. 1 only

140. [M29/2/7]
An AIRMET with the phonetic name “TANGO” indicates that this AIRMET refers to _____.
A. turbulence, strong surface winds, and low level wind shear
B. icing and freezing
C. IFR and mountain obscurations

141. [M30/1/2]
An AIRMET with the phonetic name “SIERRA” indicates that this AIRMET refers to _____.
A. turbulence, strong surface winds, and low level wind shear
B. icing and freezing
C. IFR and mountain obscurations

142. [M30/1/4]
An AIRMET with the phonetic name “ZULU” indicates that this AIRMET refers to _____.
A. turbulence, strong surface winds, and low level wind shear
B. icing and freezing
C. IFR and mountain obscurations

Convective SIGMETs (WST)

143. [M30/1/5]
A convective SIGMET is an advisory associated with convective activity such as severe _____.
A. cumulus clouds
B. thunderstorms
C. fog

144. [M30/1/5]
What information is contained in a convective SIGMET?
A. Tornadoes, embedded thunderstorms, and hail 3/4 inch or greater in diameter.
B. Severe icing, severe turbulence, or widespread dust storms lowering visibility to less than 3 miles.
C. Surface winds greater than 40 knots or thunderstorms equal to or greater than VIP (Video Integrator Processor) level 4.

145. [M30/1/5]
What is indicated when a current convective SIGMET forecasts thunderstorms?
A. Moderate thunderstorms covering 30 percent of the area.
B. Moderate or severe turbulence.
C. Thunderstorms obscured by massive cloud layers.

Pilot Reports (PIREPS)

A TYPICAL PILOT REPORT (PIREP)

KTUL UA /OV KOKC-KTUL /TM 1800 /FL 120
/TP BE90 /SK BKN018-TOP055/OVC072-TOP089 /CLR ABV
/WX RA /TA M9 /WV 08021KT /TB LGT 055-072
/IC LGT-MDT RIME 072-089.

146. [M31/1/1]
Referring to the PIREP above, if the terrain elevation is 1,295 feet MSL, what is the height above ground level of the base of the ceiling?
A. 505 feet AGL.
B. 1,295 feet AGL.
C. 6,586 feet AGL.

147. [M31/1/1]
Referring to the PIREP above, the base and tops of the overcast layer reported by a pilot are
A. 1,800 feet MSL and 5,500 feet MSL.
B. 5,500 feet AGL and 7,200 feet MSL.
C. 7,200 feet MSL and 8,900 feet MSL.

148. [M31/1/1]
Referring to the PIREP above, the wind and temperature at 12,000 feet MSL as reported by a pilot are
A. 008 degrees at 121 MPH and 90 degrees F.
B. 080 degrees at 21 knots and 9 degrees F.
C. 080 degrees at 21 knots and -9 degrees C.

149. [M31/1/1]
Referring to the PIREP above, the intensity of the turbulence reported at a specific altitude is
A. moderate at 5,500 feet and at 7,200 feet.
B. light from 5,500 feet to 7,200 feet.
C. light to moderate from 7,200 feet to 8,900 feet.

150. [M31/1/1]
Referring to the PIREP above, the intensity and type of icing reported by a pilot is
A. light to moderate rime.
B. light to moderate clear.
C. moderate rime.

Putting It All Together

Surface Map and Weather Depiction Chart

151. [M32/1/2]
Looking at the surface weather map tells you where the fronts and pressure systems are. Comparing this with the Weather Depiction Chart provides you with a picture of what effect these fronts and pressure centers_____.
A. will have at some time in the future
B. are having on surface weather
C. are having on the potential for thunderstorms

Weather Depiction Chart and Radar Summary Chart

152. [M32/1/3]
By comparing the Weather Depiction Chart and the Radar Summary Chart, you get an idea of the _____ activity associated with areas of IFR, MVFR and VFR conditions.
A. convective
B. forecast
C. predicted

Area Forecast and Prog Charts

153. [M32/1/4]
Since the Area Forecast is a textual description and the prog chart is a picture, a comparison of the two provides you with a _____.
A. forecast of convective weather only
B. present moment complete description of the weather
C. complete, big picture description of forecast weather

Weather Depiction and Prog Charts

154. [M33/1/1]
Since the Weather Depiction Chart shows the location of fronts and pressure systems, you can see _____ by comparing it to the 12 and 24 surface prog chart.
A. the present speeds of high altitude winds
B. which way these systems will move
C. the precise position of a frontal system and how these systems will move

METARs/Surface Map

155. [M33/1/2]
Comparing individual METARs with the surface weather map provides you with an understanding of _____ weather around and between your departure and destination airports.
A. convective
B. surface
C. high altitude

Chapter Thirteen Answers

1. A
2. C
3. C
4. C
5. A
6. A
7. A
8. C
9. B
10. A
11. B
12. B
13. A
14. B
15. A
16. C
17. A
18. C
19. A
20. A
21. B
22. C
23. **S**hould/**T**ina/**W**alk/**V**era's/**R**abbit/**W**ithout/**C**hecking **T**he/**D**og's/**A**ppetite
24. A/station identifier, B/time (UTC), C/wind direction & velocity, D/visibility, E/runway visual range, F/weather, G/cloud amount and type, H/temperature & dewpoint I/altimeter, J/optional remarks
25. A
26. A
27. B
28. A
29. C
30. B
31. B
32. C
33. C
34. A
35. C
36. B
37. A
38. A
39.

Qualifiers	
Intensity or proximity 1	Descriptor 2
- Light Moderate (no qualifier) + Heavy VC means: in the vicinity	MI Shallow BC Patches PR Partial DR Drifting BL Blowing SH Shower(s) TS Thunderstorm FZ Freezing

Chapter Thirteen Answers

40.

Weather Phenomena		
Precipitation	Obscuration	Other
3	4	5
DZ Drizzle	BR Mist (≥5/8sm)	PO Well developed dust/ sand whirls
RA Rain	FG Fog (<5/8sm)	SQ Squalls
SN Snow	FU Smoke	FC Funnel cloud(s)/
SG Snow grains	VA Volcanic ash	+FC Tornado/ waterspout
IC Ice crystals	DU Widespread dust	SS Sandstorm
PL Ice pellets	SA Sand	DS Duststorm
GR Hail	HZ Haze	
GS Small hail &/or snow pellets	PY Spray	
UP Unknown Precip		

41.

Remarks appended to METARS

Remarks	Definition
Sky and Ceiling	
FEW CU	Few cumulus clouds.
BINOVC	Breaks in overcast.
LWR CLDS NE	Lower clouds northeast.
CIG 14V19	Ceiling variable between 1,400 feet and 1,900 feet.
Obscuring Phenomena	
FG7	Fog obscuring 7/10 of the
BLSA3	Blowing sand obscuring 3/10 of the sky.
THN FG NE	Thin fog northeast from reporting station.
Visibility	
VSBY S1W1/4	Visibility south is 1 mile, west is 1/4 mile.
SFC VSBY 1/2	Surface visibility is 1/2 mile.

42.

Remarks appended to METARS

Remarks	Definition
Weather and Obstruction to Vision	
RAB30	Rain began 30 minutes after the hour.
RAE30	Rain ended 30 minutes after the hour.
OCNL DST LTG NW	Occasional distant lightning northwest of reporting station.
T OVHD MOVG NE	Thunderstorm overhead, moving northeast.
Wind	
WND 20V26	Wind variable between 200 degrees and 260 degrees.
PK WND 15027/35 ("PK WND" is used whenever the peak winds exceed 25 knots)	Peak wind within the past hour from 150 degrees at 27 knots occurred 35 minutes past the hour.
Pressure	
PRESSR	Pressure rising rapidly.
PRESFR	Pressure falling rapidly.

43.

METAR WEATHER REPORTING FORMAT

METAR KINK 100655Z 35016G22KT 1/2SM R30R/2000FT -RA BKN010 09/06 A3000
SPECI KMKC 100655Z 25014G28KT 3/4SM R34/2600FT -SHRA OVC003 03/02 A2998
METAR KBOI 100655Z 30010KT 4SM SCT025 18/11 A3004 RMK SLP158 T01820114
METAR KLAX 100655Z 01011G18KT 3SM FUHZ SKC 18/11 A2995

Above METAR Decoded

METAR KINK 100655Z 35016G22KT 1/2SM R16L/2000FT -RA BKN010 09/06 A3000

- KINK: Station: Wink, TX
- 100655Z: Issued on 10th day at 0655 Zulu
- 35016G22KT: Wind is 350° at 16 knots gusting to 22 knots
- 1/2SM: Reported visibility 1/2 statute mile
- R16L/2000FT: Runway 16 Left, visual range is 2,000 feet
- -RA: Light rain showers
- BKN010: (Ceiling) Broken clouds at 1000 feet AGL
- 09/06: Temperature 09°C Dewpoint 6°C
- A3000: Altimeter 30.00" Hg

SPECI KMKC 100655Z 25014G28KT 3/4SM R34/2600FT +SHRA OVC003 06/02 A2998

- SPECI: Special, unscheduled report
- KMKC: Station: Kansas city, MO
- 100655Z: Issued on 10th day at 0655 Zulu
- 25014G28KT: Wind is 250° at 14 knots gusting to 28 knots
- 3/4SM: Reported visibility ¾ statute mile
- R34/2600FT: Runway 34 visual range is 2,600 feet
- +SHRA: Heavy rain showers
- OVC003: (Ceiling) Overcast clouds at 300 feet AGL
- 06/02: Temperature 6°C Dewpoint 2°C
- A2998: Altimeter 29.98" Hg

METAR KBOI 100655Z 30010KT 4SM SCT025 18/11 A3004 RMK SLP158 T01820114

- KBOI: Station: Boise, Idaho
- 100655Z: Issued on 10th day at 0655 Zulu
- 30010KT: Wind is 300° at 10 knots
- 4SM: Reported visibility 4 statute miles
- SCT025: Scattered clouds at 2,500 feet AGL
- 18/11: Temperature 18°C Dewpoint 11°C
- A3004: Altimeter 30.04" Hg
- RMK: Remarks (follow)
- SLP158: Sea Level Pressure in millibars (Don't worry about this. You'll never use it!)
- T01820114: Temp 18.2°C./Dewpoint 11.4°C.

METAR KLAX 100655Z 01011G18KT FUHZ 3SM SKC 18/11 A2995

- KLAX: Station: Los Angeles, CA
- 100655Z: Issued on 10th day at 0655 Zulu
- 01011G18KT: Wind is 010° at 11 knots gusting to 18 knots
- FUHZ: Smoke & Haze
- 3SM: Reported visibility 3 statute miles
- SKC: Sky is clear - no clouds
- 18/11: Temp 18°C Dewpoint 11°C
- A2995: Altimeter 29.95" Hg

Chapter Thirteen Answers

44. B
45. A
46. B
47. C
48. A
49. B
50. B
51. A

52.

DECODED TERMINAL AERODROME WEATHER FORECAST (TAF)

TAF
KLAX 101740Z 101818 12020KT 4SM -RA OVC016
FM1920 30015G25KT 3SM SHRA OVC015 PROB40 2022 1/2SM TSRA OVC008CB
FM2300 27008KT 5SM -SHRA BKN020 OVC040 TEMPO 0205 00000KT 1SM -RA FG
FM0800 VRB04KT 5SM -SHRA OVC020 BECMG 1214 20010KT P6SM NSW SKC

Above TAF Decoded

KLAX 201740Z 201818 24018KT 3SM RA BKN012

- KLAX — Station: Los Angeles
- 201740Z — Issued on 20th day at 1740 Zulu
- 201818 — Forecast time from 18Z on the 20th to 18Z on the 21st
- 24018KT — Wind forecast 240° at 18 knots
- 3SM — 3 statute miles visibility
- RA — Moderate rain
- BKN012 — (Ceiling) Broken clouds at 1,200 feet AGL

FM2030 32018G25KT 3SM +SHRA OVC015 PROB40 2223 1/2SM TSRA OVC006CB

- FM2030 — A significant change in weather expected at 2030Z
- 32018G25KT — Winds of 320° at 18 knots gusting to 25 knots
- 3SM — 3 statute miles visibility
- +SHRA — Heavy rain showers
- OVC015 — Overcast (ceiling) at 1,500 feet AGL
- PROB40 2223 1/2SM TSRA OVC006CB — 40-49% probability between 2200Z & 2300Z of 1/2 statute mile visibility with thunderstorms & rain with an overcast (ceiling) of 600 foot AGL & cumulonimbus clouds

FM2400 28011KT 5SM SHRA BKN015 OVC030 TEMPO 0306 00000KT 1/2SM -RA FG

- FM2400 — A significant change in weather expected at 2400Z
- 28011KT — Wind of 280° at 11 knots
- 5SM — 5 statute miles visibility
- SHRA — Moderate rain showers
- BKN015 OVC030 — Broken clouds (ceiling) at 1,500 feet AGL & an overcast at 3,000 feet AGL
- TEMPO 0306 00000KT 1/2SM -RA FG — A temporary change between 0300Z & 0600Z is expected with calm winds and one-half statute mile visibility due to light rain & fog

FM0800 VRB05KT 5SM -SHRA OVC035 BECMG 1315 25015KT P6SM NSW SKC

- FM0800 — A significant change in weather expected at 0800Z
- VRB05KT — Wind variable in direction at 5 knots
- 5SM — 5 statute miles visibility
- -SHRA — Light rain showers
- OVC035 — Overcast clouds (ceiling) at 3,500 feet AGL
- BECMG 1315 25015KT P6SM NSW SKC — Between 1300Z to 1500Z the weather is forecast to become: winds of 250° at 15 knots, visibility > 6 miles and no significant weather & a sky that's clear

Chapter Thirteen Answers

53. A
54. C
55. B
56. A
57. C
58. A
59. B
60. BECMG
61. remain, same
62. beginning
63. TEMPO
64. 30% to 39%, 40% to 49%, 50%
65. A
66. C
67. states
68. area
69. 3, 6
70. 12, 6
71. MSL
72. 4
73. Zulu
74. 1/Communications & Header Section
    2/Precautionary Statements Section
    3/Synopsis Section
    4/VFR Clouds & Weather Section
75. A
76. A
77. B
78. C
79. A
80. 6, 20
81. B
82. B
83. A
84. B
85. twice, 6, 12, 24
86. knots, true, celsius
87. A
88. A
89. B
90. B
91. C
92. C
93. 9900, 0000
94. B
95. A
96. B
97. A
98. B
99. C
100. C
101. B
102. B
103. C
104. A
105. B

106.

SYMBOL	TOTAL SKY COVER
	Sky Clear
	FEW - (more than 1/8 to 2/8)
	SCT - Scattered (3/8 to 4/8)
	BKN - Broken (5/8 to 7/8)
	OVC - Overcast (8/8)

WX DEPICTION PLOTTING SYMBOLS

107.

PLOTTING SYMBOL	MEANING
8	Few clouds, base 800 feet, visibility more than 6
12	Broken sky cover, ceiling 1,200 feet, rain shower, visibility more than 6
5	Thin overcast with breaks, visibility 5 in haze
30	Scattered at 3,000 feet, clouds topping ridges, visibility more than 6
2=	Sky clear, visibility 2, ground fog or fog
1/2	Sky partially obscured, visibility 1/2 blowing snow, no cloud layers observed
2= 200	Sky partially obscured, visibility 2, fog, cloud layer at 20,000 feet.
1/4* 5	Sky obscured, ceiling 500 feet, visibility 1/4 snow
1 12	Overcast, ceiling 1,200 feet thunderstorms, rain showers, visibility 1

Chapter Thirteen Answers

108. B
109. A
110. C
111. A
112. B
113. C
114. B
115. C
116. B
117. B
118. C
119. B
120. A
121. B
122. A
123. A
124. A
125. A
126. B
127. A
128. C
129. A
130. B
131. C
132. C
133. C
134. B
135. C
136. C
137. C
138. A
139. C
140. A
141. C
142. B
143. B
144. A
145. C
146. A
147. C
148. C
149. B
150. A
151. B
152. A
153. C
154. B
155. B

Note: To ensure that you have the most current answers to these questions, please check the *Book & Slide Updates* section at Rod Machado's web site: www.rodmachado.com

Chapter Fourteen

Flight Planning: Getting There From Here

What is Flight Planning?

1. [N2/1/1]
Another way to think about flight planning is to consider it as _____ management.
A. wind
B. information
C. fuel

Measuring Direction

2. [N2/3/2]
We can describe the location of any point on earth by an invisible grid of lines known as _____ and _____.
A. x coordinates, y coordinates
B. width, height
C. latitude, longitude

3. [N2/3/2]
Lines of longitude run _____.
A. east and west
B. north and south
C. north and west

4. [N3/1/1]
Lines of longitude are also called _____.
A. grids
B. meridians
C. equatorial lines

5. [N3/1/2]
Lines of latitude run _____.
A. east and west
B. north and south
C. north and west

6. [N3/1/3]
The longitude line running directly through Greenwich, England is known as the _____ (the 0 degree longitude line).
A. zulu line
B. equator
C. prime meridian

7. [N3/1/3]
To the left (west) of the prime meridian are _____ lines of longitude and to the right (east) of the prime meridian are another _____ lines of longitude, also including the prime meridian.
A. 180, 180
B. 90, 90
C. 360, 360

8. [N3/1/3]
Longitude is stated as being a certain number of degrees _____ or _____ of the prime meridian.
A. right, left
B. north, south
C. east, west

9. [N3/2/2]
The zero degree line of latitude, otherwise known as the _____, is located at the midpoint of the earth.
A. north pole
B. equator
C. prime meridian

10. [N3/2/2]
Spaced one degree apart, lines of latitude run _____ to the equator, distributed north and south.
A. parallel
B. perpendicular
C. vertical

11. [N3/2/2]
Latitude lines north of the equator are called _____ and south of the equator are called _____.
A. verticals, perpendiculars
B. north latitude, south latitude
C. north longitude, south longitude

12. [N3/2/2]
Lines of latitude are sometimes called _____.
A. parallels
B. meridians
C. equatorial lines

Time Measurement

13. [N4/1/2]
It's standard practice to establish time zones in the United States at _____ degree intervals.
A. 30
B. 45
C. 15

14. [N4/1/2]
There are ____ time zones in the continental United States.
A. seven
B. four
C. three

15. [N4/1/4]
When the sun is directly above the 120th, 105th, 90th and 75th longitude lines, it's _____ in these time zones.
A. noon
B. twilight
C. midnight

16. [N4/1/4]
Under Daylight Savings Time, noon occurs when the sun is shifted _____ degrees to the left as an earthling looks up at it while facing south.
A. 20
B. 15
C. 45

17. [N4 & N5 /All]
Refer to the figure at the top of the next page. An aircraft departs an airport in the Eastern Daylight Time zone at 0945 EDT for a two-hour flight to an airport located in the Central Daylight Time zone. The landing should be at what coordinated universal time?
A. 1345Z.
B. 1445Z.
C. 1545Z.

18. [N4 & N5 /All]
Refer to the figure at the top of the next page. An aircraft departs an airport in the Central Standard Time zone at 0930 CST for a 2-hour flight to an airport located in the Mountain Standard Time zone. The landing should be at what time?
A. 0930 MST.
B. 1030 MST.
C. 1130 MST.

19. [N4 & N5 /All]
Refer to the figure at the top of the next page. An aircraft departs an airport in the Central Standard Time zone at 0845 CST for a two-hour flight to an airport located in the Mountain Standard Time zone. The landing should be at what coordinated universal time?
A. 1345Z.
B. 1445Z.
C. 1645Z.

20. [N4 & N5 /All]
Refer to the figure at the top of the next page. An aircraft departs an airport in the Mountain Standard Time zone at 1615 MST for a two-hour 15 minute flight to an airport located in the Pacific Standard Time zone. The estimated time of arrival at the destination airport should be
A. 1630 PST.
B. 1730 PST.
C. 1830 PST.

21. [N4 & N5 /All]
Refer to the figure at the top of the next page. An aircraft departs an airport in the Pacific Standard Time zone at 1030 PST for a four-hour flight to an airport located in the Central Standard Time zone. The landing should be at what coordinated universal time?
A. 2030Z.
B. 2130Z.
C. 2230Z.

22. [N4 & N5 /All]
Refer to the figure at the top of the next page. An aircraft departs an airport in the Mountain Standard Time zone at 1515 MST for a two-hour 30 minute flight to an airport located in the Pacific Standard Time zone. What is the estimated time of arrival at the destination airport?
A. 1645 PST.
B. 1745 PST.
C. 1845 PST.

23. [N4 & N5 /All]
Refer to the figure at the top of the next page. An aircraft departs an airport in the Central Standard Time zone at 1030 CST for a two-hour flight to an airport located in the Mountain Standard Time zone. The landing should be at what time?
A. 1330 MST.
B. 1130 MST.
C. 1230 MST.

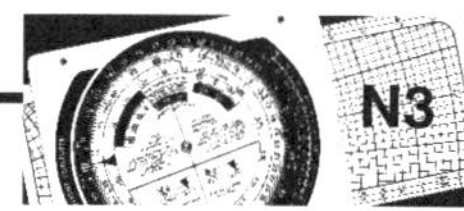

TO CONVERT FROM:	TO COORDINATED UNIVERSAL TIME
Eastern Standard Time	Add 5 hours
Eastern Daylight Time	Add 4 hours
Central Standard Time	Add 6 hours
Central Daylight Time	Add 5 hours
Mountain Standard Time	Add 7 hours
Mountain Daylight Time	Add 6 hours
Pacific Standard Time	Add 8 hours
Pacific Daylight Time	Add 7 hours
Yukon Standard Time	Add 9 hours
Alaska, Hawaii Standard Time	Add 10 hours
Bering Standard Time	Add 11 hours

TIME ZONES

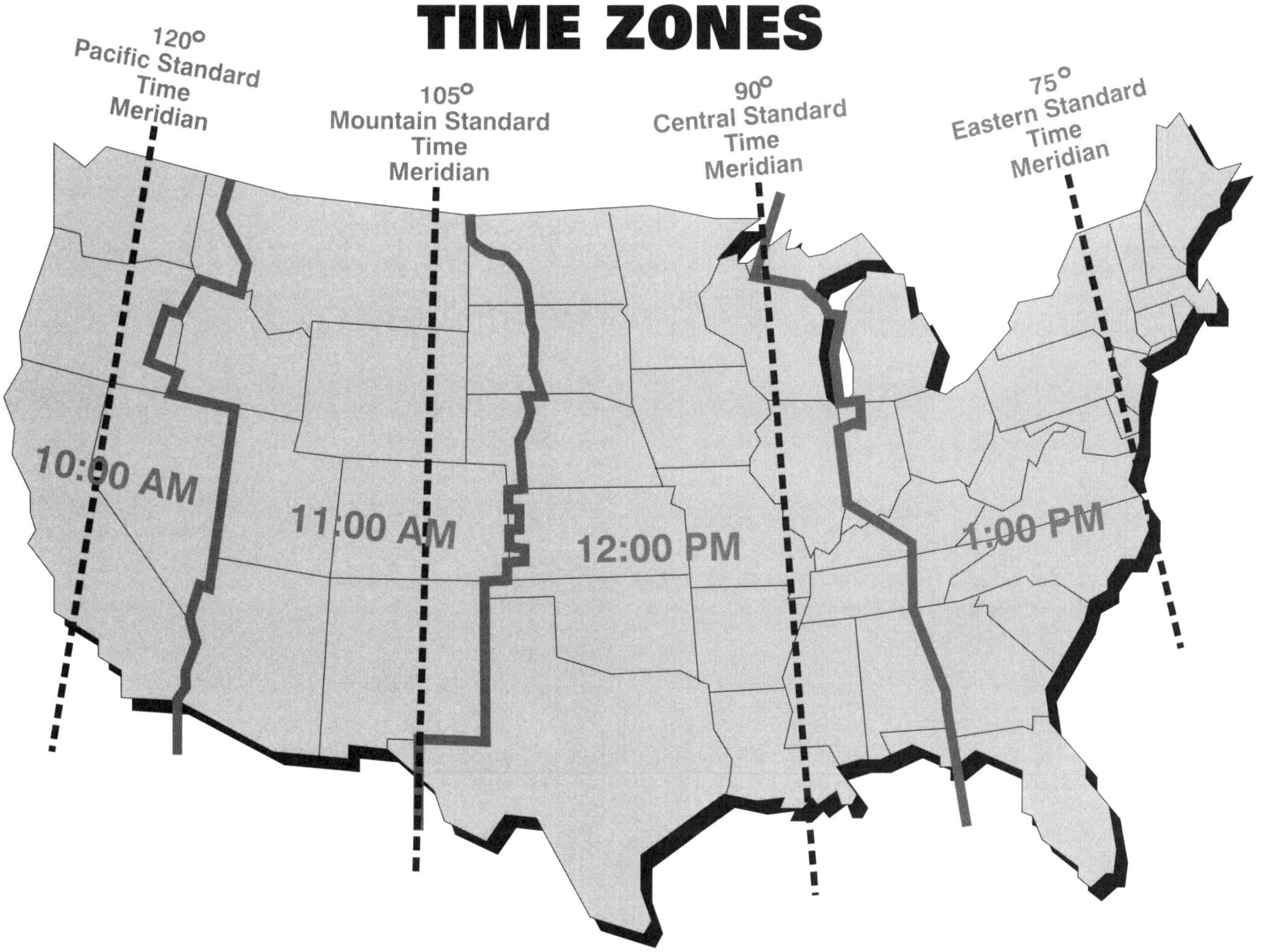

24. [N6/1/2]
Every degree of longitude and latitude can be broken down into 60 smaller segments called _____.
A. seconds
B. minutes
C. microns

25. [N7/1/2]
For even greater accuracy, each minute of distance is broken down into 60 _____.
A. seconds
B. minutes
C. microns

A Matter of Degree: Longitude and Latitude on Sectional Charts

26. [Pages N6 through N9]
Referring to the chart on the opposite page, determine the approximate latitude and longitude of General Pershing Airport (area B).
A. 39 degrees 46'N - 94 degrees 54' W.
B. 39 degrees 46'N - 93 degrees 06' W.
C. 40 degrees 14'N - 94 degrees 54' W.

27. [Pages N6 through N9]
Referring to the chart on the opposite page, which airport is located at approximately 39 degrees 04' 50" N latitude and 93 degrees 40' 55" W longitude?
A. Lexington.
B. Marshal Memorial.
C. Higginsville.

28. [Pages N6 through N9]
Referring to the chart on the opposite page, which airport is located at approximately 39 degrees 18' N latitude and 93 degrees 22' 30" W longitude?
A. Famuliner Farms.
B. Carrollton Memorial.
C. Chillicothe.

29. [Pages N6 through N9]
Referring to the chart on the opposite page, what is the approximate latitude and longitude of Marshal Memorial (area C)?
A. 39 degrees 06'N - 94 degrees 12'W.
B. 39 degrees 06'N - 93 degrees 12'W.
C. 39 degrees 54'N - 93 degrees 48'W.

30. [Pages N6 through N9]
Referring to the chart on the opposite page, determine the approximate latitude and longitude of Longwood Airport (area D).
A. 39 degrees 06'N - 93 degrees 52'W.
B. 39 degrees 54'N - 93 degrees 52'W.
C. 39 degrees 54'N - 94 degrees 52'W.

Cross Country Navigation

31. [N9/3/4]
Navigation by reference to landmarks is known as _____.
A. dead reckoning
B. pilotage
C. electronic navigation

32. [N9/3/4]
Navigation that involves calculating the effects of wind and compass error on a plotted course is known as _____.
A. dead reckoning
B. pilotage
C. electronic navigation

33. [N10/1/3]
List the steps shown below in the order you would use them to plot a cross country course.
U. Determine the true course.
V. Draw a line between airports or checkpoints.
W. Determine the compass heading.
X. Determine the magnetic heading.
Y. Determine the true heading.
Z. Determine the wind correction angle.

A. U, V, Y, W, Z, X
B. V, U, Z, Y, X, W
C. V, U, X, Z, Y, W

Our Trip

Step 1:Draw a line between airports or checkpoints

34. [N10/2/3]
The first step in cross country flight planning is to draw a line on the chart between the _____ and the _____.
A. longitude line, departure point
B. route midpoint, destination
C. departure, destination

Step 2: Determine the true course

35. [N11/Figure 12]
The true course is the angle your flight path makes with the _____.
A. magnetic north pole
B. true north pole
C. compass north pole

36. [N11/Figure 12]
The true course is measured by drawing a line on a chart and measuring the angle that line makes with a line of _____.
A. latitude
B. longitude
C. deviation

37. [N11/Figure 12]
Referring to the chart on the opposite page, determine the true course from Lexington Airport (area A) to General Pershing Airport (area B).
A. 049 degrees.
B. 229 degrees.
C. 051 degrees.

38. [N11/Figure 12]
Referring to the chart on the opposite page, determine the true course from General Pershing Airport (area B) to Lexington Airport (area A).
A. 049 degrees.
B. 229 degrees.
C. 231 degrees.

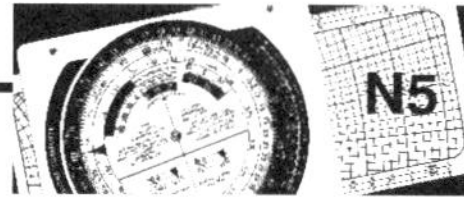

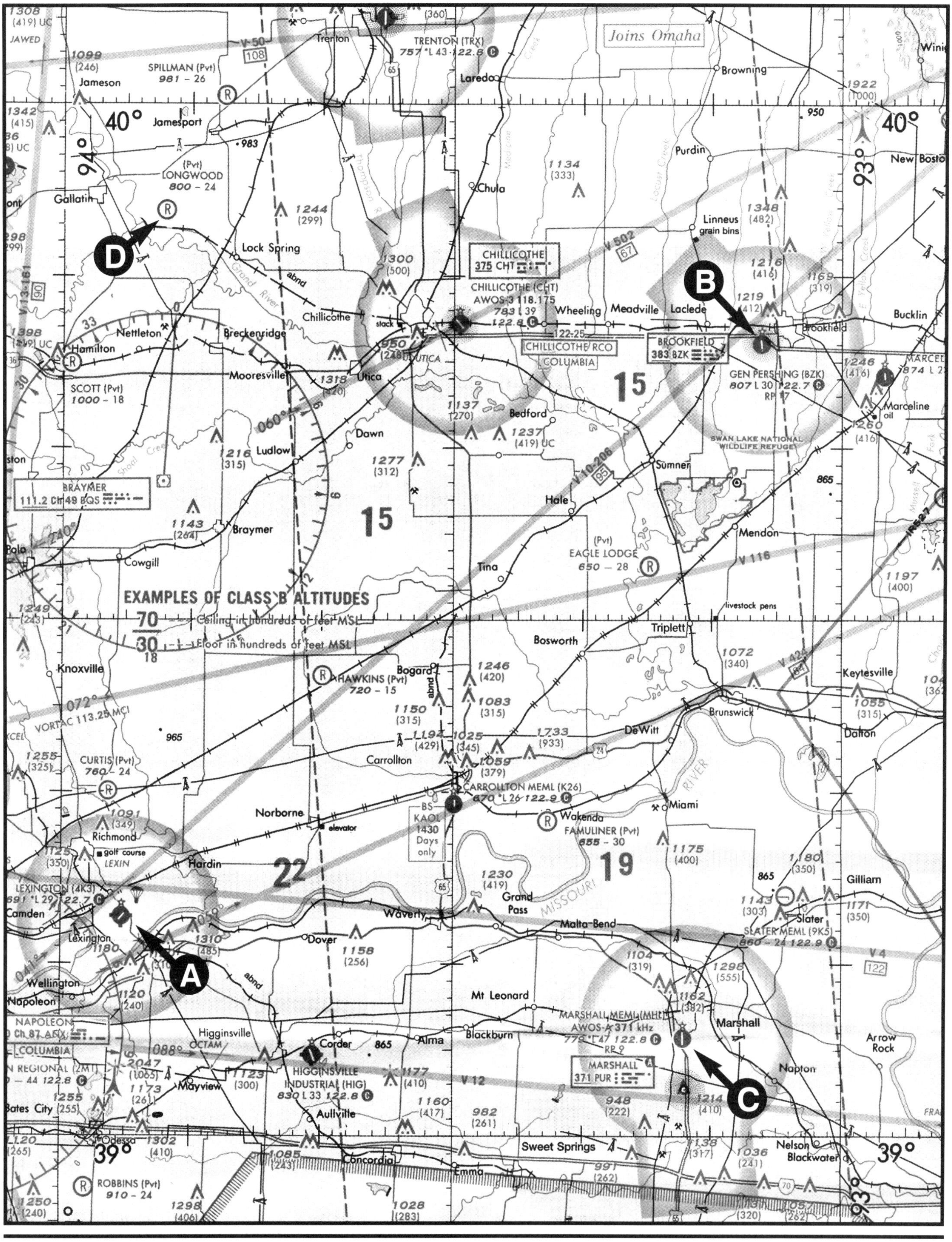

Joins Omaha

Trenton
TRENTON (TRX)
757 *L 43 122.8
Laredo
Browning
Jameson
SPILLMAN (Pvt)
981 – 26
Jamesport
40°
94°
93°
(Pvt)
LONGWOOD
800 – 24
Chula
Purdin
New Boston
Gallatin
Linneus
grain bins
Lock Spring
D
CHILLICOTHE
375 CHT
CHILLICOTHE (CHT)
AWOS-3 118.175
783 L 39
122.8
Wheeling
Meadville
Laclede
B
Bucklin
Brookfield
Chillicothe
Nettleton
Breckenridge
Hamilton
CHILLICOTHE RCO
COLUMBIA
BROOKFIELD
383 BZK
GEN PERSHING (BZK)
807 L 30 122.7
RP 17
Marceline
SCOTT (Pvt)
1000 – 18
Mooresville
Utica
Bedford
SWAN LAKE NATIONAL WILDLIFE REFUGE
Dawn
Ludlow
Sumner
BRAYMER
111.2 Ch 49 BQS
Hale
15
Braymer
Mendon
Polo
Cowgill
(Pvt)
EAGLE LODGE
650 – 28
Tina
V 116

EXAMPLES OF CLASS B ALTITUDES

70 — Ceiling in hundreds of feet MSL

30 — Floor in hundreds of feet MSL

Triplett
Bosworth
Knoxville
HAWKINS (Pvt)
720 – 15
Bogard
Keytesville
072°
VORTAC 113.25 MCI
Brunswick
DeWitt
Dalton
CURTIS (Pvt)
760 – 24
Carrollton
CARROLLTON MEML (K26)
670 *L 26 122.9
Miami
Norborne
elevator
BS-KAOL 1430 Days only
Wakenda
FAMULINER (Pvt)
655 – 30
Richmond
golf course
LEXIN
Hardin
22
19
MISSOURI
RIVER
Gilliam
LEXINGTON (4K3)
Camden
Lexington
Waverly
Grand Pass
Malta Bend
Slater
SLATER MEML (9K5)
860 – 24 122.9
Dover
A
Wellington
Napoleon
Mt Leonard
MARSHALL MEML (MHL)
AWOS-A 371 kHz
779 *L 47 122.8
RP 9
Marshall
NAPOLEON
Ch 87 ANX
COLUMBIA
Higginsville
OCTAM
088°
Corder
Alma
Blackburn
Arrow Rock
MARSHALL
371 PUR
Napton
C
Mayview
HIGGINSVILLE INDUSTRIAL (HIG)
830 L 33 122.8
V 12
Bates City
Aullville
Odessa
39°
Concordia
Sweet Springs
Nelson
Blackwater
Emma
ROBBINS (Pvt)
910 – 24

Step 3: Determine the wind correction angle

39. [N14/1/3]
If the wind is blowing from the north at 20 knots, we're saying the atmosphere is moving across the ground a distance of _____ nautical miles for every hour that passes.
A. 20
B. 10
C. 40

The Effect of Water on a Swimmer

40. [N15/Figure19]
To avoid being carried downstream, a swimmer must swim at some angle _____ the flow of water.
A. into
B. away from
C. perpendicular to

The Effects of Wind on an Airplane

41. [N15/1/2]
To maintain your ground track when flying with a wind from your left, you need to turn the airplane to the _____ at some angle sufficient to correct for drift.
A. north
B. right
C. left

42. [N15/1/2]
An airplane's track over the ground is a combination of the motion of the airplane and the motion of the _____.
A. true course
B. ground
C. air

43. [N15/1/2]
The angle between the desired course and the actual ground track is called the _____ angle.
A. gust
B. wind
C. drift

Using the Wind Side of the Slide Computer

44. [N18/Figure 29]
The E6-B computer solves a wind problem by constructing a _____.
A. pentagram
B. wind triangle
C. wind chart

Step 4: Determine the true heading

45. [N18/1/4]
A heading is technically a course that's been corrected for the effects of _____.
A. wind
B. deviation
C. variation

46. [N18/1/5]
Left WCAs (wind correction angles) are _____ and right WCAs are _____ to the true course to obtain the true heading.
A. subtracted, added
B. added, subtracted
C. subtracted, subtracted

Step 5: Determining the magnetic heading

47. [N19/See *Is There an Untrue North Pole*]
The airplane's compass points to what pole?
A. The earth's axis pole.
B. The magnetic north pole.
C. The true north pole.

48. [N20/1/2]
Lying just a little north of Hudson Bay, the _____ north pole is separated by a distance of approximately 1300 miles from the _____ north pole.
A. magnetic, true
B. true, magnetic
C. magnetic, magnetic

49. [N20/Figure 32]
According to the figure below, what type of variation would airplanes A, B and C experience, respectively?
A. Zero, east, west.
B. West, zero, east.
C. East, zero, west.

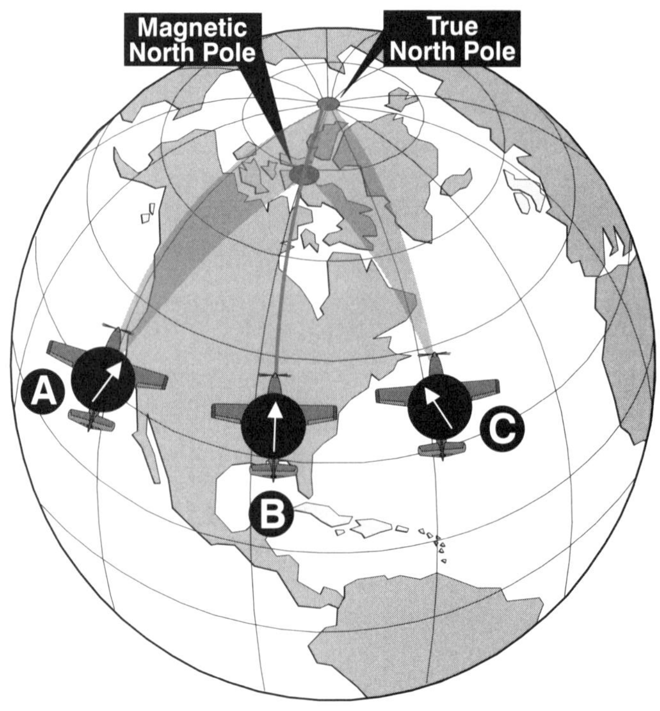

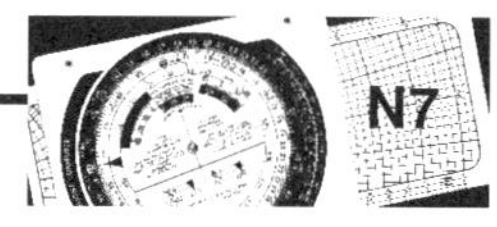

50. [N21/1/1]
An agonic line of variation represents _____ variation.
A. 10 degrees
B. zero degrees
C. maximum degrees

51. [N21/Figure 34]
Referring to the chart below, what is the magnetic variation rounded up to the nearest whole degree?
A. 14 degrees west.
B. 1.4 degrees east.
C. 15 degrees east.

52. [N22/1/2]
To convert a true heading to a magnetic heading, you must _____ easterly variation and _____ westerly variation.
A. subtract, add
B. add, add
C. add, subtract

53. [N19-N22]
Referring to the chart below, determine the magnetic heading for a flight from Paramount airport (area A) to Shafter-Minter airport (area B). The wind is from 330 degrees at 25 knots, the true airspeed is 100 knots, and the magnetic variation is 15 degrees east.
A. 080 degrees.
B. 105 degrees.
C. 265 degrees.

54. [N19-N22]
Referring to the chart below, what is the magnetic heading for a flight from Shafter-Minter airport (area B) to Paramount airport (area A)? The wind is from 030 degrees at 12 knots and the true airspeed is 95 knots (use the magnetic variation shown on the chart rounded up to the nearest whole degree).
A. 263 degrees.
B. 292 degrees.
C. 277 degrees.

55. [N19-N22]
Referring to the chart below, determine the magnetic heading for a flight from Delano Airport (area C) to Paramount airport (area A). The wind is from 300 degrees at 14 knots and the true airspeed is 90 knots (use the magnetic variation shown on the chart rounded up to the nearest whole degree).
A. 239 degrees.
B. 281 degrees.
C. 252 degrees.

56. [N19-N22]
Referring to the chart below, determine the magnetic heading for a flight from Paramount airport (area A) to Delano (area C). The wind is from 215 degrees at 25 knots and the true airspeed is 125 knots (use the magnetic variation shown on the chart rounded up to the nearest whole degree).
A. 057 degrees.
B. 074 degrees.
C. 088 degrees.

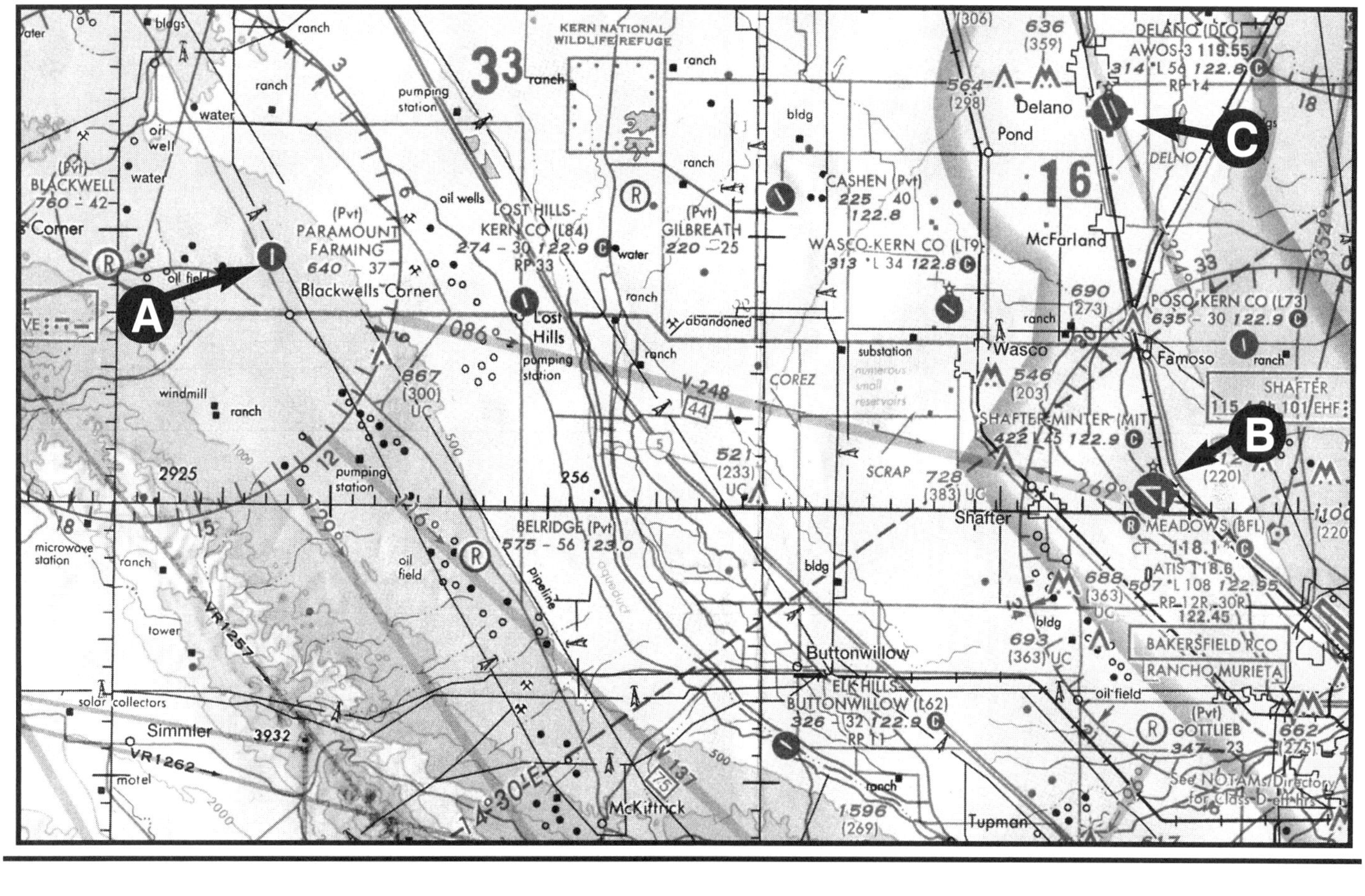

57. [N19-N22]
Referring to the chart at the bottom of page N9, determine the magnetic heading for a flight from Cold Spring airport (area A) to Wurtsboro-Sullivan County airport (area B). The wind is from 330 degrees at 25 knots, the true airspeed is 110 knots (use the magnetic variation shown on the chart).
A. 003 degrees.
B. 122 degrees.
C. 102 degrees.

58. [N19-N22]
Referring to the chart at the bottom of page N9, determine the magnetic heading for a flight from Wurtsboro-Sullivan County airport (area B) to Cold Spring airport (area A). The wind is from 340 degrees at 12 knots, the true airspeed is 136 knots (use the magnetic variation shown on the chart).
A. 289 degrees.
B. 270 degrees.
C. 296 degrees.

Step 6: Determine your compass heading

59. [N22/1/5]
Deviation in a magnetic compass is caused by the
A. presence of flaws in the permanent magnets of the compass.
B. difference in the location between true north and magnetic north.
C. magnetic fields within the aircraft distorting the lines of magnetic force.

60. [N23/1/4]
Based on the compass deviation card shown below, what compass heading should you fly if your magnetic heading is 127 degrees?
A. 124 degrees.
B. 130 degrees.
C. 120 degrees.

61. [N23/1/4]
Based on the compass deviation card shown below, what compass heading should you fly if your magnetic heading is 208 degrees?
A. 210 degrees.
B. 206 degrees.
C. 212 degrees.

TYPICAL COMPASS DEVIATION CARD

FOR (MAGNETIC)	N	30	60	E	120	150
STEER (COMPASS)	O	28	57	86	117	148
FOR (MAGNETIC)	S	210	240	W	300	330
STEER (COMPASS)	180	212	243	274	303	332

62. [N22-N24]
Referring to the chart at the bottom of page N9 (and the compass deviation chart in the top right-hand corner of that chart), determine the compass heading for a flight from Cold Spring airport (area A) to Wurtsboro-Sullivan County airport (area B). The wind is from 090 degrees at 16 knots and the true airspeed is 90 knots.
A. 110 degrees.
B. 113 degrees.
C. 107 degrees.

63. [N22-N24]
Referring to the chart at the bottom of page N9 (and the compass deviation chart in the top right-hand corner of that chart), determine the compass heading for a flight from Wurtsboro-Sullivan County airport (area B) to Cold Spring airport (area A). The wind is from 290 degrees at 18 knots and the true airspeed is 85 knots.
A. 297 degrees.
B. 294 degrees.
C. 271 degrees.

64. [N22-N24]
Referring to the chart at the bottom of page N9 (and the compass deviation chart in the top right-hand corner of that chart), determine the compass heading for a flight from Cold Spring airport (area A) to Resnick airport (area D). The wind is from 150 degrees at 26 knots and the true airspeed is 94 knots.
A. 114 degrees.
B. 111 degrees.
C. 85 degrees.

65. [N22-N24]
Based on the information below, plan a flight from Spring Hill airport (area C) to Resnick airport (area D) using the chart on page N9. Use the compass deviation card shown below. Fill out the flight log listed below. Assume the wind is from 170 degrees at 32 knots and the true airspeed is 109 knots. Fill out the appropriate parts of the basic flight log shown below.

DETERMINE THE COMPASS HEADING

Check Points	VOR Ident. Freq.	VOR Course (if applicable)	Altitude	Wind Dir. / Vel. / Temp.	True Airspeed	TC	WCA R+ L-	TH	V E- W+	MH	D	CH	Distance	GS Ground speed Est. / Act.
Spring Hill	X X	X	X	170° / 32 / X	109 Kts	63°							X	X
Resnick	X X													

TYPICAL COMPASS DEVIATION CARD

FOR (MAGNETIC)	N	30	60	E	120	150
STEER (COMPASS)	O	31	64	92	122	151
FOR (MAGNETIC)	S	210	240	W	300	330
STEER (COMPASS)	180	208	236	265	300	329

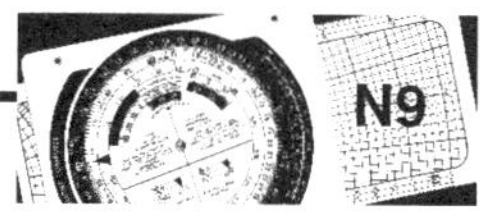

66. [N24/1/4]
What is the appropriate acronym to use as a memory aid when converting true headings to compass headings?
A. CVMDT.
B. TVMDC.
C. MDVWT.

Planning an Actual Flight

XC Planning: Selecting a Route

67. [N28/1/1]
When possible, try and select a _____ route between departure and destination airports.
A. out of the way
B. circular
C. direct

XC Planning: Measuring the Distance

68. [N28/2/2]
Referring to the chart below, what is the distance between Cold Spring airport (area A) and Wurtsboro-Sullivan County airport (area B)?
A. 37.5 statute miles.
B. 42.5 nautical miles.
C. 37.5 nautical miles.

69. [N28/2/2]
Referring to the chart below, what is the distance between Cold Spring airport (area A) and Resnick airport (area D)?
A. 41 nautical miles.
B. 41 statute miles.
C. 82 nautical miles.

70. [N28/2/2]
Referring to the chart below, what is the distance between Spring Hill airport (area C) and Resnick airport (area D)?
A. 104 nautical miles.
B. 52 nautical miles.
C. 52 statute miles.

XC Planning: Selecting Checkpoints

71. [N29/1/1] Fill in the blank:
When possible, try to pick VFR checkpoints that are separated by _____________ to _____________ miles

72. [N28/3/2]
Referring to the chart below, which of the following would make good VFR checkpoints for a flight from Spring Hill airport (area C) to Resnick airport (area D)?
A. The town of White Mills (area J) and the town resort (area K).
B. The railroad (area E) and Monticello airport (area H).
C. Boehm airport (area G) and Monticello airport (area H).

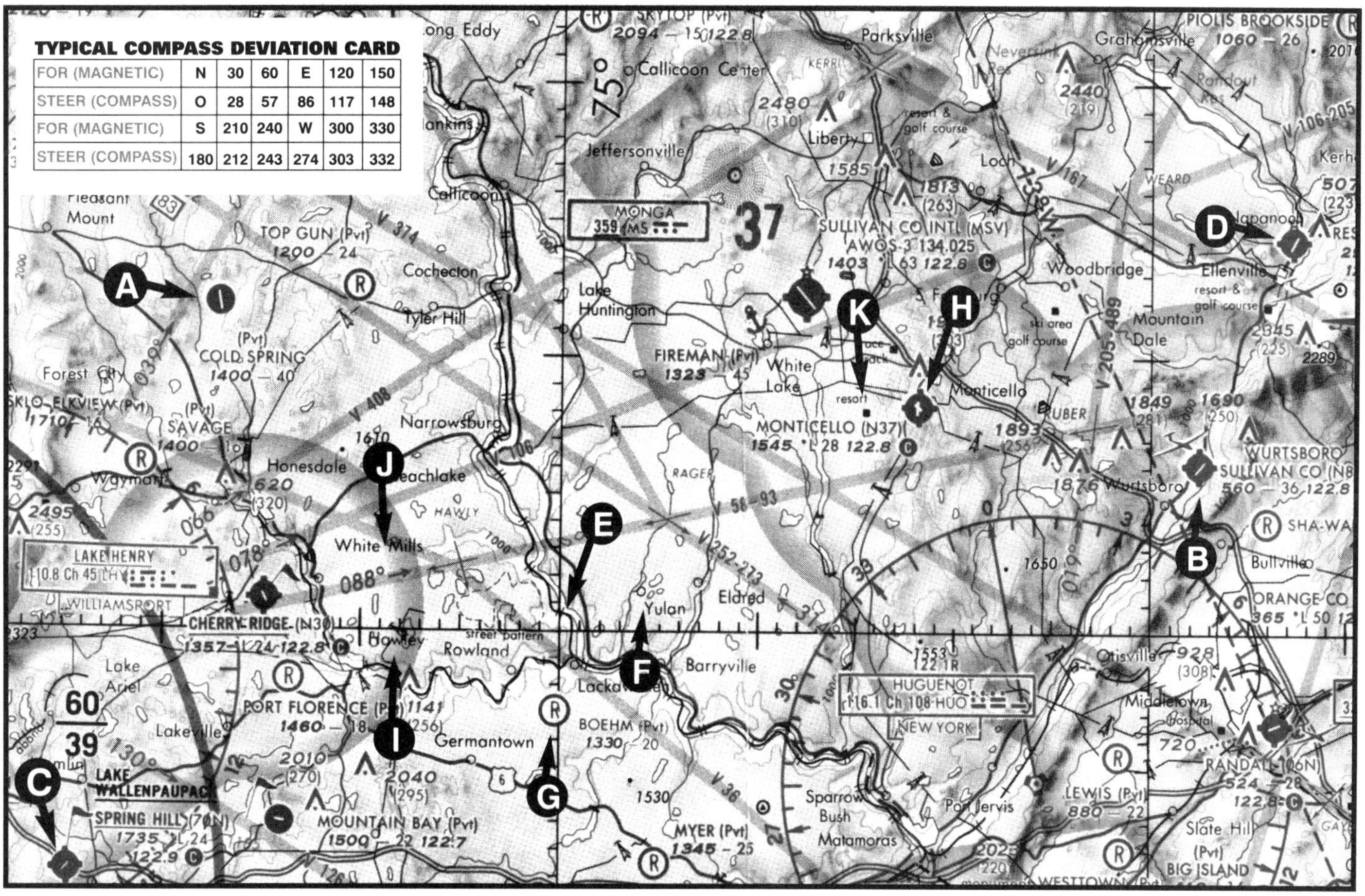

TYPICAL COMPASS DEVIATION CARD

FOR (MAGNETIC)	N	30	60	E	120	150
STEER (COMPASS)	0	28	57	86	117	148
FOR (MAGNETIC)	S	210	240	W	300	330
STEER (COMPASS)	180	212	243	274	303	332

XC Planning: Input Information

73. [N29/Figure 53]
Refer to the chart excerpt below and the basic flight log form above it. Three VFR checkpoints have been chosen for a flight from Spring Hill airport (area A) to Resnick airport (area E) and are listed in the flight plan. Measure the distance between checkpoints and place these into the flight planning form.

XC Planning: Chose an Altitude for the Flight

74. [N29/2/3]
Referring to the sectional chart excerpt below, which of the following altitudes listed would be a good cruise altitude on a flight from from Spring Hill airport (area A) to Resnick airport (area E)?
A. 5,500 feet MSL.
B. 4,500 feet MSL.
C. 2,500 feet MSL.

Basic flight log for a flight from Spring Hill airport to Resnick airport

Check Points	VOR Ident. Freq.	VOR Course (if applicable)	Altitude	Wind Dir. Vel. Temp.	True Airspeed	TC	WCA R+ L-	TH	V E- W+	MH	D	CH	Distance	GS Ground speed Est. Act.	Time Airborne ETE	ATE	GPH Gallons Per Hour Used Remain
Spring Hill Airport					110 Kt.									X	X	X	X X
Hawley					110 Kt.									X	X	X	X X
Railroad					110 Kt.									X	X	X	X X
Monticello Airport					110 Kt.									X	X	X	X X
Resnick Airport																	
Totals																	

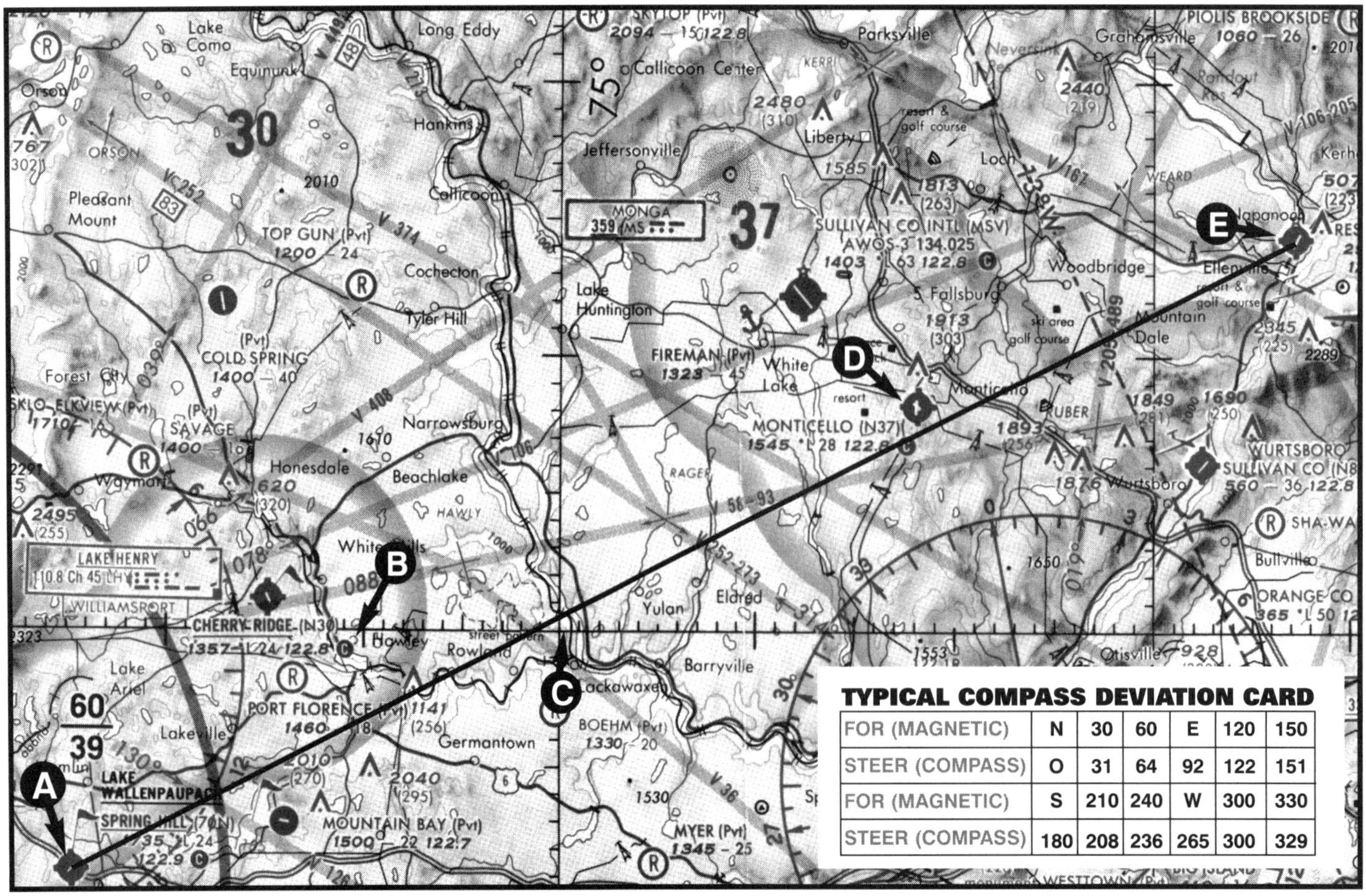

TYPICAL COMPASS DEVIATION CARD

FOR (MAGNETIC)	N	30	60	E	120	150
STEER (COMPASS)	O	31	64	92	122	151
FOR (MAGNETIC)	S	210	240	W	300	330
STEER (COMPASS)	180	208	236	265	300	329

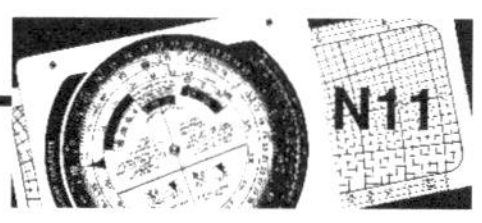

XC Flight Planning: Finding the Compass Heading

75. [N30/1/4]

Winds Aloft
3,000 feet 190 degrees at 10 knots/15 C
6,000 feet 220 degrees at 40 knots/06 C

Based on the winds aloft above and the sectional chart excerpt shown on the bottom of page N12, determine the compass heading for a flight from Spring Hill airport (area A) to Resnick airport (area E). Use the compass deviation chart shown at the bottom right corner of the sectional chart on page N10. Make sure to fill out the rest of the basic flight log.
A. 62 degrees.
B. 87 degrees.
C. 84 degrees.

XC Flight Planning: Estimating Time Enroute and Fuel Consumed

Postflight Briefing 14-1: Miles on the Menu: Converting Nautical and Statute Miles

76. [N36/2/2]
One hundred and thirty nautical miles is equal to approximately how many statute miles?
A. 150 sm.
B. 140 sm.
C. 113 sm.

77. [N36/2/2]
Eighteen statute miles is equal to how many nautical miles?
A. 156 nm.
B. 15.6 nm.
C. 20.8 nm.

78. [N36/2/2]
Twenty-three knots is equal to how many mph?
A. 2.6 mph.
B. 20 mph.
C. 26.5 mph.

79. [N36/2/2]
One hundred and ninety mph is equal to how many knots?
A. 165 knots.
B. 220 knots.
C. 143 knots.

Postflight Briefing 14-1: Time, Distance and Speed Computations

80. [N39/2/4]
At a groundspeed of 140 knots, how many miles will the airplane travel in 37 minutes?
A. 16 nautical miles.
B. 86 nautical miles.
C. 80 nautical miles.

81. [N39/2/4]
At a groundspeed of 114 knots, how many miles will the airplane travel in 1 hour and 15 minutes?
A. 143 nautical miles.
B. 79 nautical miles.
C. 91 nautical miles.

82. [N39/2/4]
At a groundspeed of 138 knots, how many miles will the airplane travel in 2 hours and 11 minutes?
A. 92 nautical miles.
B. 490 nautical miles.
C. 302 nautical miles.

83. [N39/2/4]
At a groundspeed of 95 knots, how many miles will the airplane travel in 11 minutes?
A. 17.4 nautical miles.
B. 7 nautical miles.
C. 52 nautical miles.

84. [N40/1/2]
At 146 knots groundspeed, how long will it take to travel 43 miles?
A. 10. 5 minutes.
B. 17.5 minutes.
C. 1.7 minutes.

85. [N40/1/2]
At 92 knots groundspeed, how long will it take to travel 115 miles?
A. 75 minutes.
B. 98 minutes.
C. 48 minutes.

86. [N40/1/2]
At 119 knots groundspeed, how long will it take to travel 7 miles?
A. 3.5 minutes.
B. 35 minutes.
C. .35 minutes.

87. [N40/1/2]
At 164 knots groundspeed, how long will it take to travel 33 miles?
A. 12 minutes.
B. 9 minutes.
C. 120 minutes.

88. [N40/1/1]
If you traveled 4 nautical miles in 2 minutes, what is your groundspeed?
A. 130 knots.
B. 120 knots.
C. 300 knots.

89. [N40/1/1]
If you traveled 10 nautical miles in 6 minutes, what is your groundspeed?
A. 100 knots.
B. 60 knots.
C. 36 knots.

90. [N40/1/1]
If you traveled 25 nautical miles in 13 minutes, what is your groundspeed?
A. 96 knots.
B. 315 knots.
C. 115 knots.

91. [N40/1/1]
If you traveled 21 nautical miles in 9 minutes, what is your groundspeed?
A. 140 knots.
B. 193 knots.
C. 146 knots.

92. [N31/1/5 & Pages N16 through N19]
Refer to the chart below. Assume you're enroute from Resnick Airport (area E) to Spring Hill (area A). Your flight passes over Monticello Airport (area D) at 1456 and then over the railroad (area C) at 1504. At what time should your flight arrive at Spring Hill?
A. 1515
B. 1509
C. 1526

93. [N31/1/5 & Pages N16 through N19]
Refer to the chart below. What is the estimated time enroute from Cold Spring Airport (area F) to Wurtsboro-Sullivan Airport (area G)? The wind is from 030 degrees at 25 knots and the true airspeed is 100 knots. Add 3 1/2 minutes for departure and climbout.
A. 25 minutes.
B. 14 minutes.
C. 29 minutes.

94. [N31/1/5 & Pages N16 through N19]
Refer to the chart below. Determine the estimated time enroute for a flight from Cold Spring Airport (area F) to Mountain Bay Airport (area H). The wind is from 290 at 12 knots and the true airspeed is 95 knots. Add 2 minutes for climbout.
A. 20 minutes.
B. 13.5 minutes.
C. 27 minutes.

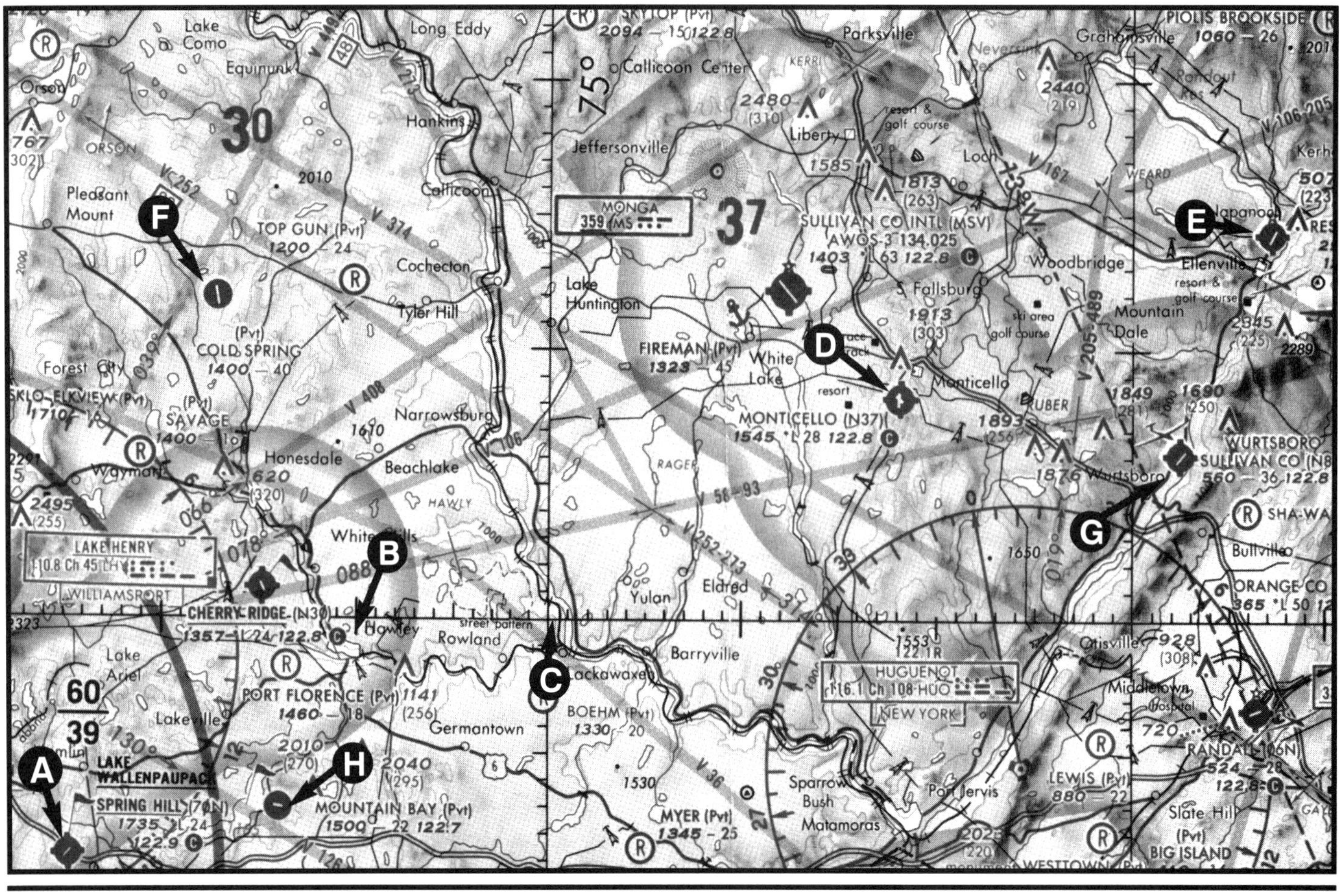

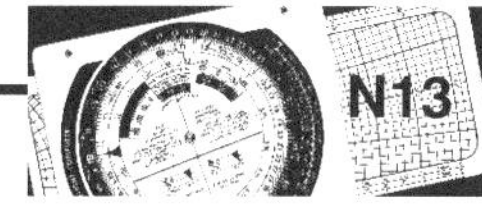

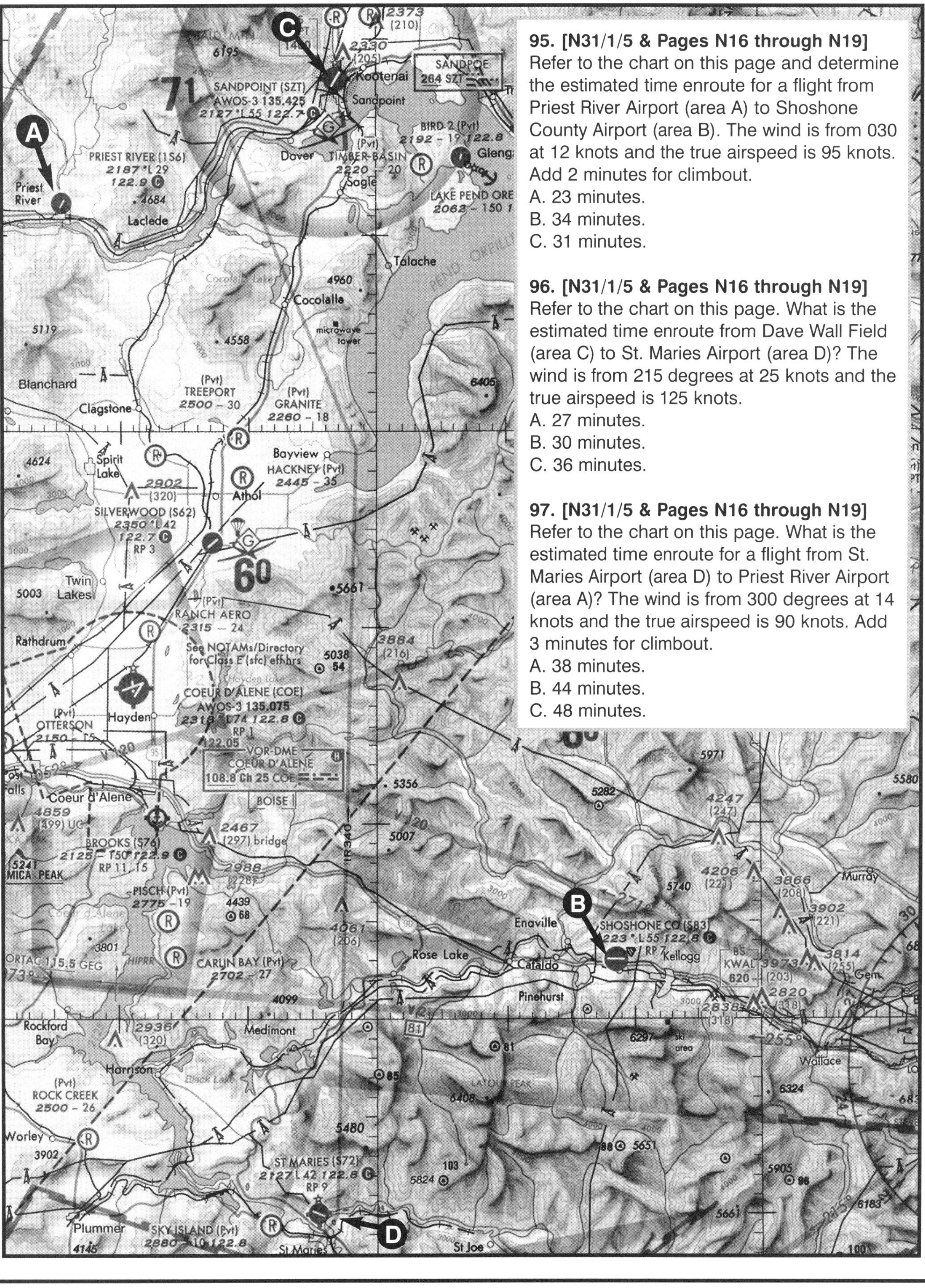

95. [N31/1/5 & Pages N16 through N19] Refer to the chart on this page and determine the estimated time enroute for a flight from Priest River Airport (area A) to Shoshone County Airport (area B). The wind is from 030 at 12 knots and the true airspeed is 95 knots. Add 2 minutes for climbout.
A. 23 minutes.
B. 34 minutes.
C. 31 minutes.

96. [N31/1/5 & Pages N16 through N19] Refer to the chart on this page. What is the estimated time enroute from Dave Wall Field (area C) to St. Maries Airport (area D)? The wind is from 215 degrees at 25 knots and the true airspeed is 125 knots.
A. 27 minutes.
B. 30 minutes.
C. 36 minutes.

97. [N31/1/5 & Pages N16 through N19] Refer to the chart on this page. What is the estimated time enroute for a flight from St. Maries Airport (area D) to Priest River Airport (area A)? The wind is from 300 degrees at 14 knots and the true airspeed is 90 knots. Add 3 minutes for climbout.
A. 38 minutes.
B. 44 minutes.
C. 48 minutes.

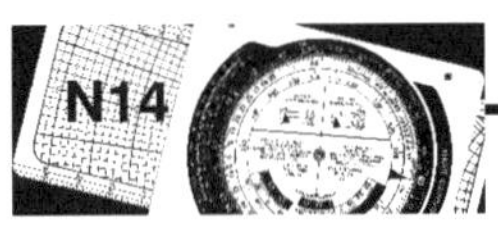

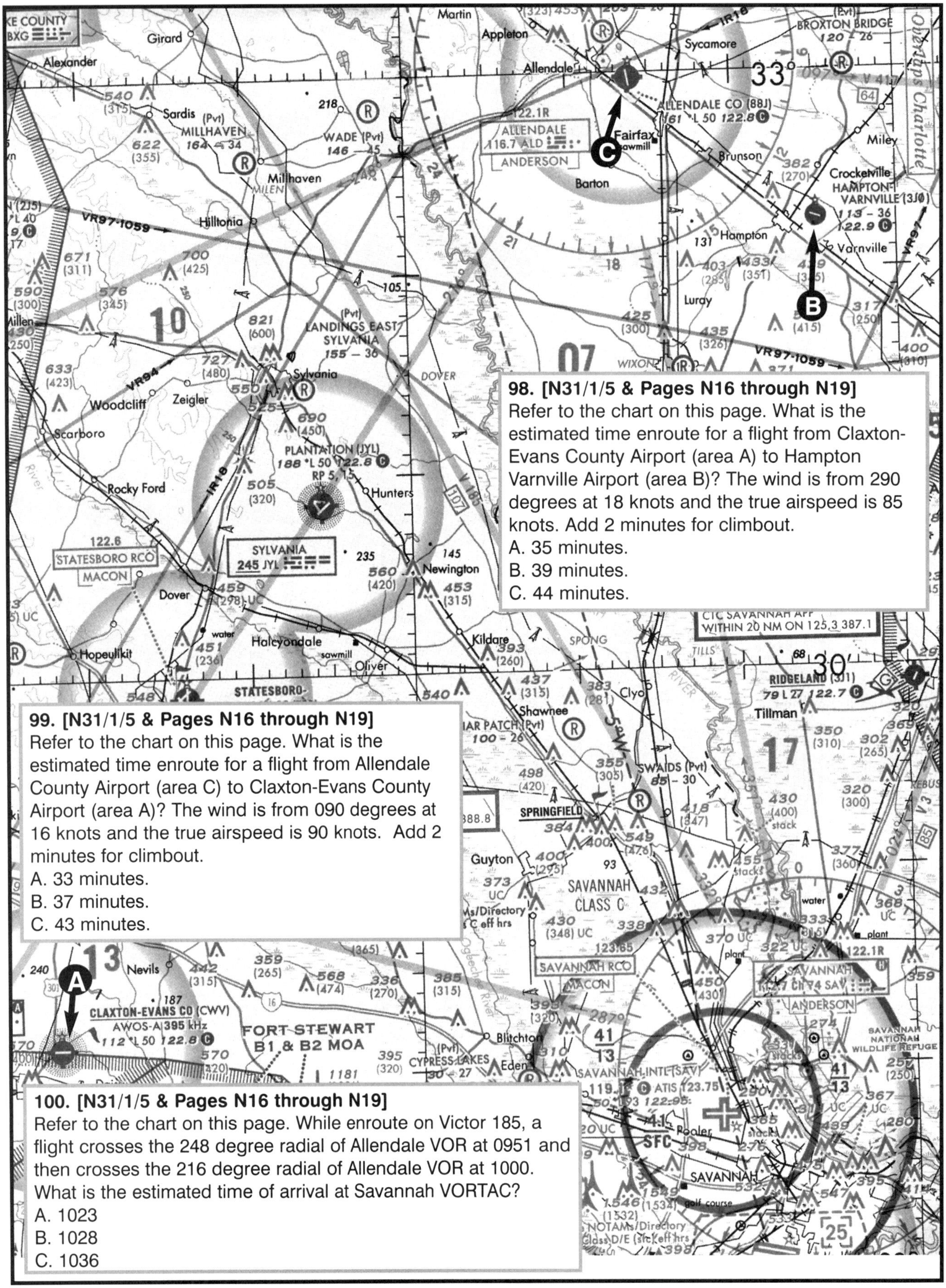

98. [N31/1/5 & Pages N16 through N19]
Refer to the chart on this page. What is the estimated time enroute for a flight from Claxton-Evans County Airport (area A) to Hampton Varnville Airport (area B)? The wind is from 290 degrees at 18 knots and the true airspeed is 85 knots. Add 2 minutes for climbout.
A. 35 minutes.
B. 39 minutes.
C. 44 minutes.

99. [N31/1/5 & Pages N16 through N19]
Refer to the chart on this page. What is the estimated time enroute for a flight from Allendale County Airport (area C) to Claxton-Evans County Airport (area A)? The wind is from 090 degrees at 16 knots and the true airspeed is 90 knots. Add 2 minutes for climbout.
A. 33 minutes.
B. 37 minutes.
C. 43 minutes.

100. [N31/1/5 & Pages N16 through N19]
Refer to the chart on this page. While enroute on Victor 185, a flight crosses the 248 degree radial of Allendale VOR at 0951 and then crosses the 216 degree radial of Allendale VOR at 1000. What is the estimated time of arrival at Savannah VORTAC?
A. 1023
B. 1028
C. 1036

101. **[N31/1/5 & Pages N16 through N19]**
Referring to the chart above, what is the estimated time enroute for a flight from Denton Muni (area A) to Addison (area B)? The wind is from 200 degrees at 20 knots, the true airspeed is 110 knots, and the magnetic variation is 7 degrees east.
A. 12 minutes.
B. 16 minutes.
C. 19 minutes.

102. **[N31/1/5 & Pages N16 through N19]**
Referring to the chart above, estimate the time enroute from Addison (area B) to Redbird (area C). The wind is from 300 degrees at 15 knots, the true airspeed is 120 knots, and the magnetic variation is 7 degrees east.
A. 8 minutes.
B. 11 minutes.
C. 14 minutes.

Fuel Consumption Problems

103. [N41/2/1]
If your fuel consumption is 8 gallons per hour, how much fuel will the airplane consume in 1:30?
A. 12 gallons.
B. 120 gallons.
C. 17.3 gallons.

104. [N41/2/1]
If your fuel consumption is 14.3 gallons per hour, how much fuel will the airplane consume in 37 minutes?
A. 15.5 gallons.
B. 8.8 gallons.
C. 9.4 gallons.

105. [N41/2/1]
If your fuel consumption is 9.4 gallons per hour, how much fuel will the airplane consume in 2:32?
A. 238 gallons.
B. 36.5 gallons.
C. 23.8 gallons.

106. [N41/2/1]
If your fuel consumption is 7.6 gallons per hour, how much fuel will the airplane consume in 12 minutes?
A. .95 gallons.
B. 1.5 gallons.
C. 9.5 gallons.

107. [N41/3/2]
If your airplane consumes 33 gallons in 3:06, what is your fuel consumption?
A. 58.5 GPH.
B. 10.6 GPH.
C. 14.5 GPH.

108. [N41/3/2]
If your airplane consumes 1.5 gallons in 9.5 minutes, what is your fuel consumption?
A. 9.5 GPH.
B. 15 GPH.
C. 8.4 GPH.

109. [N41/3/2]
If your airplane consumes 10.6 gallons in 1:12, what is your fuel consumption?
A.12.2 GPH.
B. 9.7 GPH.
C. 8.8 GPH.

110. [N41/3/2]
If your airplane consumes 45 gallons in 2:48, what is your fuel consumption?
A. 13.4 GPH.
B. 16.1 GPH.
C. 15.1 GPH.

111. [N41/3/3]
With a fuel consumption rate of 13.4 GPH, how long can you fly with 58 gallons of fuel on board?
A. 3:34
B. 1:30
C. 4:19

112. [N41/3/3]
With a fuel consumption rate of 8.2 GPH, how long can you fly with 28 gallons of fuel on board?
A. 3:58
B. 3:25
C. 3:05

113. [N41/3/3]
With a fuel consumption rate of 11.3 GPH, how long can you fly with 24 gallons of fuel on board?
A. :45
B. 2:07
C. 1:27

114. [N41/3/3]
With a fuel consumption rate of 15.7 GPH, how long can you fly with 43 gallons of fuel on board?
A. 1:39
B. 2:44
C. 3:59

Postflight Briefing #14-2: Finding Density Altitude, True Airspeed and Your True Altitude

Finding Density Altitude

115. [N42/1/1]
You're flying at a pressure altitude of 10,000 feet with an outside air temperature of -20 degrees Celsius. What is the density altitude?
A. 8,000 feet.
B. 10,000 feet.
C. 5,000 feet.

116. [N42/1/1]
You're flying at a pressure altitude of 15,000 feet with an outside air temperature of -30 degrees Celsius. What is the density altitude?
A. 15,000 feet.
B. 13,000 feet.
C. 3,000 feet.

117. [N42/1/1]
You're flying at a pressure altitude of 4,000 feet with an outside air temperature of -25 degrees Celsius. What is the density altitude?
A. 8,000 feet.
B. 10,000 feet.
C. Sea level.

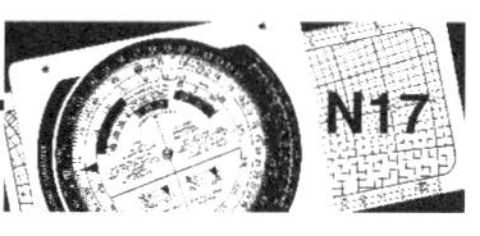

Finding True Airspeed

118. [N42/2/1]
What is your true airspeed if the pressure altitude is 10,000 feet, the temperature is -10 degrees Celsius and the indicated airspeed is 130 knots?
A. 112 knots.
B. 150 knots.
C. 130 knots.

119. [N42/2/1]
What is your true airspeed if the pressure altitude is 7,500 feet, the temperature is 15 degrees Celsius and the indicated airspeed is 120 knots?
A. 114 knots.
B. 138 knots.
C. 130 knots.

120. [N42/2/1]
What is your true airspeed if the pressure altitude is 10,000 feet, the temperature is 5 degrees Celsius and the indicated airspeed is 122 knots?
A. 124 knots.
B. 153 knots.
C. 145 knots.

Finding Your True Altitude

121. [N42/3/2]
What is your true altitude if the pressure altitude is 7,000 feet, the indicated altitude is 6,500 feet and the outside air temperature is -10 degrees Celsius?
A. 6,800 feet.
B. 6,250 feet.
C. 5,800 feet.

122. [N42/3/2]
What is your true altitude if the pressure altitude is 9,000 feet, the indicated altitude is 10,000 feet and the outside air temperature is -20 degrees Celsius?
A. 9,400 feet.
B. 9,700 feet.
C. 9,000 feet.

123. [N42/3/2]
What is your true altitude if the pressure altitude is 6,000 feet, the indicated altitude is 5,500 feet and the outside air temperature is -10 degrees Celsius?
A. 4,850 feet.
B. 5,750 feet.
C. 5,250 feet.

Postflight Briefing 14-4:

124. [N46/Postflight Briefing #14-4]
Referring to the chart below, determine the magnetic course from Paramount-Farming airport (area A) to Wasco-Kern airport (area B).
A. 078 degrees.
B. 021 degrees.
C. 181 degrees.

125. [N46/Postflight Briefing #14-4]
Referring to the chart below, determine the magnetic course from Wasco-Kern airport (area B) to Buttonwillow airport (area C).
A. 186 degrees.
B. 006 degrees.
C. 191 degrees.

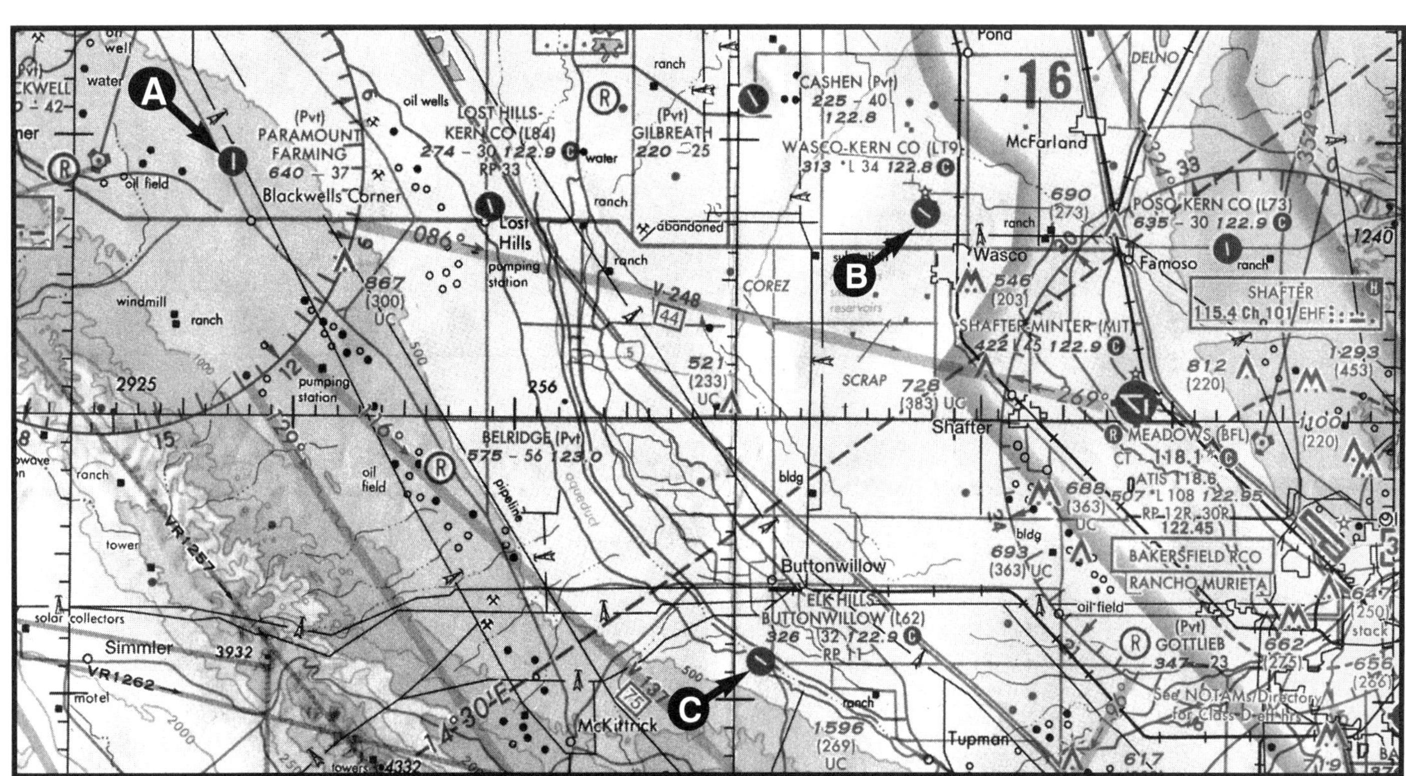

Chapter Fourteen Answers

1. B
2. C
3. B
4. B
5. A
6. C
7. A
8. C
9. B
10. A
11. B
12. A
13. C
14. B
15. A
16. B
17. C
18. B
19. C
20. B
21. C
22. A
23. B
24. B
25. A
26. B
27. C
28. A
29. B
30. B
31. B
32. A
33. B
34. C
35. B
36. B
37. A
38. B
39. A
40. A
41. C
42. C
43. C
44. B
45. A
46. A
47. B
48. A
49. C
50. B
51. C
52. A
53. A
54. C
55. C
56. B
57. C
58. C
59. C
60. A
61. A
62. C
63. A
64. A
65.

Check Points	VOR Ident. Freq.	VOR Course (if applicable)	Altitude	Wind Dir. / Vel. / Temp.	True Airspeed	TC	WCA R+ L-	TH	V E- W+	MH	D	CH	Distance	GS Ground speed Est. / Act.
Spring Hill	X X													
		X	X	170° / 32 / X	109 Kts	63°	+16	79	+13	92	+2	94	X	X
Resnick	X X													

66. B
67. C
68. C
69. A
70. B
71. 10, 15
72. B
73.

Check Points	VOR Ident. Freq.	VOR Course (if applicable)	Altitude	Wind Dir. / Vel. / Temp.	True Airspeed	TC	WCA R+ L-	TH	V E- W+	MH	D	CH	Distance	GS Ground speed Est. / Act.	Time Airborne ETE	Time Airborne ATE	GPH Gallons Per Hour Used / Remain
Spring Hill Airport																	
			5,500	215° / 35 Kts. / +7 C	110 Kt.	63°	+9°	72°	+13°	85°	+2°	87°	13.5 nm.	140K / X	X	X	X / X
Hawley																	
			5,500	215° / 35 Kts. / +7 C	110 Kt.	63°	+9°	72°	+13°	85°	+2°	87°	7.5 nm.	140K / X	X	X	X / X
Railroad																	
			5,500	215° / 35 Kts. / +7 C	110 Kt.	63°	+9°	72°	+13°	85°	+2°	87°	15.5 nm.	140K / X	X	X	X / X
Monticello Airport																	
			5,500	215° / 35 Kts. / +7 C	110 Kt.	63°	+9°	72°	+13°	85°	+2°	87°	15.5 nm.	140K / X	X	X	X / X
Resnick Airport																	
Totals													51.5 nm.				

74. A
75. B
76. A
77. B
78. C
79. A
80. B
81. A
82. C
83. A
84. B
85. A
86. A
87. A
88. B
89. A
90. C
91. A
92. A
93. C
94. B
95. B
96. C
97. B
98. B
99. B
100. C
101. A
102. A
103. A
104. B
105. C
106. B
107. B
108. A
109. C
110. B
111. C
112. B
113. B
114. B
115. A
116. B
117. C
118. B
119. B
120. C
121. B
122. A
123. C
124. A
125. A

Note: To ensure that you have the most current answers to these questions, please check the *Book & Slide Updates* section at Rod Machado's web site: www.rodmachado.com

Chapter Fifteen

Airplane Performance: Know Before You Go

Air Density

1. [O1/3/3]
Anything reducing the amount of air the engine swallows diminishes its _____ output.
A. hydrocarbon
B. fuel
C. power

Height

2. [O2/1/3]
The higher you go the _____ dense the air is.
A. less
B. more or less
C. more

Heat

3. [O2/1/4]
Heated air is much _____ dense than colder air.
A. more or less
B. more
C. less

Humidity

4. [O2/1/5]
What effect, if any, does high humidity have on aircraft performance?
A. It increases performance.
B. It decreases performance.
C. It has no effect on performance.

5. [O3/1/1]
Which combination of atmospheric conditions will reduce aircraft takeoff and climb performance?
A. Low temperature, low relative humidity, and low density altitude.
B. High temperature, low relative humidity, and low density altitude.
C. High temperature, high relative humidity, and high density altitude.

Density Altitude

6. [O3/All]
Density altitude is
A. the height above the standard datum plane.
B. pressure altitude corrected for nonstandard temperature.
C. the altitude read directly from the altimeter.

7. [O3/1/5]
The term _____ describes how dense the air feels to the airplane, regardless of the airplane's present height above sea level.
A. density altitude
B. pressure altitude
C. true altitude

8. [O3/1/5]
Which factor would tend to increase the density altitude at a given airport?
A. An increase in barometric pressure.
B. An increase in ambient temperature.
C. A decrease in relative humidity.

9. [O3/2/3]
What are the standard temperature and pressure values for sea level?
A. 15 degrees C and 29.92" Hg.
B. 59 degrees C and 1013.2 millibars.
C. 59 degrees F and 29.92 millibars.

10. [O4/1/2]
What effect does high density altitude, as compared to low density altitude, have on propeller efficiency and why?
A. Efficiency is increased due to less friction on the propeller blades.
B. Efficiency is reduced because the propeller exerts less force at high density altitudes than at low density altitudes.
C. Efficiency is reduced due to the increased force of the propeller in the thinner air.

11. [O4/1/2]
What effect does high density altitude have on aircraft performance?
A. It increases engine performance.
B. It reduces climb performance.
C. It increases takeoff performance.

12. [O4/1/2]
If the outside air temperature (OAT) at a given altitude is warmer than standard, the density altitude is
A. equal to pressure altitude.
B. lower than pressure altitude.
C. higher than pressure altitude.

Service Ceiling

13. [O5/1/1]
An airplane's service ceiling is the height at which the climb rate drops to less than _____ feet per minute.
A. 25
B. 50
C. 100

14. [O5/1/1]
If the density altitude is 13,000 feet and the airplane's service ceiling is 13,000 feet, the airplane will most likely climb at a rate of _____ at this altitude.
A. more than 100 FPM
B. less than 100 FPM
C. 500 fpm

Performance Charts
Best Rate and Best Angle of Climb Speeds

15. [O7/1/5]
After takeoff, which airspeed would the pilot use to gain the most altitude in a given period of time?
A. V_y
B. V_x
C. V_a

16. [O7/1/6]
Which would provide the greatest gain in altitude in the shortest distance during climb after takeoff?
A. V_y
B. V_a
C. V_x

Vx and Vy Change With Altitude

17. [O8/1/8]
The best rate of climb indicated airspeed _____ with an increase in altitude, while the best angle of climb indicated airspeed _____ with an increase in altitude.
A. decreases, increases
B. increases, decreases
C. increases, remains constant

Cruise Climb Speed

18. [O8/1/9]
Most of the time it's preferable to climb at some speed slightly above _____.
A. V_x
B. V_y
C. V_a

19. [O8/Figure 9] Fill in the blanks:
The graph below belongs to a particular model single-engine airplane. Estimate this airplane's best angle and best rate of climb speeds.
V_x is _____________.
V_y is _____________.

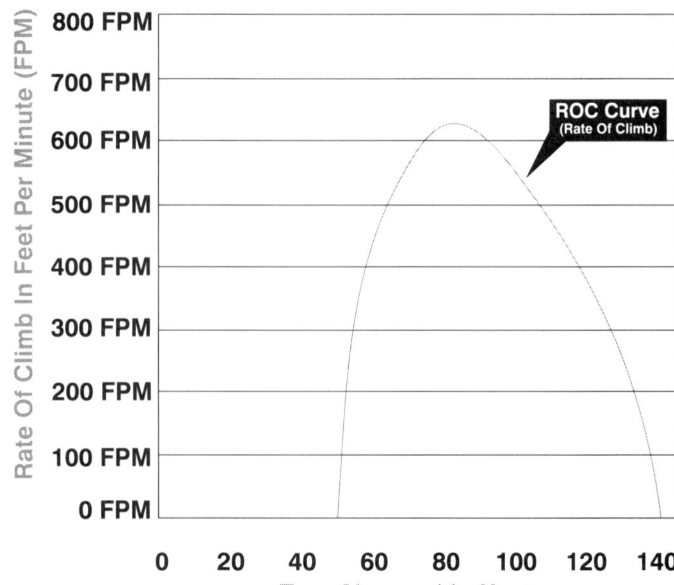

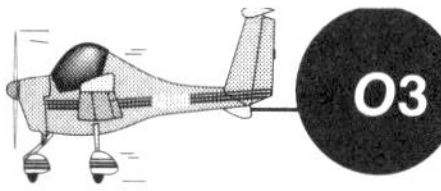

Takeoff Distance Chart

TAKEOFF DISTANCE

ASSOCIATED CONDITIONS:

POWER	FULL THROTTLE 2600 RPM
MIXTURE	LEAN TO APPROPRIATE FUEL PRESSURE
FLAPS	UP
LANDING GEAR	RETRACT AFTER POSITIVE CLIMB ESTABLISHED
COWL FLAPS	OPEN

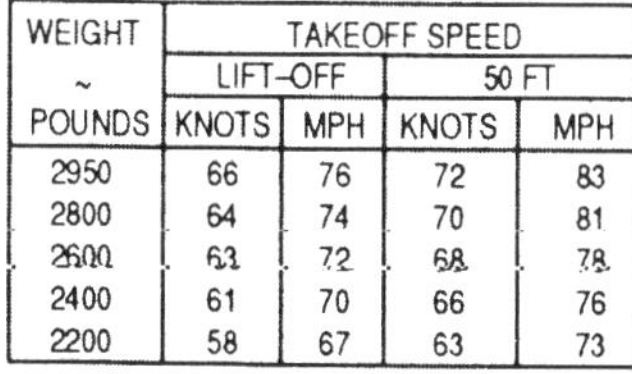

WEIGHT ~ POUNDS	TAKEOFF SPEED LIFT-OFF KNOTS	LIFT-OFF MPH	50 FT KNOTS	50 FT MPH
2950	66	76	72	83
2800	64	74	70	81
2600	63	72	68	78
2400	61	70	66	76
2200	58	67	63	73

EXAMPLE:

OAT	15 °C (59 °F)
PRESSURE ALTITUDE	5650 FT
TAKEOFF WEIGHT	2950 LB
HEADWIND COMP.	9.0 KNOTS
GROUND ROLL	1375 FT
TOTAL DISTANCE OVER A 50 FT OBSTACLE	2300 FT
TAKEOFF SPEED AT LIFT-OFF	66 KNOTS (76 MPH)
50 FT	72 KNOTS (83 MPH)

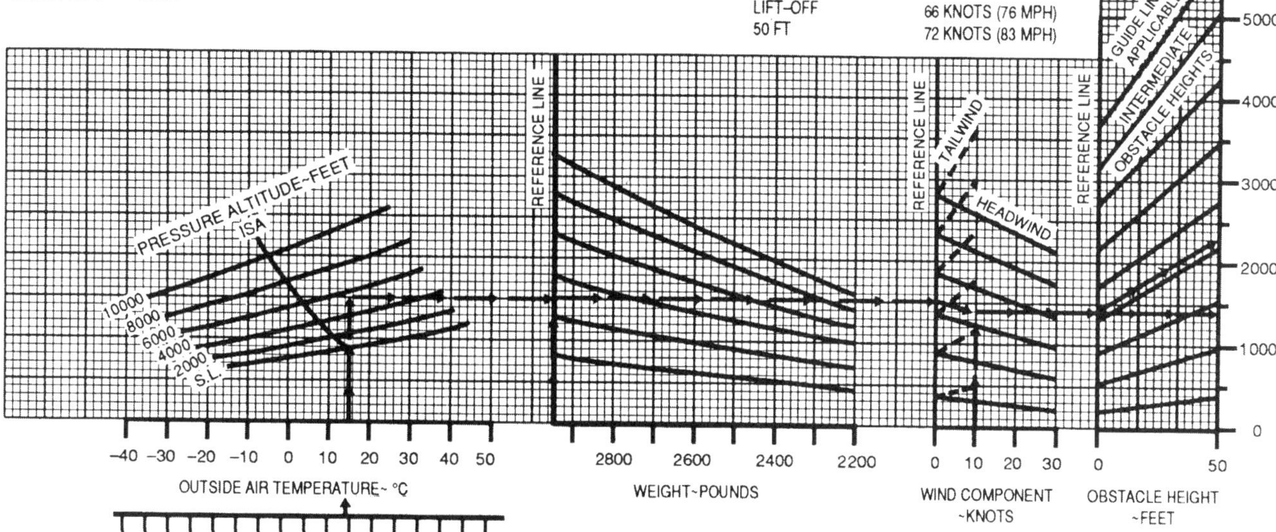

20. [O10/1/1/Entire section]
Referring to the performance chart above, determine the total distance required for takeoff to clear a 50 foot obstacle.

OAT: Standard temperature
Pressure altitude: 5,000 ft
Takeoff weight: 2,800 lb
Headwind component: calm
A. 1,150 feet.
B. 1,800 feet.
C. 2,000 feet.

21. [O10/1/1/Entire section]
Referring to the performance chart above, determine the approximate ground roll distance required for takeoff.

OAT: 90 degrees F
Pressure altitude: 2,000 ft
Takeoff weight: 2,500 lb
Headwind component: 20 kts
A. 700 feet.
B. 850 feet.
C. 1,000 feet.

22. [O10/1/1/Entire section]
Referring to the performance chart above, determine the total distance required for takeoff to clear a 50 foot obstacle.

OAT: Standard temperature
Pressure altitude: Sea level
Takeoff weight: 2,700 lb
Tailwind component: calm
A. 1,000 feet.
B. 1,300 feet.
C. 1,700 feet.

23. [O10/1/1/Entire section]
Referring to the performance chart above, determine the approximate ground roll distance required for takeoff.

OAT: 80 degrees F
Pressure altitude: 2,500 ft
Takeoff weight: 2,250 lb
Headwind component: 20 kts
A. 900 feet.
B. 500 feet.
C. 700 feet.

Takeoff Distance Chart

TAKEOFF DISTANCE

SHORT FIELD

CONDITIONS:
Flaps 10°
Full Throttle Prior to Brake Release
Paved, Level, Dry Runway
Zero Wind

NOTES:
1. Short field technique as specified in Section 4.
2. Prior to takeoff from fields above 3000 feet elevation, the mixture should be leaned to give maximum RPM in a full throttle, static runup.
3. Decrease distances 10% for each 9 knots headwind. For operation with tailwinds up to 10 knots, increase distances by 10% for each 2 knots.
4. For operation on a dry, grass runway, increase distances by 15% of the "ground roll" figure.

WEIGHT LBS	TAKEOFF SPEED KIAS		PRESS ALT FT	0°C		10°C		20°C		30°C		40°C	
	LIFT OFF	AT 50 FT		GRND ROLL	TOTAL TO CLEAR 50 FT OBS	GRND ROLL	TOTAL TO CLEAR 50 FT OBS	GRND ROLL	TOTAL TO CLEAR 50 FT OBS	GRND ROLL	TOTAL TO CLEAR 50 FT OBS	GRND ROLL	TOTAL TO CLEAR 50 FT OBS
1670	50	54	S.L.	640	1190	695	1290	755	1390	810	1495	875	1605
			1000	705	1310	765	1420	825	1530	890	1645	960	1770
			2000	775	1445	840	1565	910	1690	980	1820	1055	1960
			3000	855	1600	925	1730	1000	1870	1080	2020	1165	2185
			4000	940	1775	1020	1920	1100	2080	1190	2250	1285	2440
			5000	1040	1970	1125	2140	1215	2320	1315	2525	1420	2750
			6000	1145	2200	1245	2395	1345	2610	1455	2855	1570	3125
			7000	1270	2470	1375	2705	1490	2960	1615	3255	1745	3590
			8000	1405	2800	1525	3080	1655	3395	1795	3765	1940	4195

24. [O12/2/2/Entire section]
Referring to the performance chart above, determine the total distance required for takeoff to clear a 50 foot obstacle.
OAT: 20 degrees C
Pressure altitude: 4000 ft
Takeoff weight: 1,670 lb
Headwind component: 9 kts

A. 2,288 feet.
B. 2080 feet.
C. 1,872 feet.

25. [O12/2/2/Entire section]
Referring to the performance chart above, determine the total distance required for takeoff to clear a 50 foot obstacle.
OAT: 10 degrees C
Pressure altitude: 2,000 ft
Takeoff weight: 1,670 lb
Headwind component: 0 kts
Runway: dry, grass

A. 966 feet.
B. 1,439 feet.
C. 1,691 feet.

26. [O12/2/2/Entire section]
Referring to the performance chart above, determine the approximate ground roll distance required for takeoff.
OAT: 10 degrees C
Pressure altitude : 4,500 ft
Takeoff weight: 1,670 lb
Headwind component: 9 kts

A. 1,073 feet.
B. 965 feet.
C. 1,180 feet.

27. [O12/2/2/Entire section]
Referring to the performance chart above, determine the total distance required for takeoff to clear a 50 foot obstacle.
OAT: 20 degrees C
Pressure altitude: 5,500 ft
Takeoff weight: 1,670 lb
Tailwind component: 4 kts

A. 2,465 feet.
B. 1,972 feet.
C. 2,958 feet.

Landing Distance Performance Charts

LANDING DISTANCE

ASSOCIATED CONDITIONS:

POWER	RETARDED TO MAINTAIN 900 FT/on FINAL APPROACH
FLAPS	DOWN
LANDING GEAR	DOWN
RUNWAY	PAVED, LEVEL, DRY SURFACE
APPROACH SPEED	IAS AS TABULATED
BRAKING	MAXIMUM

WEIGHT ~ POUNDS	SPEED AT 50 FT KNOTS	MPH
2950	70	80
2800	68	78
2600	65	75
2400	63	72
2200	60	69

EXAMPLE:

OAT	25 °C (77 °F)
PRESSURE ALTITUDE	3965 FT
WEIGHT	2814 LB
WIND COMPONENT	9.0 KNOTS (HEADWIND)
GROUND ROLL	1080 FT
TOTAL OVER 50 FT OBSTACLE	1700 FT
APPROACH SPEED	68 KNOTS (78 MPH)

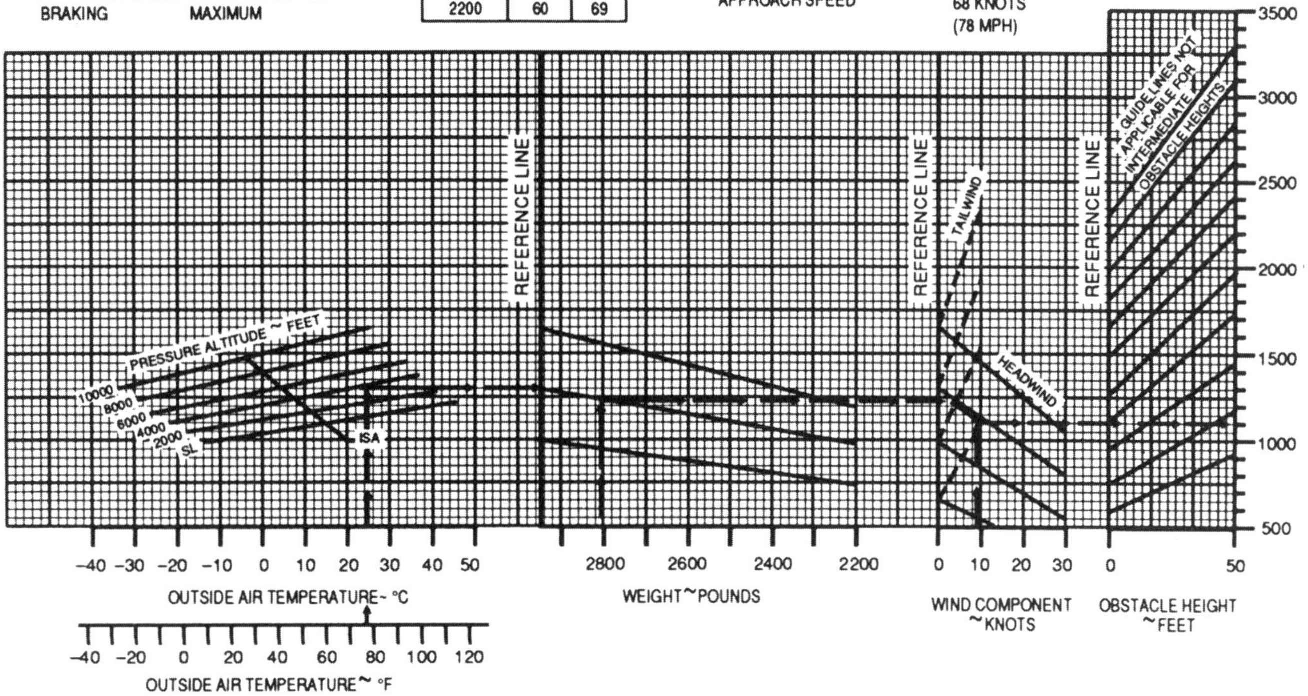

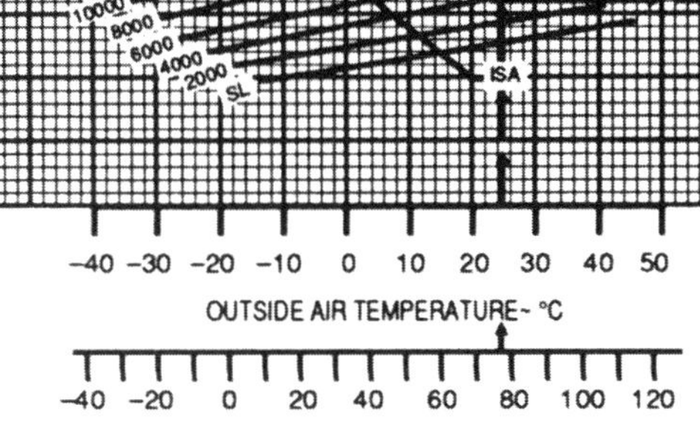

28. [O14/1/1/Entire section]
Referring to the performance chart above, determine the total distance required to land over a 50 foot obstacle.

OAT: standard
Pressure altitude: 10,000 ft
Weight: 2,400 lb
Wind component: calm
Obstacle: 50 ft
A. 750 feet.
B. 1,925 feet.
C. 1,450 feet.

29. [O14/1/1/Entire section]
Referring to the performance chart above, determine the approximate ground roll distance after landing.

OAT: 80 degrees F
Pressure altitude: 5,000 ft
Weight: 2,600 lb
Tailwind component: 10 kts
A. 1,750 feet.
B. 1,200 feet.
C. 1,050 feet.

30. [O14/1/1/Entire section]
Referring to the performance chart above, determine the total distance required to land over a 50 foot obstacle.

OAT: 90 degrees F
Pressure altitude: 4,000 ft
Weight 2,800 lb
Headwind component: 10 kts
Obstacle: 50 ft
A. 1,525 feet.
B. 1,775 feet.
C. 1,950 feet.

31. [O14/1/1/Entire section]
Referring to the performance chart above, determine the total distance required to land over a 50 foot obstacle.

OAT: 90 degrees F
Pressure altitude: 3,000 ft
Weight 2,900 lb
Headwind component: 10 kts
Obstacle: 50 ft
A. 1,450 feet.
B. 1,550 feet.
C. 1,725 feet.

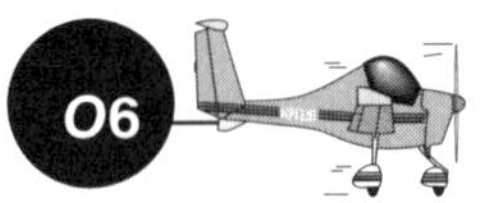

A Different Landing Distance Chart

LANDING DISTANCE — FLAPS LOWERED TO 40° - POWER OFF, HARD SURFACE RUNWAY - ZERO WIND

GROSS WEIGHT LB	APPROACH SPEED, IAS, MPH	AT SEA LEVEL & 59 °F		AT 2500 FT & 50 °F		AT 5000 FT & 41 °F		AT 7500 FT & 32 °F	
		GROUND ROLL	TOTAL TO CLEAR 50 FT OBS	GROUND ROLL	TOTAL TO CLEAR 50 FT OBS	GROUND ROLL	TOTAL TO CLEAR 50 FT OBS	GROUND ROLL	TOTAL TO CLEAR 50 FT OBS
1600	60	445	1075	470	1135	495	1195	520	1255

NOTES:
1. Decrease the distances shown by 10% for each 4 knots of headwind.
2. Increase the distance by 10% for each 60 °F temperature increase above standard.
3. For operation on a dry, grass runway, increase distances (both "ground roll" and "total to clear 50 ft obstacle") by 20% of the "total to clear 50 ft obstacle" figure.

32. [O16/1/1/Entire Section]
Referring to the performance chart above, determine the approximate landing ground roll distance.

Pressure altitude: sea level
Headwind: 4 kts
Temperature: standard
A. 356 feet.
B. 401 feet.
C. 490 feet.

33. [O16/1/1/Entire Section]
Referring to the performance chart above, determine the approximate landing ground roll distance.

Pressure altitude: 3,750 ft
Headwind: 12 kts
Temperature: standard
A. 338 feet.
B. 425 feet.
C. 483 feet.

34. [O16/1/1/Entire Section]
Referring to the performance chart above, determine the total distance required to land over a 50 foot obstacle.

Pressure altitude: 2,500 ft
Headwind: calm
Temperature: 80 degrees F
A. 1,135 feet.
B. 1,192 feet.
C. 1,078 feet.

35. [O16/1/1/Entire Section]
Referring to the performance chart above, determine the approximate landing ground roll distance.

Pressure altitude: 1,250 ft
Headwind: 8 kts
Temperature: standard
A. 275 feet.
B. 366 feet.
C. 470 feet.

36. [O16/1/1/Entire Section]
Referring to the performance chart above, determine the total distance required to land over a 50 foot obstacle.

Pressure altitude: 7,500 ft
Headwind: 8 kts
Temperature: standard
Runway: dry grass
A. 1,255 feet.
B. 1,004 feet.
C. 1,205 feet.

37. [O16/1/1/Entire Section]
Referring to the performance chart above, determine the total distance required to land over a 50 foot obstacle.

Pressure altitude: 5,000 ft
Headwind: 8 kts
Temperature: 41 degrees F
Runway: hard surface
A. 837 feet.
B. 956 feet.
C. 1,076 feet.

Time, Fuel and Distance to Climb Chart

CESSNA
MODEL 152

SECTION 5
PERFORMANCE

TIME, FUEL, AND DISTANCE TO CLIMB

MAXIMUM RATE OF CLIMB

CONDITIONS:
Flaps Up
Full Throttle
Standard Temperature

NOTES:
1. Add 0.8 of a gallon of fuel for engine start, taxi and takeoff allowance.
2. Mixture leaned above 3000 feet for maximum RPM.
3. Increase time, fuel and distance by 10% for each 10°C above standard temperature.
4. Distances shown are based on zero wind.

WEIGHT LBS	PRESSURE ALTITUDE FT	TEMP °C	CLIMB SPEED KIAS	RATE OF CLIMB FPM	FROM SEA LEVEL		
					TIME MIN	FUEL USED GALLONS	DISTANCE NM
1670	S.L.	15	67	715	0	0	0
	1000	13	66	675	1	0.2	2
	2000	11	66	630	3	0.4	3
	3000	9	65	590	5	0.7	5
	4000	7	65	550	6	0.9	7
	5000	5	64	505	8	1.2	9
	6000	3	63	465	10	1.4	12
	7000	1	63	425	13	1.7	14
	8000	-1	62	380	15	2.0	17
	9000	-3	62	340	18	2.3	21
	10,000	-5	61	300	21	2.6	25
	11,000	-7	61	255	25	3.0	29
	12,000	-9	60	215	29	3.4	34

38. [O17/1/1]
Referring to the performance chart above, estimate the amount of time and fuel consumed to climb from sea level to 7,000 feet pressure altitude under standard temperature conditions.
A. 13 minutes, 14 gallons.
B. 13 minutes, 1.7 gallons.
C. 13 minutes, 2.5 gallons.

39. [O17/1/1]
Referring to the performance chart above, estimate the amount of time and fuel consumed to climb from sea level to 5,500 feet pressure altitude at a temperature 10 degrees Celsius above standard.
A. 10 minutes, 1.4 gallons.
B. 9 minutes, 1.3 gallons.
C. 9 minutes, 2.1 gallons.

40. [O17/1/1]
Referring to the performance chart above, estimate the amount of time and fuel consumed to climb from 2,000 feet to 9,000 feet pressure altitude under standard temperature conditions.
A. 21 minutes, 2.3 gallons.
B. 25 minutes, 2.7 gallons.
C. 15 minutes, 1.9 gallons.

41. [O17/1/1]
Referring to the performance chart above, estimate the amount of time and fuel consumed to climb from 3,000 feet to 8,000 feet pressure altitude at 10 degrees Celsius above standard temperature conditions.
A. 16.5 minutes, 2.2 gallons.
B. 5.5 minutes, .7 gallons.
C. 11 minutes, 1.4 gallons.

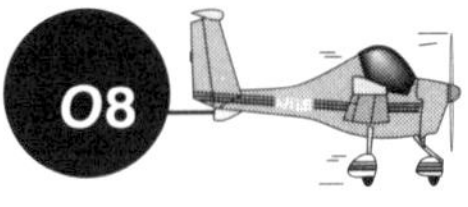

Cruise Performance Chart

CRUISE POWER SETTINGS

65% MAXIMUM CONTINUOUS POWER (OR FULL THROTTLE)
2800 POUNDS

PRESS ALT.	ISA −20 °C (−36 °F)								STANDARD DAY (ISA)								ISA +20 °C (+36 °F)							
	IOAT		ENGINE SPEED	MAN. PRESS	FUEL FLOW		TAS		IOAT		ENGINE SPEED	MAN. PRESS	FUEL FLOW		TAS		IOAT		ENGINE SPEED	MAN. PRESS	FUEL FLOW		TAS	
FEET	°F	°C	RPM	IN HG	PSI	GPH	KTS	MPH	°F	°C	RPM	IN HG	PSI	GPH	KTS	MPH	°F	°C	RPM	IN HG	PSI	GPH	KTS	MPH
SL	27	-3	2450	20.7	6.6	11.5	147	169	63	17	2450	21.2	6.6	11.5	150	173	99	37	2450	21.8	6.6	11.5	153	176
2000	19	-7	2450	20.4	6.6	11.5	149	171	55	13	2450	21.0	6.6	11.5	153	176	91	33	2450	21.5	6.6	11.5	156	180
4000	12	-11	2450	20.1	6.6	11.5	152	175	48	9	2450	20.7	6.6	11.5	156	180	84	29	2450	21.3	6.6	11.5	159	183
6000	5	-15	2450	19.8	6.6	11.5	155	178	41	5	2450	20.4	6.6	11.5	158	182	79	26	2450	21.0	6.6	11.5	161	185
8000	-2	-19	2450	19.5	6.6	11.5	157	181	36	2	2450	20.2	6.6	11.5	161	185	72	22	2450	20.8	6.6	11.5	164	189
10000	-8	-22	2450	19.2	6.6	11.5	160	184	28	-2	2450	19.9	6.6	11.5	163	188	64	18	2450	20.3	6.5	11.4	166	191
12000	-15	-26	2450	18.8	6.4	11.3	162	186	21	-6	2450	18.8	6.1	10.9	163	188	57	14	2450	18.8	5.9	10.6	163	188
14000	-22	-30	2450	17.4	5.8	10.5	159	183	14	-10	2450	17.4	5.6	10.1	160	184	50	10	2450	17.4	5.4	9.8	160	184
16000	-29	-34	2450	16.1	5.3	9.7	156	180	7	-14	2450	16.1	5.1	9.4	156	180	43	6	2450	16.1	4.9	9.1	155	178

NOTES: 1. Full throttle manifold pressure settings are approximate.
2. Shaded area represents operation with full throttle.

42. [O18/1/1/Entire section]
Referring to the performance chart above, what fuel flow should a pilot expect at 10,000 feet on a standard day with 65 percent maximum continuous power?
A. 19.9 gallons per hour.
B. 11.5 gallons per hour.
C. 6.6 gallons per hour.

43. [O18/1/1/Entire section]
Referring to the performance chart above, what is the expected fuel consumption for a 1,000 nautical mile flight under the following conditions?

Pressure altitude: 6,000 ft
Temperature: 22 degrees C
Manifold pressure: 20.8" Hg
Wind: calm
A. 72.3 gallons.
B. 41.5 gallons.
C. 56.0 gallons.

44. [O18/1/1/Entire section]
Referring to the performance chart above, what is the expected fuel consumption for a 500 nautical mile flight under the following conditions?

Pressure altitude: 4,000 ft
Temperature: +29 degrees C
Manifold pressure: 21.3" Hg
Wind: calm
A. 31.4 gallons.
B. 36.1 gallons.
C. 40.1 gallons.

45. [O18/1/1/Entire section]
Referring to the performance chart above, determine the approximate manifold pressure setting with 2,450 RPM to achieve 65 percent maximum continuous power at 6,500 feet with a temperature 36 degrees F higher than standard.
A. 19.8" Hg.
B. 20.8" Hg.
C. 21.0" Hg.

46. [O18/1/1/Entire section]
Referring to the performance chart above, approximately what true airspeed should a pilot expect with 65 percent maximum continuous power at 9,500 feet with a temperature 36 degrees F below standard?
A. 178 MPH.
B. 181 MPH.
C. 183 MPH.

47. [O18/1/1/Entire section]
Referring to the performance chart above, what is the expected fuel consumption for an 850 nautical mile flight under the following conditions?

Pressure altitude: 9,500 ft
Temperature: standard conditions
Manifold pressure: 20.0" Hg
Wind: calm
A. 60.3 gallons.
B. 31.4 gallons.
C. 11.5 gallons.

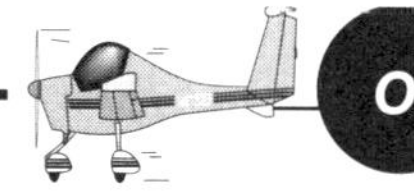

Another Variety of Cruise Performance Charts

CRUISE PERFORMANCE

CONDITIONS:
1670 Pounds
Recommended Lean Mixture (See Section 4, Cruise)

NOTE:
Cruise speeds are shown for an airplane equipped with speed fairings which increase the speeds by approximately two knots.

PRESSURE ALTITUDE FT	RPM	20°C BELOW STANDARD TEMP			STANDARD TEMPERATURE			20°C ABOVE STANDARD TEMP		
		% BHP	KTAS	GPH	% BHP	KTAS	GPH	% BHP	KTAS	GPH
2000	2400	- - -	- - -	- - -	75	101	6.1	70	101	5.7
	2300	71	97	5.7	66	96	5.4	63	95	5.1
	2200	62	92	5.1	59	91	4.8	56	90	4.6
	2100	55	87	4.5	53	86	4.3	51	85	4.2
	2000	49	81	4.1	47	80	3.9	46	79	3.8
4000	2450	- - -	- - -	- - -	75	103	6.1	70	102	5.7
	2400	76	102	6.1	71	101	5.7	67	100	5.4
	2300	67	96	5.4	63	95	5.1	60	95	4.9
	2200	60	91	4.8	56	90	4.6	54	89	4.4
	2100	53	86	4.4	51	85	4.2	49	84	4.0
	2000	48	81	3.9	46	80	3.8	45	78	3.7
6000	2500	- - -	- - -	- - -	75	105	6.1	71	104	5.7
	2400	72	101	5.8	67	100	5.4	64	99	5.2
	2300	64	96	5.2	60	95	4.9	57	94	4.7
	2200	57	90	4.6	54	89	4.4	52	88	4.3
	2100	51	85	4.2	49	84	4.0	48	83	3.9
	2000	46	80	3.8	45	79	3.7	44	77	3.6
8000	2550	- - -	- - -	- - -	75	107	6.1	71	106	5.7
	2500	76	105	6.2	71	104	5.8	67	103	5.4
	2400	68	100	5.5	64	99	5.2	61	98	4.9
	2300	61	95	5.0	58	94	4.7	55	93	4.5
	2200	55	90	4.5	52	89	4.3	51	87	4.2
	2100	49	84	4.1	48	83	3.9	46	82	3.8
10,000	2500	72	105	5.8	68	103	5.5	64	103	5.2
	2400	65	99	5.3	61	98	5.0	58	97	4.8
	2300	58	94	4.7	56	93	4.5	53	92	4.4
	2200	53	89	4.3	51	88	4.2	49	86	4.0
	2100	48	83	4.0	46	82	3.9	45	81	3.8
12,000	2450	65	101	5.3	62	100	5.0	59	99	4.8
	2400	62	99	5.0	59	97	4.8	56	96	4.6
	2300	56	93	4.6	54	92	4.4	52	91	4.3
	2200	51	88	4.2	49	87	4.1	48	85	4.0
	2100	47	82	3.9	45	81	3.8	44	79	3.7

48. [O19/1/4]
Referring to the performance chart above, determine the expected fuel consumption and true airspeed for a flight at a pressure altitude of 8,000 feet at 2,300 RPM at 20 degrees Celsius below standard conditions.
A. 4.5 GPH, 93 knots.
B. 5.5 GPH, 90 knots.
C. 5.0 GPH, 95 knots.

49. [O19/1/4]
Referring to the performance chart above, determine the expected fuel consumption and true airspeed for a flight at a pressure altitude of 5,000 feet at 2,400 RPM under standard conditions.
A. 5.4 GPH, 101 knots.
B. 5.5 GPH, 100 knots.
C. 4.7 GPH, 105 knots.

50. [O19/1/4]
Referring to the performance chart above, determine the expected fuel consumption and true airspeed for a flight at a pressure altitude of 3,000 feet at 2,200 RPM at 20 degrees Celsius above standard conditions.
A. 4.5 GPH, 90 knots.
B. 4.6 GPH, 83 knots.
C. 4.5 GPH, 95 knots.

Endurance and Range Profile Charts

ENDURANCE PROFILE

CESSNA
MODEL 152

45 MINUTES RESERVE
24.5 GALLONS USABLE FUEL

CONDITIONS:
1670 Pounds
Recommended Lean Mixture for Cruise
Standard Temperature

NOTE:
This chart allows for the fuel used for engine start, taxi, takeoff and climb, and the time during climb as shown in figure 5-6.

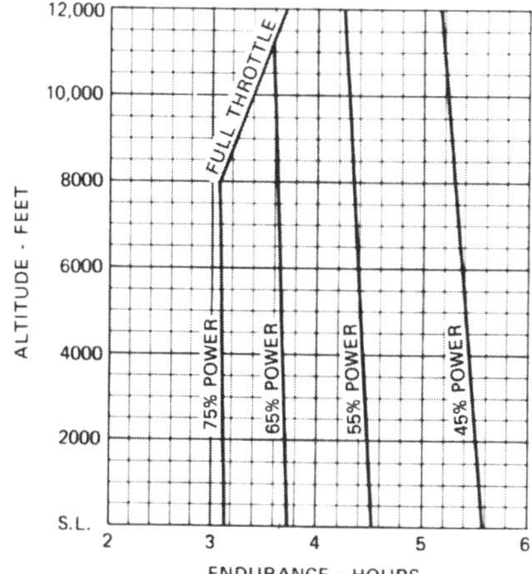

51. [O20/Figure 27]
Referring to the performance chart above, determine the airplane's endurance at 6,000 feet at 65% power with 24.5 gallons of useable fuel and a 45 minute reserve.
A. 3:38
B. 3:58
C. 3:00

52. [O20/Figure 27]
Referring to the performance chart above, determine the airplane's endurance at 4,000 feet at 75% power with 24.5 gallons of useable fuel and a 45 minute reserve.
A. 3:10
B. 3:14
C. 3:25

53. [O20/Figure 27]
Referring to the performance chart above, determine the airplane's endurance at 10,000 feet at 65% power with 24.5 gallons of useable fuel and a 45 minute reserve.
A. 3:36
B. 3:20
C. 3:50

Endurance and Range Profile Charts

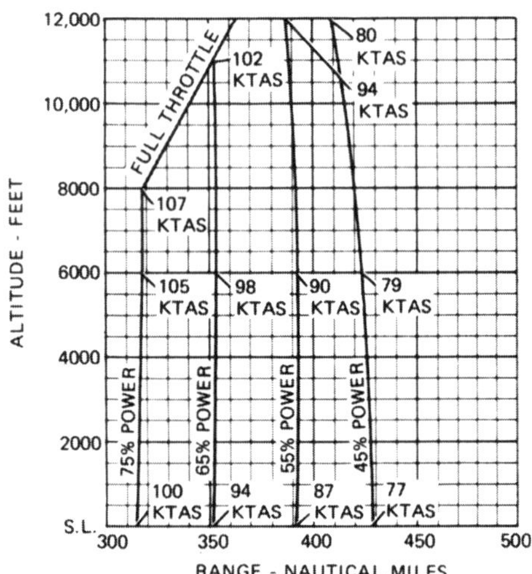

54. [O20/Figure 28]
Referring to the performance chart above, determine the airplane's range at 8,000 feet at 65% power with 24.5 gallons of useable fuel and a 45 minute reserve.
A. 350 nm.
B. 353 nm.
C. 390 nm.

55. [O20/Figure 28]
Referring to the performance chart above, determine the airplane's range at 8,000 feet at 75% power with 24.5 gallons of useable fuel and a 45 minute reserve.
A. 319 nm.
B. 340 nm.
C. 300 nm.

56. [O20/Figure 28]
Referring to the performance chart above, determine the airplane's range at 10,000 feet at 45% power with 24.5 gallons of useable fuel and a 45 minute reserve.
A. 415 nm.
B. 400 nm.
C. 450 nm.

Crosswind Component Chart

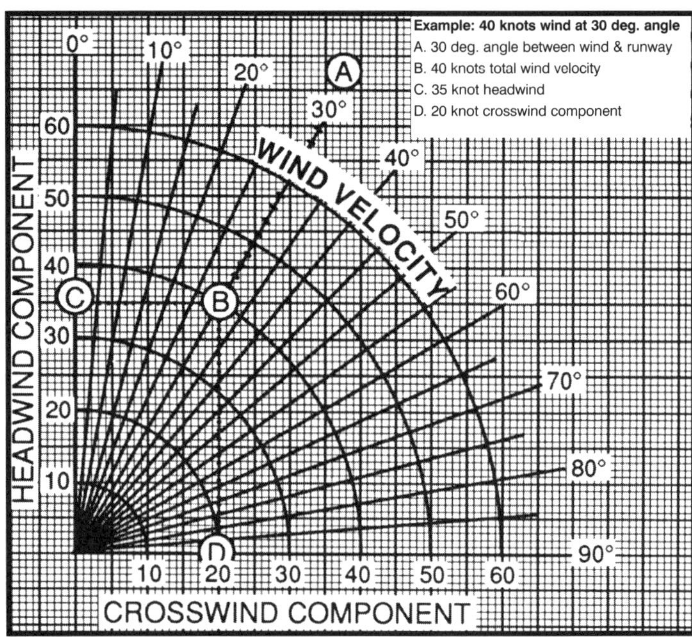

57. [O21/Entire section]
Referring to the crosswind component chart above, what is the crosswind component for a landing on Runway 18 if the tower reports the wind as 220 degrees at 30 knots?
A. 19 knots.
B. 23 knots.
C. 30 knots.

58. [O21/Entire section]
Referring to the crosswind component chart above, determine the maximum wind velocity for a 45 degree crosswind if the maximum crosswind component for the airplane is 25 knots.
A. 25 knots.
B. 29 knots.
C. 35 knots.

59. [O21/Entire section]
Referring to the crosswind component chart above, with a reported wind of north at 20 knots, which runway (6, 29, or 32) is acceptable for use for an airplane with a 13 knot maximum crosswind component?
A. Runway 6.
B. Runway 29.
C. Runway 32.

60. [O21/Entire section]
Referring to the crosswind component chart above, what is the maximum wind velocity for a 40 degree crosswind if the maximum crosswind component for the airplane is 12 knots?
A. 18 knots.
B. 22 knots.
C. 12 knots.

Postflight Briefing 15-1: Making the Forces Be With You Advanced Lessons in Density Altitude

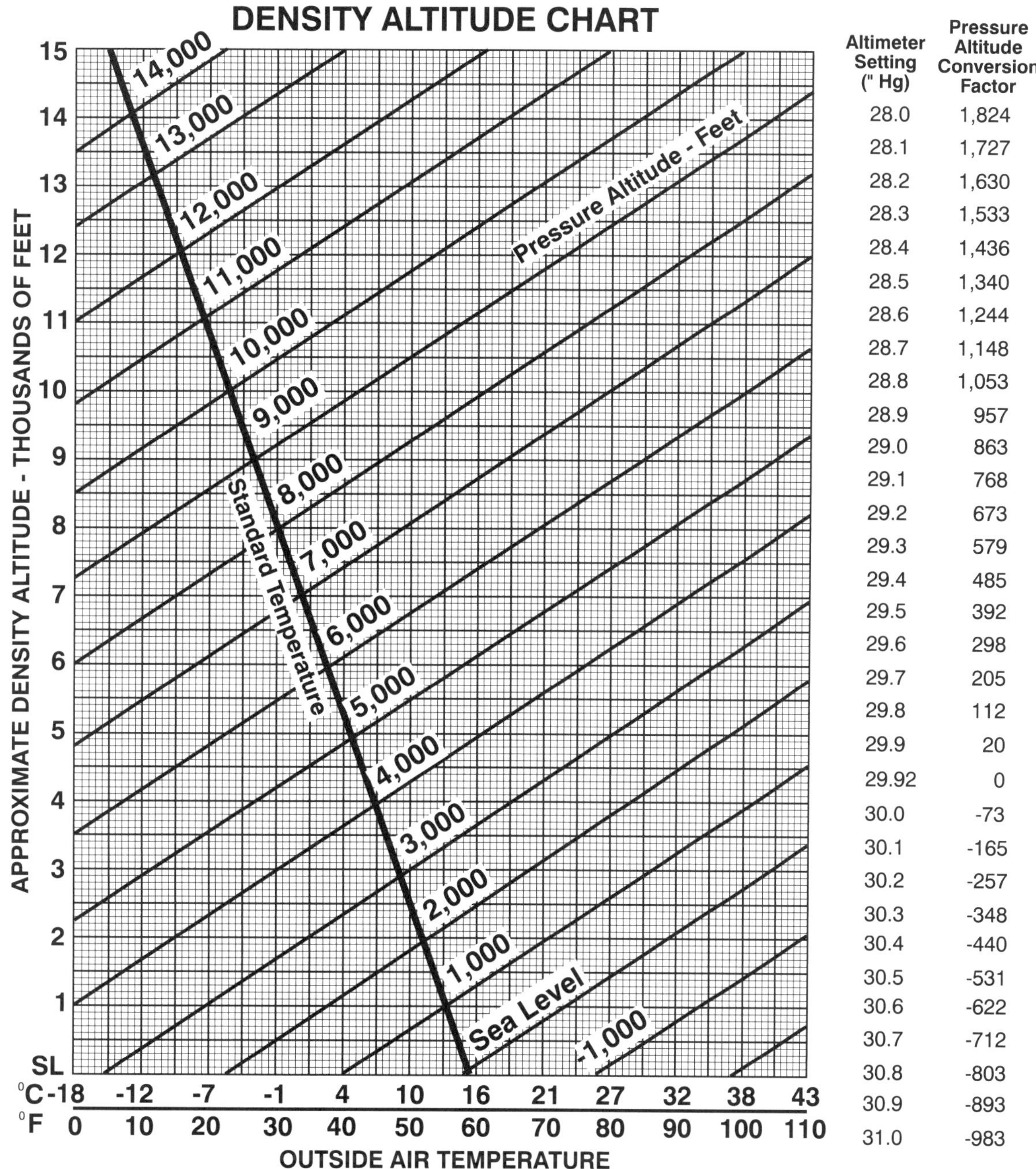

Altimeter Setting (" Hg)	Pressure Altitude Conversion Factor
28.0	1,824
28.1	1,727
28.2	1,630
28.3	1,533
28.4	1,436
28.5	1,340
28.6	1,244
28.7	1,148
28.8	1,053
28.9	957
29.0	863
29.1	768
29.2	673
29.3	579
29.4	485
29.5	392
29.6	298
29.7	205
29.8	112
29.9	20
29.92	0
30.0	-73
30.1	-165
30.2	-257
30.3	-348
30.4	-440
30.5	-531
30.6	-622
30.7	-712
30.8	-803
30.9	-893
31.0	-983

61. [O24/Postflight Briefing #15-1]
Referring to the density altitude chart above, determine the density altitude for these conditions:

Altimeter setting 30.35
Runway temperature +25 degrees F
Airport elevation 3,894 ft MSL
A. 2,000 feet MSL.
B. 2,900 feet MSL.
C. 3,500 feet MSL.

62. [O24/Postflight Briefing #15-1]
Referring to the density altitude chart above, what is the effect of a temperature decrease and a pressure altitude increase on the density altitude from 90 degrees F and 1,250 feet pressure altitude to 60 degrees F and 1,750 feet pressure altitude?
A. 500 foot increase.
B. 1,300 foot decrease.
C. 1,300 foot increase.

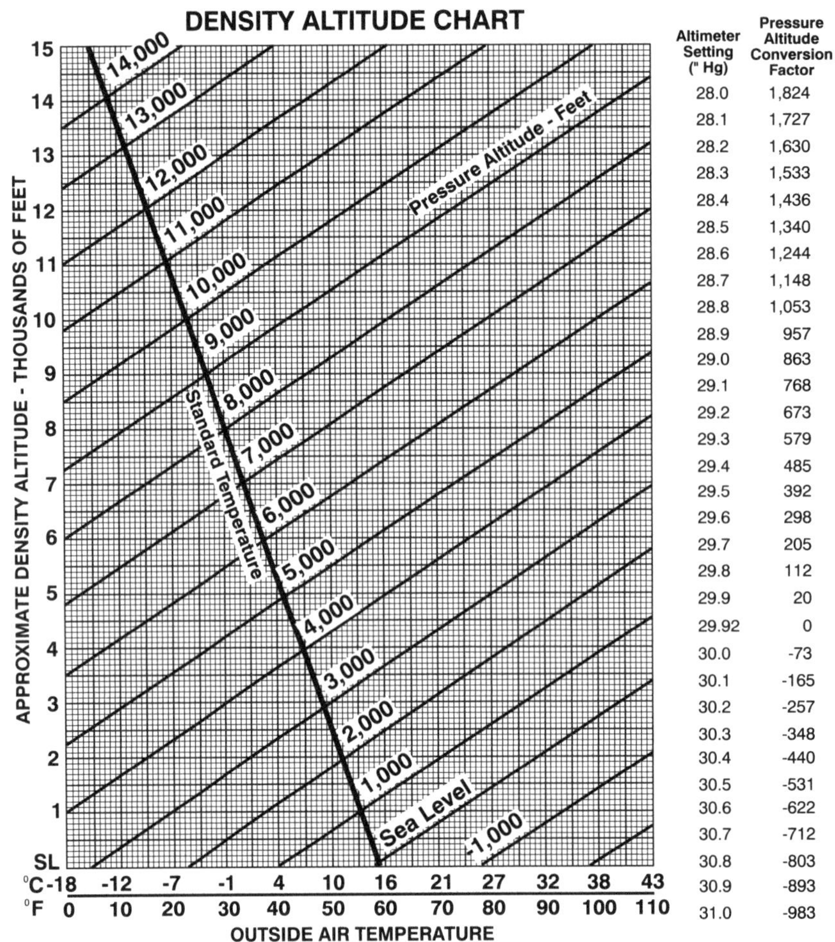

Altimeter Setting (" Hg)	Pressure Altitude Conversion Factor
28.0	1,824
28.1	1,727
28.2	1,630
28.3	1,533
28.4	1,436
28.5	1,340
28.6	1,244
28.7	1,148
28.8	1,053
28.9	957
29.0	863
29.1	768
29.2	673
29.3	579
29.4	485
29.5	392
29.6	298
29.7	205
29.8	112
29.9	20
29.92	0
30.0	-73
30.1	-165
30.2	-257
30.3	-348
30.4	-440
30.5	-531
30.6	-622
30.7	-712
30.8	-803
30.9	-893
31.0	-983

63. [O24/Postflight Briefing #15-1]
Referring to the density altitude chart above, determine the pressure altitude at an airport that is at 3,563 feet MSL with an altimeter setting of 29.96.
A. 3,527 feet MSL.
B. 3,556 feet MSL.
C. 3,639 feet MSL.

64. [O24/Postflight Briefing #15-1]
Under what condition are pressure altitude and density altitude the same value?
A. At sea level, when the temperature is 0 degrees F.
B. When the altimeter has no installation error.
C. At standard temperature.

65. [O24/Postflight Briefing #15-1]
Referring to the density altitude chart above, determine the pressure altitude at an airport that is 1,386 feet MSL with an altimeter setting of 29.97.
A. 1,341 feet MSL.
B. 1,451 feet MSL.
C. 1,562 feet MSL.

66. [O24/Postflight Briefing #15-1]
Referring to the density altitude chart above, what is the effect of a temperature increase from 25 to 50 degrees F on the density altitude if the pressure altitude remains at 5,000 feet?
A. 1,200 foot increase.
B. 1,400 foot increase.
C. 1,650 foot increase.

67. [O24/Postflight Briefing #15-1]
Referring to the density altitude chart above, determine the pressure altitude with an indicated altitude of 1,380 feet MSL and an altimeter setting of 28.22 at standard temperature.
A. 1,250 feet MSL.
B. 3,010 feet MSL.
C. 1,373 feet MSL.

68. [O24/Postflight Briefing #15-1]
Referring to the density altitude chart above, what is the effect of a temperature increase from 30 to 50 degrees F on the density altitude if the pressure altitude remains at 3,000 feet MSL?
A. 900 foot increase.
B. 1,100 foot decrease.
C. 1,300 foot increase.

Chapter Fifteen Answers

1. C
2. A
3. C
4. B
5. C
6. B
7. A
8. B
9. A
10. B
11. B
12. C
13. C
14. B
15. A
16. C
17. A
18. B
19. V_x is 71 knots
    V_y is 81 knots
20. See solution to the right
21. See solution to the right
22. See solution to the right
23. See solution to the right
24. C
25. C
26. B
27. C

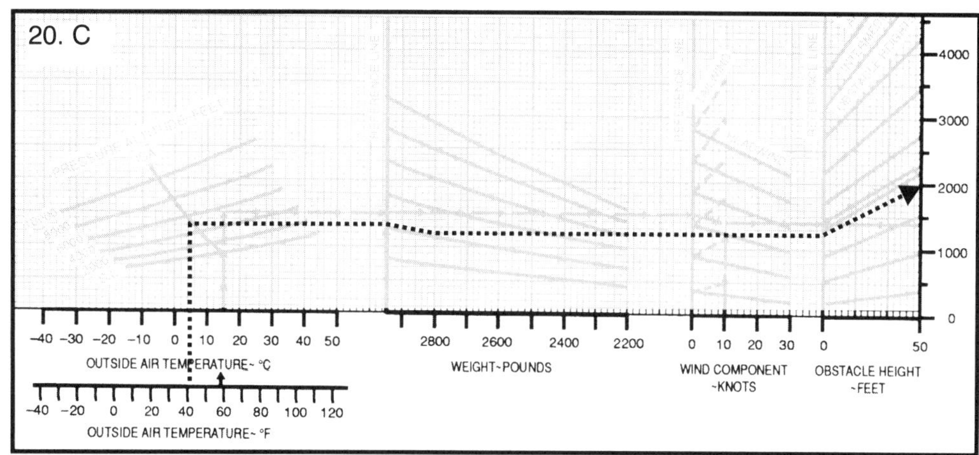

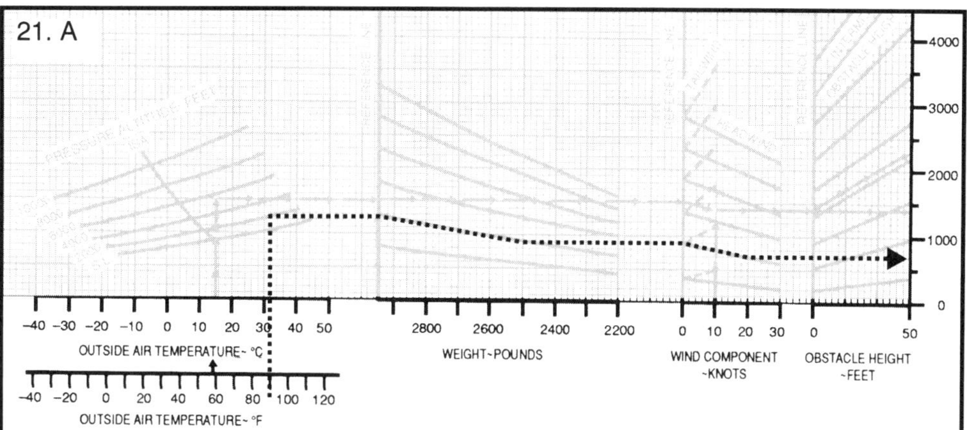

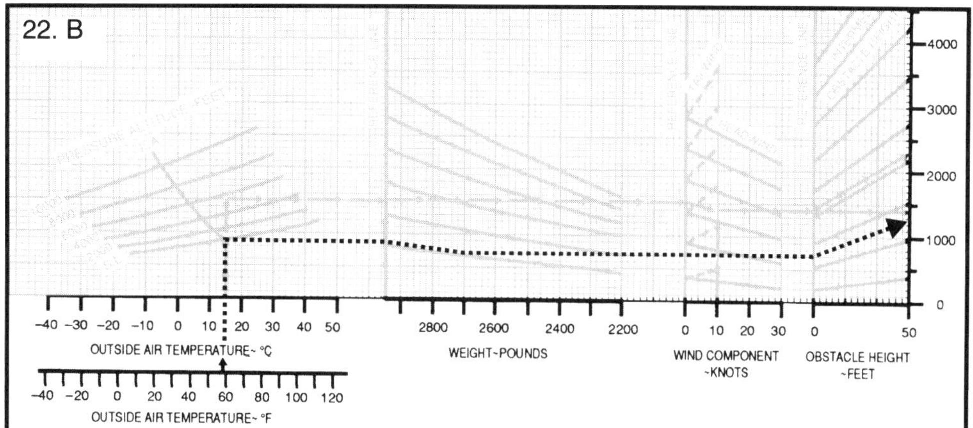

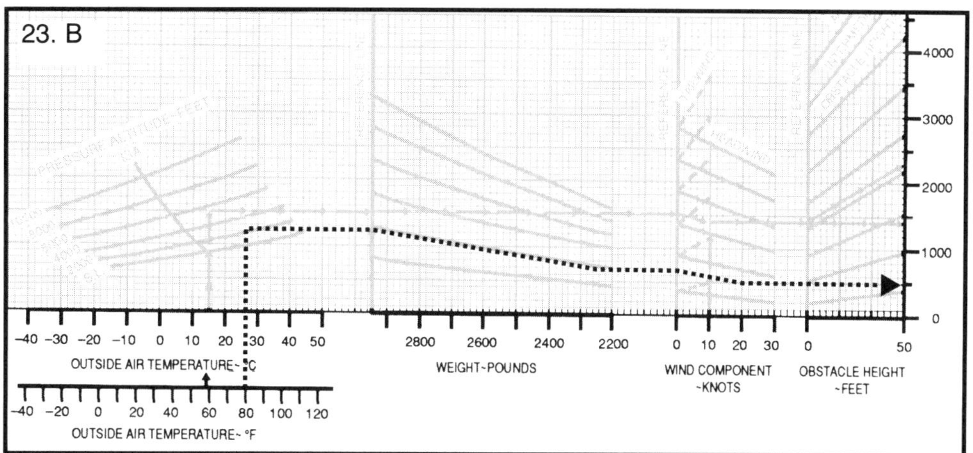

Chapter Fifteen Answers

28. See solution to the right
29. See solution to the right
30. See solution to the right
31. See solution to the right
32. B
33. A
34. B
35. B
36. C
37. B
38. B
39. A
40. C
41. C
42. B
43. A
44. B
45. C
46. C
47. A
48. C
49. B
50. A
51. A
52. A
53. A
54. B
55. A
56. A
57. A
58. C
59. C
60. A
61. A
62. B
63. A
64. C
65. A
66. C
67. B
68. C

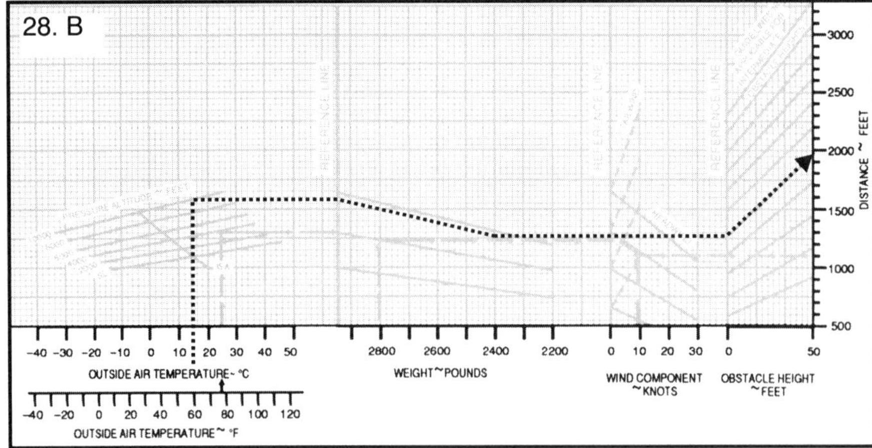

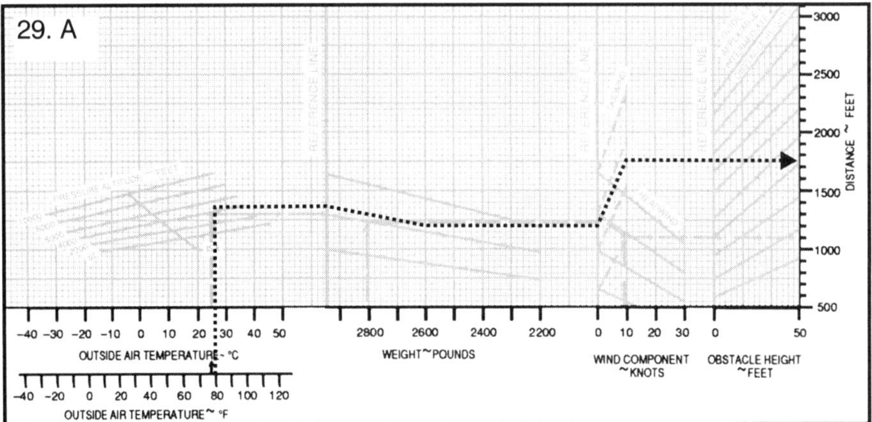

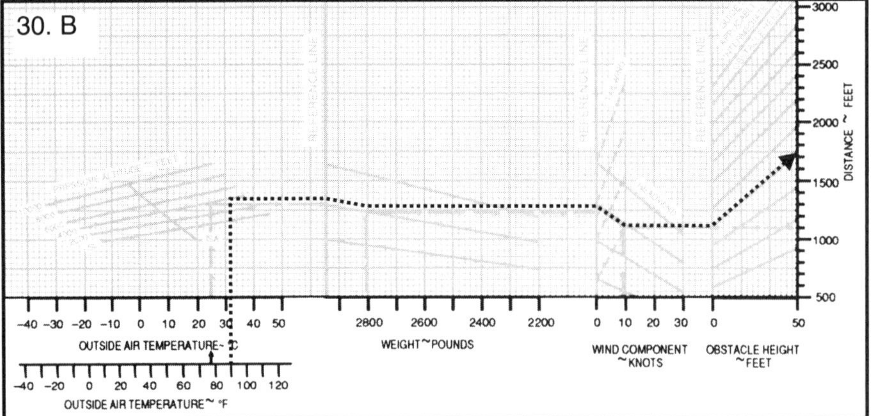

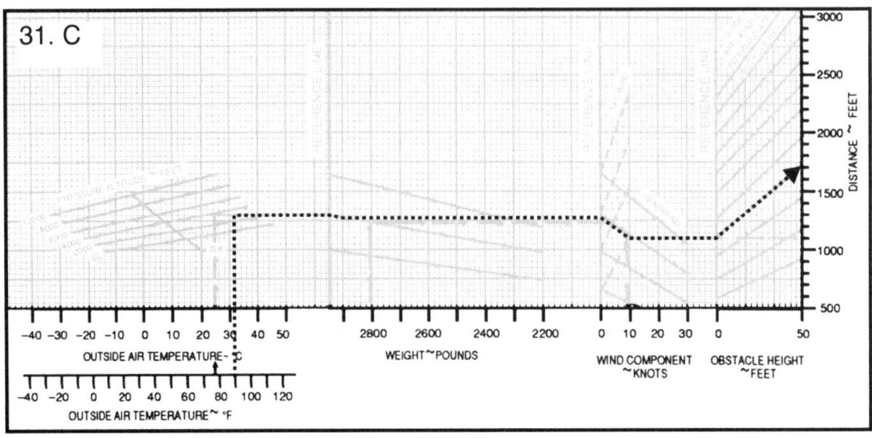

Note: To ensure that you have the most current answers to these questions, please check the *Book & Slide Updates* section at Rod Machado's web site: www.rodmachado.com

Chapter Sixteen

Weight & Balance: Let's Wait & Balance

Excessive Weight and Structural Damage

1. [P2/1/1]
Airplanes are designed to be flown up to a specific maximum _____ weight.
A. landing
B. gross
C. takeoff

2. [P2/1/1]
While it's possible to become airborne beyond the maximum gross weight, _____ problems can arise when turbulence or high-G maneuvering enters the picture.
A. magneto
B. emotional
C. structural

Center of Gravity

3. [P2/3/1]
An object's center of gravity is the place where that object would _____ if it were supported at this position.
A. balance
B. tip
C. move

4. [P3/1/1]
If an airplane will return, unassisted, to level flight after its controls are disturbed, it is said to have _____ dynamic stability.
A. negative
B. positive
C. neutral

5. [P3/1/1]
If the airplane won't return to its original flight configuration, and in fact keeps diverging farther from it in a series of oscillations, it is said to be exhibiting _____ dynamic stability.
A. negative
B. positive
C. neutral

6. [P3/1/1]
Engineers tell us that for an airplane to be positively stable, its weight must not be concentrated too far forward nor too far aft. These forward and aft weight limits are known as the _____ limits.
A. center of gravity
B. envelope
C. moment

7. [P3/Figure 3]
The term used to describe the airplane's pitching motion is known as _____ stability.
A. vertical
B. longitudinal
C. lateral

8. [P3/Figure 3]
Dihedral, weight placement and keel effect are common means of enhancing _____ stability.
A. vertical
B. longitudinal
C. lateral

9. [P3/Figure 3]
Since the airplane pitches about its lateral (sideways) axis, it's correct to say that _____ stability describes the airplane's pitching motion about its lateral axis.
A. vertical
B. longitudinal
C. lateral

10. [P3/Figure 3]
The location of the CG with respect to the center of lift determines the airplane's _____ stability.
A. vertical
B. longitudinal
C. lateral

11. [P3/Figure 3]
The term used to describe the airplane's rolling motion is known as _____ stability.
A. vertical
B. longitudinal
C. lateral

12. [P3/Figure 3]
Since the airplane rolls about its longitudinal (long) axis, it's correct to say that _____ stability describes the airplane's motion about its longitudinal axis.
A. vertical
B. longitudinal
C. lateral

13. [P3/Figure 3]
Properly loaded airplanes are _____ stable.
A. vertically
B. longitudinally
C. laterally

14. [P3/Figure 3]
Airplanes with good _____ stability tend to favor a wings-level flight condition. They resist _____ movement.
A. yaw, pitch
B. longitudinal, roll
C. lateral, roll

15. [P3/Figure 3]
What determines the longitudinal stability of an airplane?
A. The location of the CG with respect to the center of lift.
B. The effectiveness of the horizontal stabilizer, rudder and rudder trim tab.
C. The relationship of thrust and lift to weight and drag.

16. [P3/1/4]
An airplane said to be inherently stable will
A. be difficult to stall.
B. require less effort to control.
C. not spin.

Other CG Considerations

17. [P5/See *The Center of Lift*]
The location where the wing's total lifting force is concentrated is known as the _____.
A. center of gravity
B. center of lift
C. aerodynamic center

18. [P5/See *The Center of Lift*]
As the airplane's angle of attack increases, the center of lift _____.
A. moves forward
B. remain stationary
C. moves aft

19. [P5/See *The Center of Lift*]
As the airplane's angle of attack decreases, the center of lift _____.
A. moves forward
B. remain stationary
C. moves aft

20. [P4/1/4]
For an airplane to have positive longitudinal stability, the center of gravity must remain _____ the center of lift.
A. centered at
B. behind
C. in front of

21. [P4/1/5]
An airplane has been loaded in such a manner that the CG is located aft of the aft CG limit. One undesirable flight characteristic a pilot might experience with this airplane would be
A. a longer takeoff run.
B. difficulty in recovering from a stalled condition.
C. stalling at higher-than-normal airspeed.

22. [P4/2/2]
Loading an airplane to the most aft CG will cause the airplane to be
A. less stable at all speeds.
B. less stable at slow speeds, but more stable at high speeds.
C. less stable at high speeds, but more stable at low speeds.

23. [P4/3/3]
An airplane has been loaded in such a manner that the CG is located forward of the forward CG limit. One undesirable flight characteristic a pilot might experience with this airplane would be
A. lower stick (elevator) forces.
B. full rearward elevator pressure is required to flare.
C. difficulty in recovering from a stalled condition.

Just a Moment

24. [P5/1/4 & P3/Figure 6]
Referring to the figure above, how much should Block R weigh to balance the scale?
A. 10 lb.
B. 20 lb.
C. 30 lb.

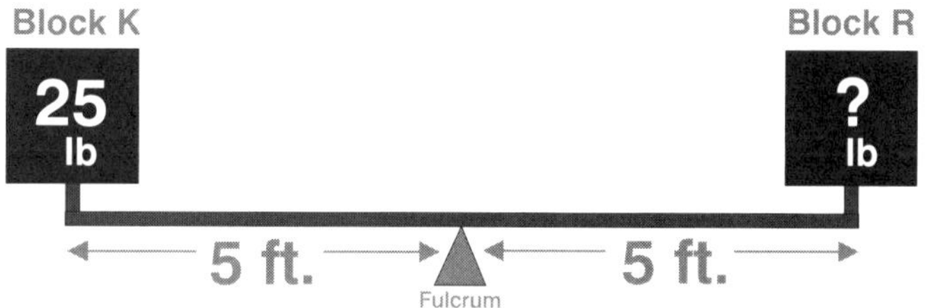

25. [P5/1/4 & P3/Figure 6]
Referring to the figure above, how much should Block R weigh to balance the scale?
A. 10 lb.
B. 25 lb.
C. 50 lb.

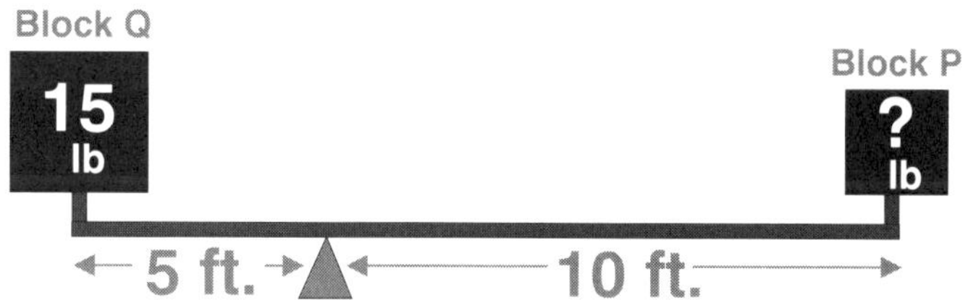

26. [P6/1/4 & P3/Figure 7]
Referring to the figure above, how much should Block P weigh to balance the scale?
A. 15 lb.
B. 7.5 lb.
C. 50 lb.

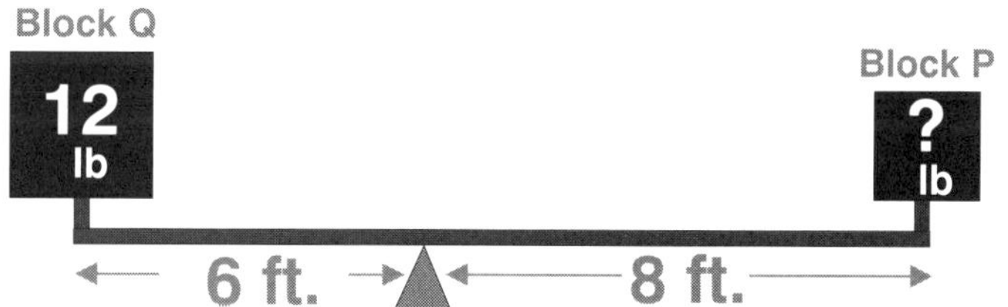

27. [P6/1/4 & P3/Figure 7]
Referring to the figure above, how much should Block P weigh to balance the scale?
A. 12 lb.
B. 9 lb.
C. 6 lb.

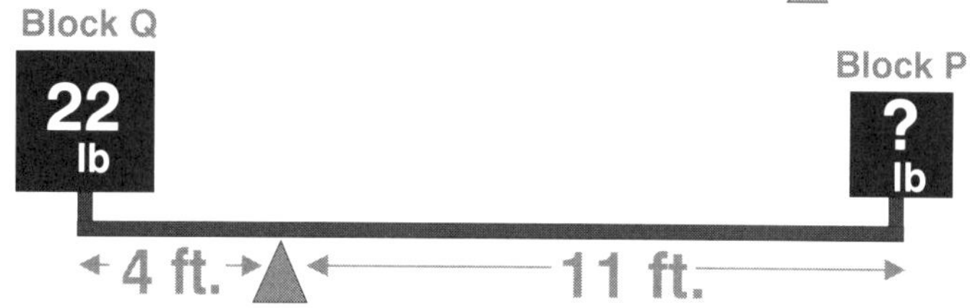

28. [P6/1/4 & P3/Figure 7]
Referring to the figure above, how much should Block P weigh to balance the scale?
A. 8 lb.
B. 22 lb.
C. 4 lb.

29. [P6/1/6]
An arbitrary vertical reference line from which all weights on an airplane are measured is known as the _____ line.
A. moment
B. CG
C. datum

30. [P8/2/2]
Which items are typically included in the empty weight of an aircraft?
A. Unusable fuel and undrainable oil.
B. Only the airframe, powerplant, and optional equipment.
C. Full fuel tanks and engine oil to capacity.

31. [P8/2/5]
The empty weight plus the useful load equals _____.
A. the gross weight
B. the datum weight
C. the minimum takeoff weight

32. [P8/2/6]
Aviation gasoline weighs _____ pounds per gallon.
A. 6
B. 8
C. 7.5

33. [P8/2/6]
Oil weighs _____ pounds per gallon.
A. 6
B. 8
C. 7.5

34. [P8/2/6]
An aircraft is loaded 110 pounds over maximum certificated gross weight. If fuel (gasoline) is drained to bring the aircraft weight within limits, how much fuel should be drained?
A. 15.7 gallons.
B. 16.2 gallons.
C. 18.4 gallons.

35. [P8/2/6]
If an aircraft is loaded 90 pounds over maximum certificated gross weight and fuel (gasoline) is drained to bring the aircraft weight within limits, how much fuel should be drained?
A. 10 gallons.
B. 12 gallons.
C. 15 gallons.

Don't Wait to Balance

36. [P9/2/2]
To find the airplane's center of gravity you must divide the airplane's total _____.
A. moment by the total weight
B. weight by the total moment
C. arm by the total moment

	Weight (lb)	x	Arm (inches)	=	Moment (lb-in)
Empty weight	1,495.0		101.4		151,593.0
Pilot & front passenger	380.0		62.0		?
Fuel (30 gal. no reserve)	180.0		93.0		?
Oil (8 qts.)	15.0		30.0		?
Total	?				?

How far aft of the datum is the CG located?

Arm (new CG) = Total Moment / Total Weight = ? / ? = ? (new CG aft datum)

37. [P9/Figure11]
According to the figure above, what is the airplane's present center of gravity in inches aft of datum?
A. 90.23 inches.
B. 92.91 inches.
C. 94.21 inches.

	Weight (lb)	x	Arm (inches)	=	Moment (lb-in)
Empty weight	1,476.0		103.4		152,618.4
Pilot & front passenger	365.0		60.5		?
Fuel (30 gal. no reserve)	179.0		92.5		?
Oil (8 qts.)	15.0		31.0		?
Total	?				?

How far aft of the datum is the CG located?

Arm (new CG) = Total Moment / Total Weight = ? / ? = ? (new CG aft datum)

38. [P9/Figure11]
According to the figure above, what is the airplane's present center of gravity in inches aft of datum?
A. 52.0 inches.
B. 91.21 inches.
C. 94.21 inches.

Basic Weight and Balance Problems

Weight & Balance Problem

Determine if the airplane's weight and balance are within safe limits in the problem below.

Pilot & front seat occupants.......	340 lb
Rear seat occupants..................	295 lb
Fuel (main & aux tanks both full)	44 gal
Baggage..................................	56 lb

39. [P14/1/2]
Based on the conditions listed above and the weight and balance charts shown on page P5, what is the airplane's weight in relation to maximum takeoff weight and is the airplane within proper CG limits?
A. 20 pounds overweight, CG aft of aft limits.
B. 20 pounds overweight, CG within limits.
C. 20 pounds underweight, CG forward of forward limits.

Weight & Balance Problem

Determine if the airplane's weight and balance are within safe limits in the problem below.

Pilot & front seat occupants.......	350 lb
Rear seat occupants..................	325 lb
Fuel (main & aux tanks both full)	27 gal
Baggage..................................	35 lb

40. [P14/1/2]
Based on the conditions listed above and the weight and balance charts shown on page P5, what is the airplane's center of gravity location and is the airplane within proper CG limits?
A. CG 81.7, out of limits forward.
B. CG 83.6, within limits.
C. CG 84.1, within limits.

Weight & Balance Problem

Determine if the airplane's weight and balance are within safe limits in the problem below.

Pilot & front seat occupants.......	387 lb
Rear seat occupants..................	293 lb
Fuel (main & aux tanks both full)	35 gal
Baggage..................................	?

41. [P14/1/2]
What is the maximum amount of baggage that can be carried when the airplane is loaded as shown above?
A. 45 pounds.
B. 63 pounds.
C. 220 pounds.

USEFUL LOAD WEIGHTS AND MOMENTS

OCCUPANTS

FRONT SEATS ARM 85		REAR SEATS ARM 121	
Weight	Moment/100	Weight	Moment/100
120	102	120	145
130	110	130	157
140	119	140	169
150	128	150	182
160	136	160	194
170	144	170	206
180	153	180	218
190	162	190	230
200	170	200	242

USABLE FUEL

MAIN WING TANKS ARM 75		
Gallons	Weight	Moment/100
5	30	22
10	60	45
15	90	68
20	120	90
25	150	112
30	180	135
35	210	158
40	240	180
44	264	198

BAGGAGE OR 5TH SEAT OCCUPANT ARM 140

Weight	Moment/100
10	14
20	28
30	42
40	56
50	70
60	84
70	98
80	112
90	126
100	140
110	154
120	168
130	182
140	196
150	210
160	224
170	238
180	252
190	266
200	280
210	294
220	308
230	322
240	336
250	350
260	364
270	378

AUXILIARY WING TANKS ARM 94

Gallons	Weight	Moment/100
5	30	28
10	60	56
15	90	85
19	114	107

*OIL

Quarts	Weight	Moment/100
10	19	5

*Included in basic Empty Weight

Basic **Empty Weight** ~ 2015

MOM / 100 ~ 1554

MOMENT LIMITS vs WEIGHT

Moment limits are based on the following weight and center of gravity limit data (landing gear down).

WEIGHT CONDITION	FORWARD CG LIMIT	AFT CG LIMIT
2950 lb (takeoff or landing)	82.1	84.7
2525 lb	77.5	85.7
2475 lb or less	77.0	85.7

MOMENT LIMITS vs WEIGHT (Continued)

Weight	Minimum Moment/100	Maximum Moment/100
2100	1617	1800
2110	1625	1808
2120	1632	1817
2130	1640	1825
2140	1648	1834
2150	1656	1843
2160	1663	1851
2170	1671	1860
2180	1679	1868
2190	1686	1877
2200	1694	1885
2210	1702	1894
2220	1709	1903
2230	1717	1911
2240	1725	1920
2250	1733	1928
2260	1740	1937
2270	1748	1945
2280	1756	1954
2290	1763	1963
2300	1771	1971
2310	1779	1980
2320	1786	1988
2330	1794	1997
2340	1802	2005
2350	1810	2014
2360	1817	2023
2370	1825	2031
2380	1833	2040
2390	1840	2048
2400	1848	2057
2410	1856	2065
2420	1863	2074
2430	1871	2083
2440	1879	2091
2450	1887	2100
2460	1894	2108
2470	1902	2117
2480	1911	2125
2490	1921	2134
2500	1932	2143
2510	1942	2151
2520	1953	2160
2530	1963	2168
2540	1974	2176
2550	1984	2184
2560	1995	2192
2570	2005	2200
2580	2016	2208
2590	2026	2216

Weight	Minimum Moment/100	Maximum Moment/100
2600	2037	2224
2610	2048	2232
2620	2058	2239
2630	2069	2247
2640	2080	2255
2650	2090	2263
2660	2101	2271
2670	2112	2279
2680	2123	2287
2690	2133	2295
2700	2144	2303
2710	2155	2311
2720	2166	2319
2730	2177	2326
2740	2188	2334
2750	2199	2342
2760	2210	2350
2770	2221	2358
2780	2232	2366
2790	2243	2374
2800	2254	2381
2810	2265	2389
2820	2276	2397
2830	2287	2405
2840	2298	2413
2850	2309	2421
2860	2320	2428
2870	2332	2436
2880	2343	2444
2890	2354	2452
2900	2365	2460
2910	2377	2468
2920	2388	2475
2930	2399	2483
2940	2411	2491
2950	2422	2499

42. [P15/1/1]
Upon landing, the front seat passenger (200 pounds) departs the airplane. A rear passenger (120 pounds) moves to the front passenger position. What effect does this have on the CG if the airplane weighed 2,800 pounds and the moment/100 was 2,270 just prior the passenger transfer? (Refer to the weight and balance charts above.)
A. The CG shifted 1.96 inches forward and is beyond the forward CG limits.
B. The CG shifted 1.96 inches aft and is within proper CG limits.
C. The CG shifted 1.96 inches forward and is within proper CG limits.

43. [P16/1/1]
What effect does a 30 gallon fuel burn (main tanks) have on the weight and balance if the airplane weighed 2,890 pounds and the moment/100 was 2,452 at takeoff? (Refer to the weight and balance charts above.)
A. The CG has shifted beyond the forward CG limit.
B. The CG is within proper CG limits.
C. The CG has shifted beyond the aft CG limit.

44. [P16/1/6]
With the airplane loaded as follows, what action must be taken to place the airplane within the proper weight and balance limits? (Refer to the weight and balance charts above.)

Front seat occupants .... 440 pounds
Rear seat occupants .... 90 pounds
Main wing tanks .... 44 gallons

A. Add 130 pounds of baggage in the baggage compartment.
B. Add 60 pounds of baggage in the baggage compartment.
C. Transfer 15 gallons of fuel from the main to the aux tanks.

45. [P16/1/6]
Which action can adjust the airplane's weight to maximum gross weight and the CG within limits for takeoff?

Front seat occupants .... 425 pounds
Rear seat occupants .... 300 pounds
Main wing tanks .... 44 gallons

A. Drain 12 gallons of fuel.
B. Drain 9 gallons of fuel.
C. Transfer 12 gallons of fuel from the main tanks to the auxiliary tanks.

	Weight (lb)	Moment (lb-in)
Empty weight	1,350.0	51.5
Pilot & front passenger	340.0	?
Fuel (Std tanks)	Capacity	?
Oil (8 qts.)	?	?
Total	?	?

46. [P18/1/1]
Using the airplane loading information shown above and the weight and balance charts shown to the right, determine the moment.
A. 69.9 pound-inches.
B. 75.0 pound-inches.
C. 77.6 pound-inches.

	Weight (lb)	Moment (lb-in)
Empty weight	1,350.0	51.5
Pilot & front passenger	380.0	?
Fuel (48 gal)	288	?
Oil (8 qts.)	?	?
Total	?	?

47. [P18/1/1]
Using the airplane loading information shown above and the weight and balance charts shown to the right, determine the aircraft loaded moment and the aircraft category.
A. 78.2 pound-inches, normal category.
B. 79.2 pound-inches, normal category.
C. 80.4 pound-inches, utility category.

	Weight (lb)	Moment (lb-in)
Empty weight	1,350.0	51.5
Pilot & front passenger	400.0	?
Fuel (38 gal)	Capacity	?
Oil (8 qts.)	?	?
Baggage	100.0	
Total	?	?

48. [P18/1/1]
Using the airplane loading information shown above and the weight and balance charts shown to the right, determine if the airplane is within its proper CG limits.
A. The CG is beyond the forward CG limit.
B. The CG is within proper CG limits.
C. The CG is beyond the aft CG limit.

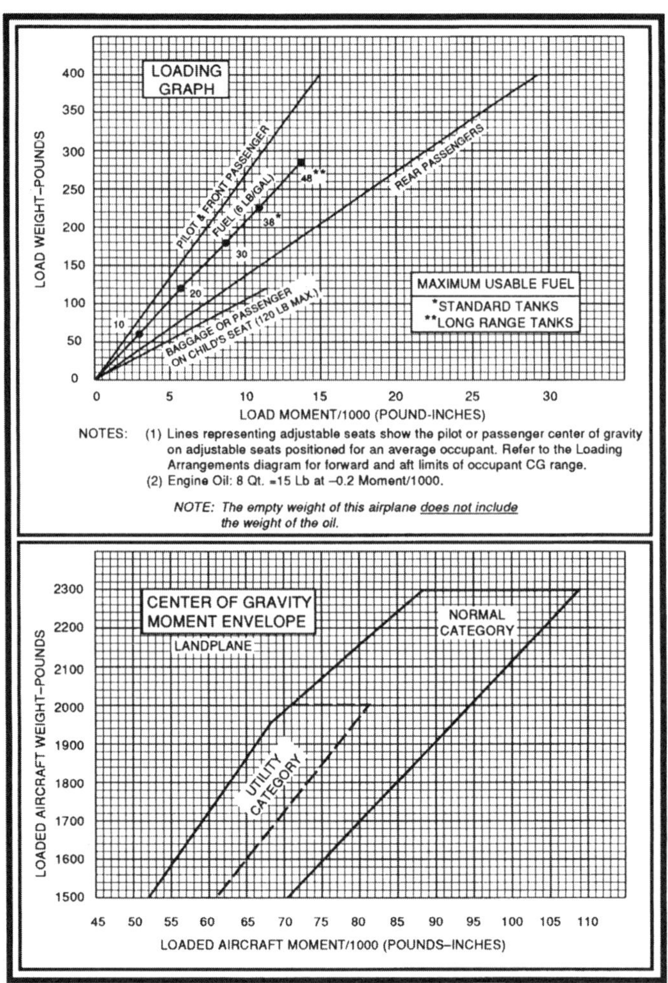

	Weight (lb)	Moment (lb-in)
Empty weight	1,350.0	51.5
Pilot & front passenger	400.0	?
Rear passengers	400.0	?
Fuel (std tanks)	?	?
Oil (8 qts.)	15.0	-.2
Baggage	17.0	1.7
Total	?	?

49. [P20/1/1]
Using the airplane loading information and the weight and balance charts shown above, determine the maximum amount of fuel that can be carried aboard the aircraft.
A. 6 gallons.
B. 19.6 gallons.
C. 10 gallons.

Chapter Sixteen Answers

1. B
2. C
3. A
4. B
5. A
6. A
7. B
8. C
9. B
10. B
11. C
12. C
13. B
14. C
15. A
16. B
17. B
18. A
19. C
20. C
21. B
22. A
23. B
24. A
25. B
26. B
27. B
28. A
29. C
30. A
31. A
32. A
33. C
34. C
35. C
36. A

37. B

	Weight (lb)	x Arm (inches) =	Moment (lb-in)
Empty weight	1,495.0	101.4	151,593.0
Pilot & front passenger	380.0	62.0	23,560
Fuel (30 gal. no reserve)	180.0	93.0	16,740
Oil (8 qts.)	15.0	30.0	450
Total	2,070		192,343

How far aft of the datum is the CG located?

Arm (new CG) = Total Moment / Total Weight = 192,343 / 2,070 = 92.91 (new CG aft datum)

38. C

	Weight (lb)	x Arm (inches) =	Moment (lb-in)
Empty weight	1,476.0	103.4	152,618.4
Pilot & front passenger	365.0	60.5	22,082.5
Fuel (30 gal. no reserve)	179.0	92.5	16,557.5
Oil (8 qts.)	15.0	31.0	465
Total	2,035		191,723.4

How far aft of the datum is the CG located?

Arm (new CG) = Total Moment / Total Weight = 191,723.4 / 2,035 = 94.21 (new CG aft datum)

39. B

	Weight (lb)	x Arm (inches) =	Moment (lb-in/100)
Empty weight	2,015.0		1,554.00
Pilot & front occupants	340.0	85.0	289.00
Rear seat occupants	295.0	121.0	357.00
Fuel (main tanks-44 gal.)	264.0	75.0	198.00
Fuel (aux tank-19 gal.)	0.0	94.0	0.00
Baggage	56.0	140.0	78.00
Oil (8 qts.)	Included in basic empty weight		
Total	2,970		2,476.00

How far aft of the datum is the CG located?

Arm (new CG) = Total Moment / Total Weight = 247,600 / 2,970 = 83.36 (new CG aft datum)

40. B

	Weight (lb)	x Arm (inches) =	Moment (lb-in/100)
Empty weight	2,015.0		1,554.00
Pilot & front occupants	350.0	85.0	298.00
Rear seat occupants	325.0	121.0	393.00
Fuel (main tanks-44 gal.)	162.0	75.0	122.00
Fuel (aux tank-19 gal.)	0.0	94.0	0.00
Baggage	35.0	140.0	48.00
Oil (8 qts.)	Included in basic empty weight		
Total	2,887		2,415.00

How far aft of the datum is the CG located?

Arm (new CG) = Total Moment / Total Weight = 241,500 / 2,887 = 83.65 (new CG aft datum)

41. A

	Weight (lb)	x Arm (inches) =	Moment (lb-in/100)
Empty weight	2,015.0		1,554.00
Pilot & front occupants	387.0	85.0	329.00
Rear seat occupants	293.0	121.0	355.00
Fuel (main tanks-44 gal.)	210.0	75.0	158.00
Fuel (aux tank-19 gal.)	0.0	94.0	0.00
Baggage	?	140.0	0.00
Oil (8 qts.)	Included in basic empty weight		
Total	2,905		

What is the maximum amount of baggage we can carry?

Gross Weight (2,950) - 2,905 = 45 pounds

42. C

	Weight (lb) x	= Moment (lb-in/100)
Airplane weight	2,800	2,270
Front seat pass departs	-200	-170
Rear seat pass leaves	-120	-145
Rear seat pass enters	+120	+102
Total	2,600	2,057

New arm (new CG) = Total Moment / Total Weight = 205,700 / 2,600 = 79.11 (new CG aft datum)

Old arm (old CG) = Total Moment / Total Weight = 227,000 / 2,800 = 81.07 (old CG aft datum)

81.07 - 79.11 = a forward CG shift of 1.96 inches

Chapter Sixteen Answers

43. C

	Weight (lb) x	= Moment (lb-in/100)
Airplane weight	2,890	2,452
Fuel burn of 30 gal.	-180	-135
Total	2,710	2,317

Old arm (old CG) = Total Moment / Total Weight = 245,200 / 2,890 = 84.84 (old CG aft datum)

New arm (new CG) = Total Moment / Total Weight = 231,700 / 2,710 = 85.49 (new CG aft datum)

The CG has shifted beyond the aft limits.

44. A

	Weight (lb) x	Arm (inches) =	Moment (lb-in/100)
Empty weight	2,015.0		1,554.00
Pilot & front occupants	440.0	85.0	374.00
Rear seat occupants	90.0	121.0	109.00
Fuel (main tanks-44 gal.)	264.0	75.0	198.00
Baggage	00.0	140.0	00.00
Oil (8 qts.)	Included in basic empty weight		
Total	2,809		2,235.00

How far aft of the datum is the CG located?

Arm (new CG) = Total Moment / Total Weight = 223,500 / 2,809 = 79.56 (new CG aft datum)

This present CG is too far forward and needs to be moved aft.

Try adding 60 pounds of baggage:

2,809 2,235.00
+60 +84 According to chart CG still not within limits
2,869 2,319

Try adding shifting 15 gal. of fuel to aux tanks:

2,809 2,235.00
-90 -68
+90 +85 According to chart CG still not within limits
2,809 2,251

Try adding 130 pounds of baggage.

2,809 2,235.00
+130 +182 The CG is at 82.23 inches and within limits
2,939 2,417

45. B

	Weight (lb) x	Arm (inches) =	Moment (lb-in/100)
Empty weight	2,015.0		1,554.00
Pilot & front occupants	425.0	85.0	361.00
Rear seat occupants	300.0	121.0	363.00
Fuel (main tanks-44 gal.)	264.0	75.0	198.00
Baggage	00.0	140.0	00.00
Oil (8 qts.)	Included in basic empty weight		
Total	3,004		2,476.00

The airplane is 54 pound over gross.

Eliminating 9 gallons of fuel puts the gross weight within limits. Let's check to see if the CG is now within limits:

3,004 2,476.00
-54 -41
2,950 2,435

Arm (new CG) = Total Moment / Total Weight = 243,500 / 2,950 = 82.5 inches (new CG aft datum)

The CG is now at 82.5 inches & within limits

Note: To ensure that you have the most current answers to these questions, please check the *Book & Slide Updates* section at Rod Machado's web site: www.rodmachado.com

46. B

	Weight (lb)	Moment (lb-in/1000)
Empty weight	1,350.0	51.5
Pilot & front passenger	340.0	12.7
Fuel (Std tanks)	228.0	11.0
Oil (8 qts.)	15.0	-.2
Total	1,933.0	75.0

47. B

	Weight (lb)	Moment (lb-in/1000)
Empty weight	1,350.0	51.5
Pilot & front passenger	380.0	14.2
Fuel (Std tanks)	288.0	13.7
Oil (8 qts.)	15.0	-.2
Total	2,033.0	79.2

48. B

	Weight (lb)	Moment (lb-in/1000)
Empty weight	1,350.0	51.5
Pilot & front passenger	400.0	15.0
Fuel (Std tanks)	228.0	11.0
Oil (8 qts.)	15.0	-.2
Baggage	100.0	9.5
Total	2,093.0	86.8

49. B

	Weight (lb)	Moment (lb-in/1000)
Empty weight	1,350.0	51.5
Pilot & front passenger	400.0	15.0
Rear passenger(s)	400.0	29.2
Fuel (Std tanks)	228.0	11.0
Oil (8 qts.)	15.0	-.2
Baggage	17.0	1.4
Total	2,410.0	107.9

The airplane is 110 pounds over gross.
Eliminate 110 pounds from 228 pounds of fuel.
This leaves 118 pounds of fuel to be carried.
118 lbs/6 lbs/gal = 19.6 gallons of allowable fuel.
Let's see if the airplane is still within CG limits:

2,410 107.9 (at 228 lbs the moment is 11.0
-110 -5.3 at 118 lbs the moment is 5.7
2,300 102.6 11.0 - 5.7 = 5.3 difference)

The CG is within limits

Chapter Seventeen

Pilot Potpourri: Neat Aeronautical Information

Fitness for Flight

1. [Q2/1/5]
The FARs prohibit you from acting as PIC if you have a _____ that would make you unable to meet the standards for a medical certificate.
A. cold
B. earache
C. known medical condition

Medication

2. [Q3/1/2]
If you're not sure whether medication you are taking might affect your ability to fly, you should contact your _____.
A. spiritual doctor
B. aviation medical examiner
C. psychiatrist

Alcohol - Don't Fly High

3. [Q3/1/4]
The effects of alcohol can last _____.
A. longer than eight hours
B. only eight hours
C. never less than eight hours

4. [Q3/2/1]
Although not required, it's usually best to wait _____ hours between bottle and throttle.
A. 4 to 8
B. 112 to 354
C. 12 to 24

Hypoxia: Low O Two

5. [Q3/2/2]
Which statement best defines hypoxia?
A. A state of oxygen deficiency in the body.
B. An abnormal increase in the volume of air breathed.
C. A condition of gas bubble formation around the joints or muscles.

6. [Q3/2/2]
What happens to the percentage of oxygen available in the atmosphere as altitude increases?
A. It decreases dramatically.
B. It actually increases slightly.
C. It remains the same.

7. [Q3/2/2]
If the percentage of oxygen remains relatively constant with altitude, then why would a pilot experience hypoxia at these higher altitudes?
A. The fast speed of the airplane creates a slight over the fuselage vacuum that depletes the cockpit of oxygen.
B. The partial pressure of oxygen decreases with altitude.
C. Carbon monoxide always increases with altitude.

8. [Q3/2/3]
The effects of hypoxia can occur at altitudes as low as _____.
A. 5,000 feet
B. 10,000 feet
C. 15,000 feet

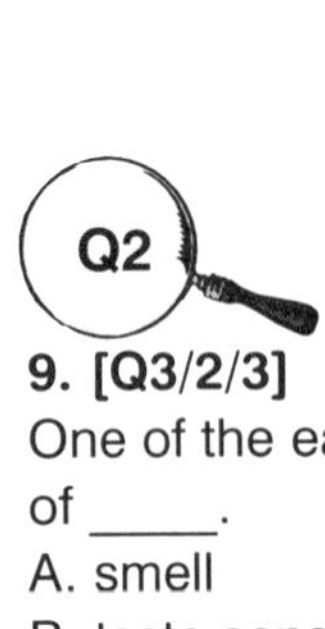

9. [Q3/2/3]
One of the early effects of hypoxia is a deterioration of _____.
A. smell
B. taste sensation
C. night vision

10. [Q3/3/2]
Above _____ feet MSL, most individuals begin to experience some decrease in their judgment, memory, alertness and coordination.
A. 10,000
B. 2,000
C. 1,000

Hyperventilation

11. [Q4/1/3]
Rapid or extra deep breathing while using oxygen can cause a condition known as
A. hyperventilation.
B. aerosinusitis.
C. aerotitis.

12. [Q4/1/3]
Which would most likely result in hyperventilation?
A. Emotional tension, anxiety, or fear.
B. The excessive consumption of alcohol.
C. An extremely slow rate of breathing and insufficient oxygen.

13. [Q4/2/2]
A pilot should be able to overcome the symptoms or avoid future occurrences of hyperventilation by
A. closely monitoring the flight instruments to control the airplane.
B. slowing the breathing rate, breathing into a bag, or talking aloud.
C. increasing the breathing rate in order to increase lung ventilation.

14. [Q4/3/2]
The early symptoms of hyperventilation and hypoxia are _____.
A. dissimilar
B. similar
C. not similar at all

CO Oh Oh

15. [Q4/3/3]
Carbon monoxide (CO) is a _____ gas.
A. colorless, odorless, tasteless
B. tasteful, visible and smelly
C. harmless

16. [Q4/3/3]
Carbon monoxide has a stronger affinity than oxygen for _____.
A. white blood cells
B. hemoglobin
C. nose molecules

17. [Q4/3/3]
If a hemoglobin molecule is occupied with a CO molecule, _____ can't get aboard and be transported.
A. oxygen
B. nitrogen
C. argon

18. [Q4/3/4]
Susceptibility to carbon monoxide poisoning increases as
A. altitude increases.
B. altitude decreases.
C. air pressure increases.

19. [Q5/1/2]
Large accumulations of carbon monoxide in the human body result in
A. tightness across the forehead.
B. loss of muscular power.
C. an increased sense of well being.

20. [Q5/1/2]
If you notice the odor of exhaust or experience symptoms of _____ or loss of muscular power when the aircraft heater is in use, immediately turn off the heater and open the air vents.
A. hunger
B. euphoria
C. headache, drowsiness, dizziness

Ear Ye, Ear Ye

21. [Q5/1/4]
Ear problems common to pilots usually involve a little flaccid tube that connects the middle ear to the back of the throat. This tube is known as the _____ tube.
A. throat
B. eustachian
C. middle ear

22. [Q5/2/4]
The first line of defense against ear block is to stop _____.
A. holding a constant altitude
B. climbing
C. descending

Spatial Disorientation

23. [Q6/1/5]
A state of temporary confusion resulting from misleading information being sent to the brain by various sensory organs is defined as
A. spatial disorientation.
B. ground school.
C. hypoxia.

24. [Q6/1/5]
Pilots are more subject to spatial disorientation if
A. they ignore the sensations of muscles and inner ear.
B. body signals are used to interpret flight attitude.
C. eyes are moved often in the process of cross checking the flight instruments.

25. [Q6/2/1]
If a pilot experiences spatial disorientation during flight in a restricted visibility condition, the best way to overcome the effect is to
A. rely upon the aircraft instrument indications.
B. concentrate on yaw, pitch, and roll sensations.
C. consciously slow the breathing rate until symptoms clear and then resume normal breathing rate.

26. [Q6/2/1]
The danger of spatial disorientation during flight in poor visual conditions may be reduced by
A. shifting the eyes quickly between the exterior visual field and the instrument panel.
B. having faith in the instruments rather than taking a chance on the sensory organs.
C. leaning the body in the opposite direction of the motion of the aircraft.

27. [Q6/2/2]
Vertigo is caused by problems associated with three of our sensory systems: _____.
A. inner ear, throat and tongue
B. visual, tactile and circulatory
C. vestibular, kinesthetic and visual

28. [Q6/2/3]
The visual system is exactly what it sounds like: information sent to our brain from our _____.
A. ears
B. tongue
C. eyes

29. [Q6/2/3]
The _____ is the sensory information sent to the brain by the seat of our pants. It's the information transmitted by sensors in our skin and from areas deeper within our bodies.
A. derrière
B. kinesthetic system
C. vestibular

30. [Q6/2/4]
The vestibular system consists of the _____ canals located in the inner ear.
A. fluidic
B. circular
C. semicircular

31. [Q6/2/4]
These canals consist of three circular tubes, each containing a fluid whose movement causes the bending of small hair filaments known as _____ organs located at the base of each canal.
A. otolith
B. internal
C. movement

32. [Q6/2/4]
Movement of the fluid within the tubes, caused by acceleration (a change in direction or velocity), stimulates the otolith organs, alerting the brain that the _____ is/are in motion.
A. hand
B. airplane
C. body

33. [Q7/1/3]
The semicircular canal system was evolved as a _____ system where gravity always pulls the body in one direction—straight downward.
A. non-motion based
B. air based
C. ground based

34. [Q7/1/4]
Abrupt head movements under instrument or instrument-like conditions can cause you to perceive maneuvers that aren't really happening. This vertigo-type illusion is called the _____ illusion.
A. canal
B. Coriolis
C. black hole

Visual Illusions

35. [Q8/1/2]
At night, a blending of the earth and sky is often responsible for creating an indiscernible _____, resulting in near-instrument flight conditions. This is most prevalent on moonless nights when stars take on the appearance of _____ and city lights appear to be stars.
A. star map, planets
B. horizon, city lights
C. horizon, the sky

Flight Vision

36. [Q9/1/1]
Light sensitive areas of the retina are made up of individual cells known as _____ and _____.
A. rods, bones
B. rods, cones
C. foveas, pupils

37. [Q9/3/1]
Cone cells are responsible for allowing you to perceive _____. Unfortunately, the cones don't work well when it's _____. This explains why it's more difficult to perceive color at night than in daylight hours.
A. color, dark
B. black and white, dark
C. motion, bright

38. [Q9/Figure 7]
Fill in the blanks:

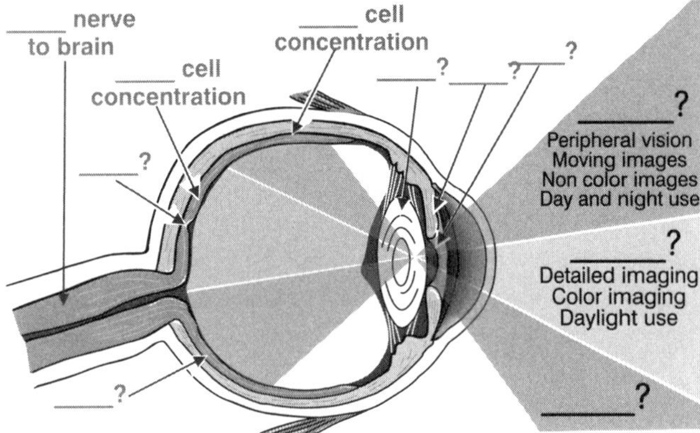

39. [Q10/1/1]
The dim light receptors in the eye are known as _____ cells.
A. pupil
B. rod
C. cone

40. [Q10/1/3]
If you want the best view of a dimly lit object you need to expose the _____ to the light. You can do this by using your _____ vision for off-center viewing.
A. cones, peripheral
B. cones, direct
C. rods, peripheral

41. [Q10/1/3]
What is the most effective way to use the eyes during night flight?
A. Look only at far away, dim lights.
B. Scan slowly to permit off-center viewing.
C. Concentrate directly on each object for a few seconds.

42. [Q10/1/3]
The best method to use when looking for other traffic at night is to
A. look to the side of the object and scan slowly.
B. scan the visual field very rapidly.
C. look to the side of the object and scan rapidly.

Night Vision

43. [Q10/1/4]
It may take at least _____ minutes for your eyes to completely adapt to the dark.
A. 60
B. 45
C. 30

44. [Q10/1/4]
Adapting to darkness is one reason you want to avoid very bright lights for at least _____ minutes before the flight if you're planning on flying at night.
A. 60
B. 45
C. 30

45. [Q10/1/4]
What preparation should a pilot make to adapt the eyes for night flying?
A. Wear sunglasses after sunset until ready for flight.
B. Avoid red lights at least 30 minutes before the flight.
C. Avoid bright white lights at least 30 minutes before the flight.

Haze and Collision Avoidance

46. [Q10/1/6]
What effect does haze have on the ability to see traffic or terrain features during flight?
A. Haze causes the eyes to focus at infinity.
B. The eyes tend to overwork in haze and do not detect relative movement easily.
C. All traffic or terrain features appear to be farther away than their actual distance.

47. [Q11/1/2]
Wearing _____ lens sunglasses is often recommended for hazy, smoggy conditions.
A. yellow
B. green
C. blue

Scanning for Traffic During the Day

48. [Q11/1/3]
A military study once determined that of a 17 second cycle, approximately _____ seconds should be spent inside the cockpit with _____ seconds spent looking outside.
A. 14, 3
B. 3, 14
C. 7, 7

49. [Q11/Figure 10]
Which technique should a pilot use to scan for traffic to the right and left during straight-and-level flight?
A. Systematically focus on different segments of the sky for short intervals.
B. Concentrate on relative movement detected in the peripheral vision area.
C. Continuous sweeping of the windshield from right to left.

50. [Q11/1/3]
Prior to starting each flight maneuver, pilots should
A. check altitude, airspeed, and heading indications.
B. visually scan the entire area for collision avoidance.
C. announce their intentions on the nearest CTAF.

51. [Q11/Figure 10]
The most effective method of scanning for other aircraft for collision avoidance during daylight hours is to use
A. regularly spaced concentration on the 3, 9, and 12-o'clock positions.
B. a series of short, regularly spaced eye movements to search each 10 degree sector.
C. peripheral vision by scanning small sectors and utilizing off-center viewing.

52. [Q12/1/2] Fill in the blanks:
Empty field myopia is the condition that causes the eyes to relax and seek a comfortable focal distance ranging from ____________ to ____________ feet.

53. [Q12/1/3 & Q13/Figure 13]
How can you determine if another aircraft is on a collision course with your aircraft?
A. The other aircraft will always appear to get larger and closer at a rapid rate.
B. The nose of each aircraft is pointed at the same point in space.
C. There will be no apparent relative motion between your aircraft and the other aircraft.

54. [Q12/1/3] Fill in the blank:
Take action ____________ if you see a target with little or no apparent motion in your windscreen.

Night Scanning For Traffic

55. [Q13/Figure 15] Fill in the blanks:

What are the colors of the lights on both wings and the tail? Fill in the blanks with the appropriate colors.

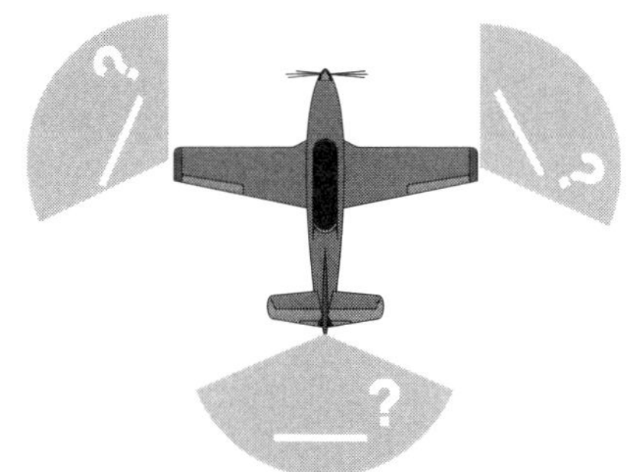

56. [Q13/1/1]
While it's easier to spot aircraft at night, that doesn't necessarily mean it's easier to identify the aircraft's _____ of movement, much less its _____ and _____.
A. direction, weight, color
B. position, size, speed
C. direction, size, shape

57. [Q13/Figure 15]
During a night flight, you observe a steady red light and a flashing red light ahead and at the same altitude. What is the general direction of movement of the other aircraft?
A. The other aircraft is crossing to the left.
B. The other aircraft is crossing to the right.
C. The other aircraft is approaching head-on.

58. [Q13/Figure 15]
During a night flight, you observe a steady white light and a flashing red light ahead and at the same altitude. What is the general direction of movement of the other aircraft?
A. The other aircraft is flying away from you.
B. The other aircraft is crossing to the left.
C. The other aircraft is crossing to the right.

59. [Q13/Figure 15]
During a night flight, you observe steady red and green lights ahead and at the same altitude. What is the general direction of movement of the other aircraft?
A. The other aircraft is crossing to the left.
B. The other aircraft is flying away from you.
C. The other aircraft is approaching head-on.

Airplane Blind Spots

60. [General Knowledge Question]
What procedure is recommended when climbing or descending VFR on an airway?
A. Execute gentle banks, left and right for continuous visual scanning of the airspace.
B. Advise the nearest FSS of the altitude changes.
C. Fly away from the centerline of the airway before changing altitude.

Filing a VFR Flight Plan

61. [Q15/1/3]
The FSS specialist will be expecting a pilot to close his or her flight plan when the _____ has expired.
A. ETE
B. ATC
C. 30 minute limit

62. [Q15/1/4]
How should a VFR flight plan be closed at the completion of the flight at a controlled airport?
A. The tower will automatically close the flight plan when the aircraft turns off the runway.
B. The pilot must close the flight plan with the nearest FSS or other FAA facility upon landing.
C. The tower will relay the instructions to the nearest FSS when the aircraft contacts the tower for landing.

63. [Q15/2/2]
Referring to the flight plan form below, if your airplane has a transponder with altitude encoding capability, what suffix should you list in block 3 (*aircraft type/special equipment*) of a flight plan?
A. /A.
B. /U.
C. /T.

Form Approved: OMB No. 2120-0026

U.S. DEPARTMENT OF TRANSPORTATION FEDERAL AVIATION ADMINISTRATION **FLIGHT PLAN**	(FAA USE ONLY) ☐ PILOT BRIEFING ☐ VNR ☐ STOPOVER	TIME STARTED	SPECIALIST INITIALS

1 TYPE ☐ VFR ☐ IFR ☐ DVFR	2 AIRCRAFT IDENTIFICATION	3 AIRCRAFT TYPE/ SPECIAL EQUIPMENT	4 TRUE AIRSPEED KTS	5 DEPARTURE POINT	6 DEPARTURE TIME PROPOSED (Z) ACTUAL (Z)	7 CRUISING ALTITUDE

8 ROUTE OF FLIGHT

9 DESTINATION (Name of airport and city)	10 EST. TIME ENROUTE HOURS MINUTES	11 REMARKS

12 FUEL ON BOARD HOURS MINUTES	13 ALTERNATE AIRPORT(S)	14 PILOT'S NAME, ADDRESS & TELEPHONE NUMBER & AIRCRAFT HOME BASE / 17 DESTINATION CONTACT/TELEPHONE (OPTIONAL)	15 NUMBER ABOARD

16 COLOR OF AIRCRAFT	CIVIL AIRCRAFT PILOTS. FAR Part 91 requires you file an IFR flight plan to operate under instrument flight rules in controlled airspace. Failure to file could result in a civil penalty not to exceed $1,000 for each violation (Section 901 of the Federal Aviation Act of 1958, as amended). Filing of a VFR flight plan is recommended as a good operating practice. See also Part 99 for requirements concerning DVFR flight plans.

FAA Form 7233-1 (8-82) CLOSE VFR FLIGHT PLAN WITH ____________ FSS ON ARRIVAL

64. [Q16/1/2]
Referring to the flight plan form above, if more than one cruising altitude is intended, which should be entered in block 7 (*cruising altitude*) of the flight plan?
A. Initial cruising altitude.
B. Highest cruising altitude.
C. Lowest cruising altitude.

65. [Q16/3/2]
Referring to the flight plan form above, what information should be entered in block 9 (*destination*) for a VFR day flight?
A. The name of the airport of first intended landing.
B. The name of destination airport if no stopover for more than 1 hour is anticipated.
C. The name of the airport where the aircraft is based.

67. [Q17/1/2]
Referring to the flight plan form above, what information should be entered in block 12 (*fuel on board*) for a VFR day flight?
A. The estimated time en route plus 30 minutes.
B. The estimated time en route plus 45 minutes.
C. The amount of usable fuel on board expressed in time.

68. [Q17/1/2]
Referring to the flight plan form to the left, what information should be entered in block 16 for a VFR day flight?
A. The predominant colors of the aircraft.
B. The color of the aircraft after it's washed.
C. Only the color of the oil on the bottom of the airplane.

Airport/Facility Directory

69. [Q17/3/2]
(Refer to the A/FD excerpt to the right and the A/FD legend in the Appendix.) When approaching Lincoln Municipal from the west at noon for the purpose of landing, initial communications should be with
A. Lincoln Approach Control on 124.0 MHz.
B. Minneapolis Center on 128.75 MHz.
C. Lincoln Tower on 118.5 MHz.

70. [Q17/3/2 & H19/3/2]
(Refer to the A/FD excerpt to the right and the A/FD legend in the Appendix.) Which type radar service is provided to VFR aircraft at Lincoln Municipal?
A. Sequencing to the primary Class C airport and standard separation.
B. Sequencing to the primary Class C airport and conflict resolution so that radar targets do not touch, or 1,000 feet vertical separation.
C. Sequencing to the primary Class C airport, traffic advisories, conflict resolution, and safety alerts.

71. [Q17/3/2 & G19/Figure 32]
(Refer to the A/FD excerpt to the right and the A/FD legend in the Appendix.) What is the recommended communications procedure for landing at Lincoln Municipal during the hours when the tower is not in operation?
A. Monitor airport traffic and announce your position and intentions on 118.5 MHz.
B. Contact UNICOM on 122.95 MHz for traffic advisories.
C. Monitor ATIS for airport conditions, then announce your position on 122.95 MHz.

72. [Q17/3/2]
(Refer to the A/FD excerpt to the right and the A/FD legend in the Appendix.) Where is Loup City Municipal located with relation to the city?
A. Northeast approximately 3 miles.
B. Northwest approximately 1 mile.
C. East approximately 10 miles.

73. [Q17/3/2]
(Refer to the A/FD excerpt to the right and the A/FD legend in the Appendix.) Traffic patterns in effect at Lincoln Municipal are
A. to the right on Runway 17L and Runway 35L; to the left on Runway 17R and Runway 35R.
B. to the left on Runway 17L and Runway 35L; to the right on Runway 17R and Runway 35R.
C. to the right on Runways 14-32.

NEBRASKA

LINCOLN MUNI (LNK) 4 NW UTC-6(-5DT) N40°51.05' W96°45.55' **OMAHA**
1218 B S4 **FUEL** 100LL, JET A TPA—2218(1000) ARFF Index B **H-1E, 3F, 4F, L-11B**
RWY 17R-35L: H12901X200 (ASPH-CONC-GRVD) S-100, D-200, DT-400 HIRL **IAP**
RWY 17R: MALSR. VASI(V4L)—GA 3.0° TCH 55'. Rgt tfc. 0.4% down.
RWY 35L: MALSR. VASI(V4L)—GA 3.0° TCH 55'.
RWY 14-32: H8620X150 (ASPH-CONC-GRVD) S-80, D-170, DT-280 MIRL
RWY 14: REIL. VASI(V4L)—GA 3.0° TCH 48'.
RWY 32: VASI(V4L)—GA 3.0° TCH 53'. Thld dsplcd 431'. Pole. 0.3% up.
RWY 17L-35R: H5400X100 (ASPH-CONC-AFSC) S-49, D-60 HIRL 0.8% up N
RWY 17L: PAPI(P4L)—GA 3.0° TCH 33'. **RWY 35R**: PAPI(P4L)—GA 3.0° TCH 40'. Pole. Rgt tfc.
AIRPORT REMARKS: Attended continuously. Birds in vicinity of arpt. Twy D clsd between taxiways S and H indef. For MALSR Rwy 17R and Rwy 35L ctc twr. When twr clsd MALSR Rwy 17R and Rwy 35L preset on med ints. and REIL Rwy 14 left on when wind favor. NOTE: See Land and Hold Short Operations Section.
WEATHER DATA SOURCES: ASOS (402) 474-9214. LLWAS
COMMUNICATIONS: **CTAF** 118.5 **ATIS** 118.05 **UNICOM** 122.95
COLUMBUS FSS (OLU) TF 1-800-WX-BRIEF. NOTAM FILE LNK.
RCO 122.65 (COLUMBUS FSS)
Ⓡ **APP/DEP CON** 124.0 (170°-349°) 124.8 (350°-169°) (1130-0630Z‡)
Ⓡ **MINNEAPOLIS CENTER APP/DEP CON** 128.75 (0630-1130Z‡)
TOWER 118.5 125.7 (1130-0630Z‡) **GND CON** 121.9 **CLNC DEL** 120.7
AIRSPACE: **CLASS C** svc 1130-0630Z‡ ctc **APP CON** other times CLASS E.
RADIO AIDS TO NAVIGATION: NOTAM FILE LNK. VHF/DF ctc FSS.
(H) VORTACW 116.1 LNK Chan 108 N40°55.43' W96°44.52' 181° 4.5 NM to fld. 1370/9E
POTTS NDB (MHW/LOM) 385 LN N40°44.83' W96°45.75' 355° 6.2 NM to fld. Unmonitored when twr clsd.
ILS 111.1 I-OCZ Rwy 17R. MM and OM unmonitored.
ILS 109.9 I-LNK Rwy 35L LOM POTTS NDB. MM unmonitored. LOM unmonitored when twr clsd.
COMM/NAVAID REMARKS: Emerg frequency 121.5 not available at tower.

LOUP CITY MUNI (NEØ3) 1 NW UTC-6(-5DT) N41°17.42' W98°59.44' **OMAHA**
2070 B **FUEL** 100LL **L-11B**
RWY 15-33: H3200X50 (ASPH) S-8 LIRL
RWY 33: Trees.
RWY 04-22: 2100X100 (TURF)
RWY 04: Tree. **RWY 22**: Road.
AIRPORT REMARKS: Unattended. For svc call 308-745-0328/1244/0664.
COMMUNICATIONS: **CTAF** 122.9
COLUMBUS FSS (OLU) TF 1-800-WX-BRIEF. NOTAM FILE OLU.
RADIO AIDS TO NAVIGATION: NOTAM FILE OLU.
WOLBACH (H) VORTAC 114.8 OBH Chan 95 N41°22.54' W98°21.22' 253° 29.3 NM to fld. 2010/7E.

MARTIN FLD (See SO SIOUX CITY)

MC COOK MUNI (MCK) 2 E UTC-6(-5DT) N40°12.36' W100°35.51' **OMAHA**
2579 B S4 **FUEL** 100LL, JET A ARFF Index Ltd. **H-2D, L-11A**
RWY 12-30: H5999X100 (CONC) S-30, D-38 MIRL 0.6% up NW **IAP**
RWY 12: MALS. VASI(V4L)—GA 3.0° TCH 33'. Tree. **RWY 30**: REIL. VASI(V4L)—GA 3.0° TCH 42'.
RWY 03-21: H3999X75 (CONC) S-30, D-38 MIRL
RWY 03: VASI(V2L)—GA 3.0° TCH 26'. Rgt tfc. **RWY 21**: VASI(V2L)—GA 3.0° TCH 26'.
RWY 17-35: 1350X200 (TURF)
AIRPORT REMARKS: Attended daylight hours. Parachute Jumping. Deer on and in vicinity of arpt. Numerous waterfowl/migratory birds invof arpt. Arpt closed to air carrier operations with more than 30 passengers except 24 hour PPR, call arpt manager 308-345-2022. Avoid McCook State (abandoned) arpt 7 miles NW on the MCK VOR/DME 313° radial at 8.3 DME. ACTIVATE VASI Rwys 12 and 30 and MALS Rwy 12—CTAF.
COMMUNICATIONS: **CTAF/UNICOM** 122.8
COLUMBUS FSS (OLU) TF 1-800-WX-BRIEF. NOTAM FILE MCK.
RCO 122.6 (COLUMBUS FSS)
DENVER CENTER APP/DEP CON 132.7
AIRSPACE: **CLASS E** svc effective 1100-0500Z‡ except holidays other times CLASS G.
RADIO AIDS TO NAVIGATION: NOTAM FILE MCK.
(H) VORW/DME 115.3 MCK Chan 100 N40°12.23' W100°35.65' at fld. 2570/8E.

---

74. [Q18/1/2]
Sectional charts are revised only once every _____ months while the A/FD is reissued every _____ weeks.
A. 6, 6
B. 8, 6
C. 6, 8

75. [Q18/1/3]
For information about parachute jumping and glider operations at Silverwood Airport, refer to
A. notes on the border of the chart.
B. the Airport/Facility Directory.
C. the Notices to Airmen (NOTAM) publication.

The Aeronautical Information Manual Notices To Airmen (NOTAMs)

76. [Q19/1/3]
NOTAM D is information that is given _____ distribution from its generating source.
A. distant
B. local
C. flight data center

77. [Q20/1/1]
NOTAMS are published once every _____ weeks in the Notices to Airmen Publication (NTAP).
A. two
B. three
C. eight

78. [Q20/1/2]
NOTAM L provides you with information on a _____ level.
A. distant
B. local
C. flight data center

79. [Q20/2/1]
A separate file of NOTAM L information is maintained at each FSS for facilities _____ only.
A. in their service area
B. outside their service area
C. within 40 miles of their service area

80. [Q20/3/2]
How might you identify the FSS having jurisdiction over your destination airport?
A. Look in the FDC NOTAMS.
B. Look in the A/FD.
C. Look in the Advisory Circulars.

81. [Q20/3/3]
FDC NOTAMS are issued when changes that are _____ in nature occur.
A. unofficial
B. advisory
C. regulatory

82. [Q21/1/2]
FDC NOTAMS are issued once and kept on file at the FSS _____.
A. until revoked
B. until published or canceled
C. until amended

83. [Q21/1/2]
FSSs are responsible for maintaining a file of current, unpublished FDC NOTAMS concerning conditions within _____ miles of their facilities.
A. 1,000
B. 200
C. 400

Advisory Circulars

84. [Q21/1/3]
FAA advisory circulars (some free, others at cost) are available to all pilots and are obtained by
A. distribution from the nearest FAA district office.
B. ordering those desired from the Government Printing Office.
C. subscribing to the Federal Register.

85. [Q21/1/4]
FAA advisory circulars containing subject matter specifically related to airmen are issued under which subject number?
A. 60
B. 70
C. 90

86. [Q21/1/4]
FAA advisory circulars containing subject matter specifically related to airspace are issued under which subject number?
A. 60
B. 70
C. 90

87. [Q21/1/4]
FAA advisory circulars containing subject matter specifically related to air traffic control and general operations are issued under which subject number?
A. 60
B. 70
C. 90

Postflight Briefing 17-1: Handpropping A New Way to Give Blood

88. [Q22/1/3]
Should it become necessary to handprop an airplane engine, it is extremely important that a competent pilot
A. call "contact" before touching the propeller.
B. be at the controls in the cockpit.
C. be in the cockpit and call out all commands.

Postflight Briefing 17-2: Aviation Decision Making: Thoughts for Life

89. [Q23/2/10]
What is one of the neglected items when a pilot relies on short and long term memory for repetitive tasks?
A. Flying outside the envelope.
B. Situational awareness.
C. Checklists.

90 [Q23/2/3]
Who is responsible for determining whether a pilot is fit to fly for a particular flight, even though he or she holds a current medical certificate?
A. The FAA.
B. The medical examiner.
C. The pilot.

91. [Q24/1/3]
What is it often called when a pilot pushes his or her capabilities and the aircraft's limits by trying to maintain visual contact with the terrain in low visibility and ceiling?
A. Peer pressure.
B. Scud running.
C. Mindset.

92. [Q24/1/4]
What often leads to spatial disorientation or collision with ground/obstacles when flying under Visual Flight Rules (VFR)?
A. Getting behind the aircraft.
B. Duck-under syndrome.
C. Continued flight into instrument conditions.

93. [See Figure on page Q25]
What is the antidote when a pilot has a hazardous attitude, such as "anti-authority"?
A. Follow the rules.
B. Rules do not apply in this situation.
C. I know what I am doing.

94. [See Figure on page Q25]
What is the antidote when a pilot has a hazardous attitude, such as "impulsivity"?
A. Do it quickly to get it over with.
B. Not so fast, think first.
C. It could happen to me.

95. [See Figure on page Q25]
What is the antidote when a pilot has a hazardous attitude, such as "invulnerability"?
A. It could happen to me.
B. It cannot be that bad.
C. It will not happen to me.

96. [See Figure on page Q25]
What is the antidote when a pilot has a hazardous attitude, such as "macho"?
A. Taking chances is foolish.
B. I can do it.
C. Nothing will happen.

97. [See Figure on page Q25]
What is the antidote when a pilot has a hazardous attitude, such as "resignation"?
A. I am not helpless.
B. Someone else is responsible.
C. What is the use.

98. [Q25/3/1]
What is the common factor in most preventable accidents?
A. Human error.
B. Mechanical difficulties.
C. Luck.

99. [Bonus Question-General Knowledge]
Prior to starting each maneuver, pilots should
A. check altitude, airspeed and heading indications.
B. visually scan the entire area for collision avoidance.
C. announce their intentions on the nearest CTAF.

100. [Bonus Question-General Knowledge]
what procedure is recommended when climbing or descending VFR on an airway?
A. Execute gentle banks, left and right for continuous visual scanning of the airspace.
B. Advise the nearest FSS of the altitude change.
C. Fly away from the centerline of the airway before changing altitude.

Chapter Seventeen Answers

1. C
2. B
3. A
4. C
5. A
6. C
7. B
8. A
9. C
10. A
11. A
12. A
13. B
14. B
15. A
16. B
17. A
18. A
19. B
20. C
21. B
22. C
23. A
24. B
25. A
26. B
27. C
28. C
29. B
30. C
31. A
32. C
33. C
34. B
35. B
36. B
37. A
38. Fill in the blanks:

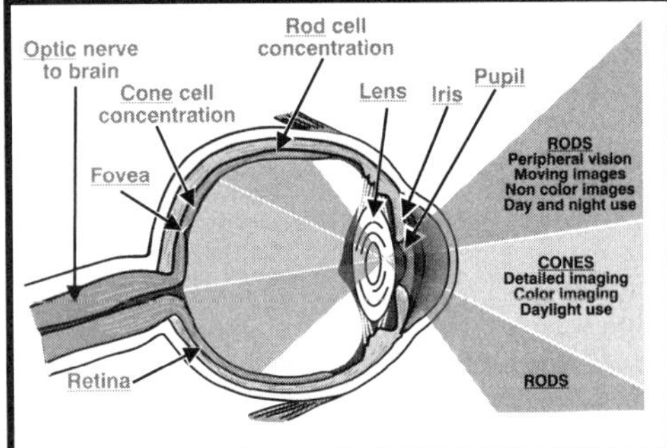

39. B
40. C
41. B
42. A
43. C
44. C
45. C
46. C
47. A
48. B
49. A
50. B
51. B
52. 10, 30
53. C
54. immediately
55. Fill in the blanks:

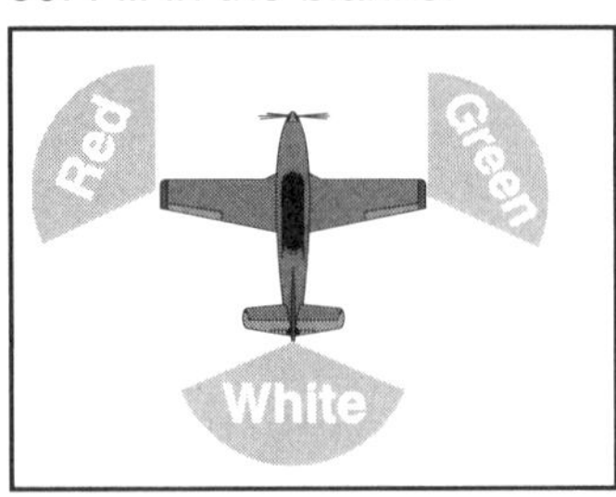

56. C
57. A
58. A
59. C
60. C
61. A
62. B
63. B
64. A
65. B
67. C
68. A
69. A
70. C
71. A
72. B
73. B
74. C
75. B
76. A
77. A
78. B
79. A
80. B
81. C
82. B
83. C
84. B
85. A
86. B
87. C
88. B
89. C
90 C.
91. B
92. C
93. A
94. B
95. A
96. A
97. A
98. A
99. B
100. A

Note: To ensure that you have the most current answers to these questions, please check the *Book & Slide Updates* section at Rod Machado's web site: www.rodmachado.com

Rod Machado's Private Pilot, FAA Approved, Part 141 Ground Training Syllabus

Presented by: Rod Machado/Certified Flight Instructor

Business address: P. O. Box 6030, San Clemente, CA. 92674

E-mail: webmail@rodmachado.com

Rod Machado's Private Pilot Ground Training Syllabus

Presented by: Rod Machado/Certified Flight Instructor
P. O. Box 6030, San Clemente, CA. 92674
E-mail: webmail@rodmachado.com

Ground School Format

The ground school presented here is a nine week class. Classes meet two times a week for three hours of instruction at each meeting (an hour consists of 50 minutes of instruction and a 10 minute break). The class provides a total of 51.5 hours of presented material including 2.5 hours for issuing two stage exams and a final exam.

Ground School Presentation Sequence

The ground school presented here consists of three stages. **Stage One** covers Chapters 1-6 of *Rod Machado's Private Pilot Handbook*. **Stage Two** covers Chapters 7-12 of *Rod Machado's Private Pilot Handbook*. **Stage Three** covers Chapters 13-17 of *Rod Machado's Private Pilot Handbook*. At the end of Stage One and Stage Two, an exam is given. A final (Stage Three) exam is scheduled at the end of the nine week class period. The Stage Three exam is a comprehensive exam given on the last official day of class.

The ground school is designed to allow a student to join the class at any time during the nine week class cycle. Each class is an individualized module of learning that is independent of the information presented in a prior ground lesson.

Joining the Class Mid-Session

A student may join the class at the beginning of any scheduled class session. The student will not be eligible to take the Stage One or Two exam or the final (Stage Three) exam until that student completes all of the lessons covered by those stages.

Example 1: A student begins class during Ground Lesson 4. The student will forgo the Stage One exam. The student is eligible for the Stage Two exam after completing Ground Lessons 7-12. When the student completes Ground Lessons 13-18, that student is still not eligible for the final (Stage Three) exam. However, once the student completes Ground Lessons 1-3, that student is now eligible for the Stage One exam and the final (Stage Three) exam. Once eligible, these exams may be given to the student by any authorized instructor.

Example 2: A student begins class during Ground Lesson 8. The student will forgo the Stage Two exam and the final (Stage Three) exam as the class progresses. The student will then be eligible for the Stage Two exam and the final (Stage Three) exam only after completing Ground Lessons 1-7 (during which time that student will have completed the Stage One exam at the end of Ground Lesson 6). Once eligible, the Stage Two exam and the final (Stage Three) exam can be given to that student by an authorized instructor.

Directed Self-Study

In the event a student misses not more than four ground lessons (i.e., four, three-hour class sessions), that student may complete the ground lessons missed by directed self-study, which will be conducted under the supervision of an authorized instructor. Once the student completes the ground lessons missed, that student is eligible for the appropriate stage or final exams.

Rod Machado's Private Pilot Ground Training Syllabus

Ground Training Objectives

This syllabus prescribes the course of training necessary for a student to obtain the required knowledge for the Private Pilot Knowledge Exam specified in CFAR Part 61/Part 141.

Ground Training Completion Standards

Students will demonstrate by written exam that the knowledge required for the Private Pilot Knowledge Exam as prescribed in CFAR Part 61/Part 141 has been met.

Stage One

Stage One Objectives

In this stage, the student is introduced to the airplane's major components and learns the basics of aerodynamics, light airplane engine operations, the airplane's electrical system, the airplane's flight instruments and the Federal Aviation Regulations, as these subjects pertain to typical light airplane operations by a private pilot.

Stage One Completion Standards

Completion of this stage will occur when the student takes the Stage One exam, completing it with a minimum passing score of 80%. The instructor will orally review each incorrect response, thus ensuring an adequate understanding of the material before proceeding to the next stage.

Stage One

Ground Lesson 1

Text Reference:

Rod Machado's Private Pilot Handbook

Rod Machado's Private Pilot Workbook

Presentation Format:

Any combination of visual or oral means may be used to present the required information.

Recommended Presentation Sequence:

Chapter One - Pages A1-8, *Airplane Components (1 hour)*
Chapter Two - Pages B1-25, *Aerodynamics (2 hours)*

Lesson Objective:

During this lesson, the student will become familiar with and develop a basic understanding of an airplane's major components and the terms used in the operation of an airplane. The student will also become familiar with and develop an understanding of the basic aerodynamic principles of flight from the four forces to ground effect.

Lesson Content:

Chapter 1:

_____ Airplane Components
_____ The Wing
_____ Stall Equipment
_____ Moving Parts
_____ The Empennage
_____ Antennas
_____ The Engine

Chapter 2:

_____ The Four Forces
_____ Climbs
_____ Descents
_____ Defining the Wing
_____ How the Wing Works
_____ Relative Wind
_____ Attacking the Air
_____ How Lift Develops
_____ Impact vs. Pressure Lift
_____ Angle of Attack
_____ Stalls & Angle of Attack
_____ Stall at Any Attitude or Airspeed
_____ Five Stall Warning Signs
_____ Stalling Speed, Gee Whiz and G-Force
_____ What a Drag
_____ Horizontal and Vertical Movement of Air
_____ Total Drag and Your Go Far Speed
_____ Stretching the Glide, Saving the Hide
_____ Ground Effect
_____ Where to Use Caution in Ground Effect
_____ Pitch Changes In and Out of Ground Effect

Completion Standards:

By the following class period the student will complete the sections in *Rod Machado's Private Pilot Workbook* that pertain to the material covered in this ground lesson.

Study Assignments (Assign the homework assignment below during the first class session):

Rod Machado's Private Pilot Handbook

Read: Chapter One - Pages A1-8, *Airplane Components*
Read: Chapter Two - Pages B1-48, *Aerodynamics*
Read: Chapter Three - Pages C1-9, *Engines*

Ground Lesson 2

Text Reference
Rod Machado's Private Pilot Handbook
Rod Machado's Private Pilot Workbook

Presentation Format
Any combination of visual or oral means may be used to present the required information.

Recommended Presentation Sequence
Chapter Two - Pages B26-48, *Aerodynamics (2 hours)*
Chapter Three - Pages C1-9, *Engines (1 hour)*

Lesson Objective
During this lesson, the student will become familiar with and develop a basic understanding of advanced aerodynamic concepts from flap operations to maneuvering speed. Additionally, the student will become familiar with and develop a basic understanding of general aviation engine operations from engine design to carburetor operation.

Lesson Content
Chapter 2:
_____ Flap Over Flaps
_____ Flap Varieties
_____ Why Use Flaps?
_____ How Airplanes Turn
_____ Flight Controls
_____ Ailerons
_____ Adverse Yaw
_____ Rudders
_____ Elevator
_____ Trim Tabs
_____ Left Turning Tendencies
_____ How a Spin Occurs
_____ Parasite Drag
_____ Induced Drag
_____ Maximum Range
_____ Maximum Endurance
_____ The Best Glide Speed and Weight Changes
_____ A Different Look at Maneuvering Speed
_____ Weight Change and V_a

Chapter 3:
_____ The Airplane Engine
_____ Four Cycle Engine
_____ The Ignition System
_____ Dual Ignition Systems
_____ Meet Mister Magneto
_____ Impulse Coupling
_____ Selecting Magnetos
_____ The P-Lead
_____ The Exhaust System
_____ The Induction System
_____ The Carburetor

Completion Standards:
By the following class period the student will complete the sections in *Rod Machado's Private Pilot Workbook* that pertain to the material covered in this ground lesson.

Study Assignments:
Rod Machado's Private Pilot Handbook:
Read: Chapter Three - Pages C9-38, *Engines*

Ground Lesson 3

Text Reference:
Rod Machado's Private Pilot Handbook
Rod Machado's Private Pilot Workbook

Presentation Format:
Any combination of visual or oral means may be used to present the required information.

Recommended Presentation Sequence:
Chapter Three - Pages C9-38, *Engines (3 hours)*

Lesson Objective:
During this lesson, the student will become familiar with and develop a basic understanding of the advanced operations of the modern general aviation airplane engine from the carburetor's idling system to fuel injection operations.

Lesson Content:
Chapter 3:
_____ The Idling System
_____ The Accelerator Pump
_____ Atomization of Fuel
_____ Your Carburetor, the Ice Maker
_____ Ice: Just Your Type
_____ The Carburetor Heater
_____ Carb Ice Symptoms
_____ Apply Carb Heat as a Precautionary Measure
_____ Carburetor Icing Potential in Different Engines
_____ The Mixture Control
_____ The Fuel/Air Mixture
_____ When to Lean
_____ How to Lean
_____ Too Rich and Too Lean
_____ Leaning & High Alt Takeoffs for Nonturbocharged Airplanes
_____ EGT Gauge Setting for Best Power or Best Economy
_____ The Fuel System
_____ Components
_____ Fuel Colors
_____ Fuel Vents
_____ Auxiliary Fuel Pumps
_____ Prime Time
_____ Fuel Gauges
_____ How Much Is Enough?
_____ The Oil System
_____ Change of Life
_____ Malfunctions in the Oil System
_____ The Engine Cooling System
_____ The Propeller
_____ Why Constant Speed Propellers?
_____ How to Make Power Changes
_____ Propeller Tips and Ideas
_____ Detonation and Preignition
_____ Fuel Injection Systems
_____ Turbocharging (optional)
_____ Pressurization (optional)

Completion Standards:
By the following class period the student will complete the sections in *Rod Machado's Private Pilot Workbook* that pertain to the material covered in this ground lesson.

Study Assignments:
Rod Machado's Private Pilot Handbook:
Read: Chapter Four - Pages D1-16, *Electrical System*
Read: Chapter Five - Pages E1-18, *Flight Instruments*

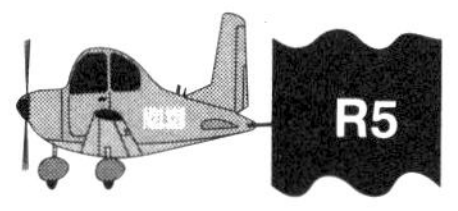

Ground Lesson 4

Text Reference:
Rod Machado's Private Pilot Handbook
Rod Machado's Private Pilot Workbook

Presentation Format:
Any combination of visual or oral means may be used to present the required information.

Recommended Presentation Sequence:
Chapter Four - Pages D1-16, *Electrical Systems (1 hour)*
Chapter Five - Pages E1-18, *Flight Instruments (2 hours)*

Lesson Objective:
During this lesson, the student will become familiar with and develop a basic understanding of the typical general aviation airplane's electrical system. Additionally, the student will become familiar with and develop a basic understanding of the workings of the non-gyro instruments from the airspeed indicator to the altimeter.

Lesson Content:
Chapter 4:
_____ Electricity and Water
_____ The Water Pump
_____ The Electrical Ground
_____ Load Meter
_____ The Battery
_____ Battery Potential
_____ The Charge-Discharge Ammeter
_____ Load Meters
_____ Electrical Drain
_____ The Voltage Regulator
_____ Problems With Brains
_____ Making Connections
_____ Drawing It All Together
_____ How the Battery Contactor Works

Chapter 5:
_____ Non-Gyro Instruments
_____ Airspeed Indicator
_____ Static Pressure
_____ Pitot Tubes
_____ The Airspeed Indicator's Face
_____ Indicated Airspeeds
_____ Calibrated Airspeed
_____ True Airspeed
_____ Dense Doings
_____ The Altimeter
_____ Pressure Variations and the Altimeter
_____ Temperature Variations and the Altimeter
_____ Sensitive Altimeters
_____ Pressure Altitude

Completion Standards:
By the following class period the student will complete the sections in *Rod Machado's Private Pilot Workbook* that pertain to the material covered in this ground lesson.

Study Assignments:
Rod Machado's Private Pilot Handbook:
Read: Chapter Five - Pages E18-36, *Flight Instruments*

Ground Lesson 5:

Text Reference:
Rod Machado's Private Pilot Handbook
Rod Machado's Private Pilot Workbook

Presentation Format:
Any combination of visual or oral means may be used to present the required information.

Recommended Presentation Sequence:
Chapter Five - Pages E18-36, *Flight Instruments (2 hours)*
Chapter Six - Pages F1-19, *FARs (1 hour)*

Lesson Objective:
During this lesson, the student will become familiar with and develop a basic understanding of the airplane's gyroscopic instruments and how they work. Additionally, the student will become familiar with and develop a basic understanding of the following sections of the Code of Federal Aviation Regulations: *Definitions* and *Part 61.*

Lesson Content:
Chapter 5:
_____ Reading the Altimeter
_____ The Vertical Speed Indicator (VSI)
_____ Alternate Static Source
_____ The Gyroscopic Instruments
_____ The Attitude Indicator
_____ The Heading Indicator
_____ The Turn Coordinator
_____ The Magnetic Compass Acceleration and Deceleration Error
_____ Northerly Turning Errors
_____ Gyroscopic Precession (optional)

Chapter 6:
_____ Definitions
_____ Aircraft
_____ Category
_____ Class
_____ Type Ratings
_____ Visual Flight Rules (VFR)
_____ Instrument Flight Rules (IFR)
_____ Night
_____ Pilot In Command (PIC)
_____ FAR 61.3 Requirements for Certificates,Ratings & Authorizations
_____ FAR 61.15 Offenses Involving Alcohol or Drugs
_____ FAR 61.23 Duration of Medical Certificates
_____ FAR 61.31 General Limitations: High performance/complex AC
_____ FAR 61.31 High Altitude Airplanes
_____ FAR 61.31 Tailwheel Airplanes
_____ FAR 61.56 Flight Reviews
_____ FAR 61.57 Recent Flight Experience - Pilot In Command
_____ FAR 61.57 Recent Experience at Night
_____ FAR 61.60 Change of Address
_____ FAR 61.81 Student and Recreational Pilots (optional)
_____ FAR 61.89 General Limitations
_____ FAR 61.93 Cross Country Flight Requirements (optional)
_____ FAR 61.103 Private Pilot Requirements (optional)
_____ FAR 61.107 Flight Experience (optional)
_____ FAR 61.118 Private Pilot Privileges and Limitations: as PIC

Completion Standards:
By the following class period the student will complete the sections in *Rod Machado's Private Pilot Workbook* that pertain to the material covered in this ground lesson.

Study Assignments:
Rod Machado's Private Pilot Handbook:
Read: Chapter Six - Pages F20-52, *FARs*

Ground Lesson 6:

Text Reference:
Rod Machado's Private Pilot Handbook
Rod Machado's Private Pilot Workbook

Presentation Format:
Any combination of visual or oral means may be used to present the required information.

Lesson Objective:
During this lesson, the student will become familiar with and develop a basic understanding of the relevant sections of Part 91 of the Code of Federal Aviation Regulations and the relevant sections of NTSB 830 rules and regulations. The Stage One written exam will be administered during class.

Recommended Presentation Sequence:
Chapter Six - Pages F20-52, *FARs (2:45 hours)*
Issue Stage One Exam (:15)

Lesson Content:
Chapter 6:

_____ FAR 91.3 Responsibility and Authority of the Pilot In Command
_____ FAR 91.7 Civil Aircraft Airworthiness
_____ FAR 91.9 Civil Aircraft Flight Manual
_____ FAR 91.15 Dropping Objects
_____ FAR 91.17 Alcohol or Drugs
_____ FAR 91.103 Preflight Action
_____ FAR 91.105 Flight Crewmembers at Stations
_____ FAR 91.107 Use of Safety Belts
_____ FAR 91.111 Operating Near Other Aircraft
_____ FAR 91.113 Right of Way Rules: Except Water
_____ FAR 91.113 Right of Way Rules: Water Operations
_____ FAR 91.117 Aircraft Speed
_____ FAR 91.119 Minimum Safe Altitudes
_____ FAR 91.121 Altimeter Settings
_____ FAR 91.123 Compliance with ATC Clearances and Instructions
_____ FAR 91.125 ATC Light Signals
_____ FAR 91.126 Operating on or in the Vicinity of an Airport in Class G Airspace (familiarize only)*
_____ FAR 91.127 Operations on or in the Vicinity of an Airport in Class E Airspace (familiarize only)*
_____ FAR 91.129 Operations in Class D Airspace (familiarize only)*
_____ FAR 91.130 Operations in Class C Airspace (familiarize only)*
_____ FAR 91.131 Operations in Class B Airspace (familiarize only)*
_____ FAR 91.133 Restricted and Prohibited Areas
_____ FAR 91.135 Operations in Class A Airspace
_____ FAR 91.151 Fuel Requirements for Flight in VFR Conditions
_____ FAR 91.155 Basic VFR Weather Minimums
_____ FAR 91.157 Special VFR Weather Minimums
_____ FAR 91.159 VFR Cruising Altitude or Flight Level
_____ FAR 91.203 Civil Aircraft: Certifications Required
_____ FAR 91.207 Emergency Locator Transmitters
_____ FAR 91.209 Aircraft Lights
_____ FAR 91.211 Use of Supplemental Oxygen
_____ FAR 91.215 ATC Transponder and Altitude Reporting Equipment and Use
_____ FAR 91.303 Aerobatic Flight
_____ FAR 91.307 Parachutes and Parachuting
_____ FAR 91.313 Restricted Cat. Civil Aircraft:Operating Limitations
_____ FAR 91.319 Aircraft Having Experimental Certificates:
_____ FAR 91.403 Aircraft Maintenance: General
_____ FAR 91.407 Operations After Maintenance, Preventive Maintenance, Rebuilding or Alteration
_____ FAR 91.409 Inspections
_____ FAR 91.413 ATC Transponder Tests and Inspections
_____ FAR 91. 417 Maintenance Records
_____ National Transportation Safety Board 830
_____ NTSB 830.2 Definitions
_____ NTSB 830.5 Immediate Notification
_____ NTSB 830.10 Preservation of Aircraft Wreckage, Mail, Cargo and Records
_____ NTSB 830.15 Reports and Statements to Be Filed

Completion Standards:
By the following class period the student will complete the sections in *Rod Machado's Private Pilot Workbook* that pertain to the material covered in this ground lesson.

Study Assignments:
Rod Machado's Private Pilot Handbook:
Read: Chapter Seven - Pages G1-29, *Airport Operations*
Read: Chapter Eight - Pages H1-11, *Radio Operations*

***Note**: The instructor should familiarize the student with the parts of these regulations that don't pertain to the construction of the associated airspace. Airspace construction will be thoroughly covered in Ground Lessons 8 & 9.

Ground Training Objectives

This syllabus prescribes the course of training necessary for a student to obtain the required knowledge for the Private Pilot Knowledge Exam specified in CFAR Part 61/Part 141.

Ground Training Completion Standards

Students will demonstrate by written exam that the knowledge required for the Private Pilot Knowledge Exam as prescribed in CFAR Part 61/Part 141 has been met.

Stage Two

Stage Two Objectives

In this stage, the student will be introduced to the basic fundamentals of airport operations at controlled and uncontrolled airports, airspace, chart symbology, radio navigation and meteorology, as these subjects pertain to typical light airplane operations by a private pilot.

Stage Two Completion Standards

Completion of this stage occurs when the student takes the Stage Two exam, completing it with a minimum passing score of 80%. The instructor will orally review each incorrect response, thus ensuring an adequate understanding of the material before proceeding to the next stage.

Ground Lesson 7:

Text Reference:
Rod Machado's Private Pilot Handbook
Rod Machado's Private Pilot Workbook

Presentation Format:
Any combination of visual or oral means may be used to present the required information.

Recommended Presentation Sequence:
Chapter Seven - Pages G1-29, *Airport Operations (2 hours)*
Chapter Eight - Pages H1-11, *Radio Operations (1 hour)*

Lesson Objective:
During this lesson, the student will become familiar with and develop a basic understanding of airport operations including signage, traffic patterns, wind designators and operations at towered and nontowered airports as well as radio telephony, HIWAS and TWEB.

Lesson Content:
Chapter 7:
_____ Runway Lighting
_____ Taxiway Markings
_____ Additional Runway Markings
_____ Airport Beacons
_____ The Traffic Pattern
_____ Traffic Pattern Components
_____ Crabbing in the Pattern
_____ Entering the Traffic Pattern
_____ The Segmented Circle
_____ Wind and Landing-Direction Indicators
_____ The 45 Degree Entry Point
_____ CTAF (Common Traffic Advisory Frequency)
_____ Using Unicom and Multicom for Information
_____ Finding Out What's Common
_____ Automatic Terminal Information Service (ATIS)
_____ Pilot Control of Airport Lighting
_____ Visual Approach Slope Indicator (VASI)
_____ Precision Approach Path Indicator (PAPI)
_____ Tricolor VASI
_____ Pulsating VASI Systems
_____ Wake Turbulence
_____ ATC Wake Turbulence Separation Requirements
_____ Taxiing in Crosswind Conditions
_____ LAHSO (Land and Hold Short Operations)

Chapter 8:
_____ Radio Technique
_____ Radio Equipment
_____ Talking the Talk
_____ Controlled Airports
_____ Control Tower Communications
_____ HIWAS and TWEB

Completion Standards:
By the following class period the student will complete the sections in *Rod Machado's Private Pilot Workbook* that pertain to the material covered in this ground lesson.

Study Assignments:
Rod Machado's Private Pilot Handbook:
Read: Chapter Eight - Pages H11-22, *Radio Operations*
Read: Chapter Nine - Pages I1-13, *Airspace*

Ground Lesson 8:

Text Reference:
Rod Machado's Private Pilot Handbook
Rod Machado's Private Pilot Workbook

Presentation Format:
Any combination of visual or oral means may be used to present the required information.

Recommended Presentation Sequence:
Chapter Eight - Pages H11-22, *Radio Operations (1 hour)*
Chapter Nine - Pages I1-13, *Airspace (2 hours)*

Lesson Objective:
During this lesson, the student will become familiar with and develop a basic understanding of radio operations, frequencies and radar services. Additionally, the student will become familiar with and develop a basic understanding of Class A and E airspace, controlled and uncontrolled airspace, surface-based controlled airspace and special VFR operations.

Lesson Content:
Chapter 8:
_____ The Emergency Frequency
_____ The Airport/Facility Directory
_____ Radar and the ATC System
_____ Transponders
_____ Airborne Cowboy: Riding a DF Steer
_____ Radar Services for Pilots
_____ Radar Assistance to VFR Aircraft
_____ Basic Radar Service
_____ Terminal Radar Service Area (TRSA) Service
_____ Class C Service
_____ Class B Service
_____ Clearance Delivery
_____ How ATC Keeps an Eye on You

Chapter 9:
_____ Controlled and Uncontrolled Airspace
_____ The Big Picture
_____ Class A Airspace Class E Airspace
_____ Class E at and Above 10,000 Feet MSL
_____ Class E Below 10,000 Feet MSL
_____ Class E Airspace Starting at 700 Feet AGL
_____ Additional Requirements in Surface-Based Controlled Airspace
_____ Special VFR Clearance
_____ Obtaining a SVFR Clearance
_____ Satellite Airports Lying Within the Primary Airport's Surface-Based Controlled Airspace

Completion Standards:
By the following class period the student will complete the sections in *Rod Machado's Private Pilot Workbook* that pertain to the material covered in this ground lesson.

Study Assignments:
Rod Machado's Private Pilot Handbook:
Read: Chapter Nine - Pages I14-36, *Airspace*

Ground Lesson 9:

Text Reference:
Rod Machado's Private Pilot Handbook
Rod Machado's Private Pilot Workbook

Presentation Format:
Any combination of visual or oral means may be used to present the required information.

Recommended Presentation Sequence:
Chapter Nine - Pages I14-36, *Airspace (3 hours)*

Lesson Objective:
During this lesson, the student will become familiar with and develop a basic understanding of Class G, D, B, and C airspace and the equipment and requirements to operate within this airspace. The student will also become familiar with and develop a basic understanding of special use airspace.

Lesson Content:
Chapter 9
_____ Class G Airspace
_____ Night Operations in Class G Airspace at 1,200 Feet AGL and Below
_____ Basic VFR Minimums in Class G Airspace
_____ Operations in Class G Airspace Above 1,200 Feet AGL
_____ Basic VFR Minimums in Class G Airspace
_____ General Conclusions About Class A, E and G Airspace
_____ Class B, C and D Airspace
_____ Class D Airspace
_____ Weather Minimums for Class D Airspace
_____ Satellite Airports Within Class D Airspace
_____ Class C Airspace
_____ Equipment Requirements to Operate Within Class C Airspace
_____ Class C Service
_____ Satellite Airports Within Class C Airspace
_____ Variations in Class C Airspace
_____ Weather Minimums for Class C Airspace
_____ Class B Airspace
_____ Requirements to Enter Class B Airspace
_____ Special VFR Within Class B Airspace
_____ Corridors and Circumnavigating Class B Airspace
_____ Transponder and Mode C Within 30 NM of Certain Airports
_____ Transponders and Mode C Above 10,000 Feet MSL
_____ Transponders in Controlled Airspace
_____ Transponder and Mode C Deviations
_____ Speed Restriction in Class C and D Airspace
_____ Terminal Radar Service Area
_____ Special Use Airspace
_____ Warning Areas
_____ Military Operations Areas
_____ Military Training Routes
_____ Variable Floors of Class E Airspace

Completion Standards:
By the following class period the student will complete the sections in *Rod Machado's Private Pilot Workbook* that pertain to the material covered in this ground lesson.

Study Assignments:
Rod Machado's Private Pilot Handbook:
Read: Chapter Ten - Pages J1-14, *Aviation Maps*
Read: Chapter Eleven - Pages K1-9, *Radio Navigation*

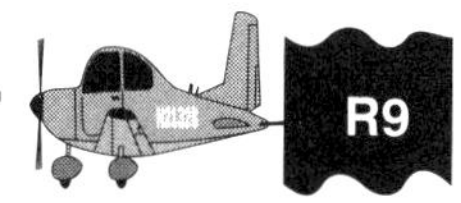

Ground Lesson 10:

Text Reference:
Rod Machado's Private Pilot Handbook
Rod Machado's Private Pilot Workbook

Presentation Format:
Any combination of visual or oral means can be used to present the required information.

Recommended Presentation Sequence:
Chapter Ten - Pages J1-14, *Aviation Maps (2 hours)*
Chapter Eleven - Pages K1-9, *Radio Navigation (1 hour)*

Lesson Objective:
During this lesson, the student will become familiar with and develop a basic understanding of aviation chart symbology, pilotage and VOR navigation.

Lesson Content:
Chapter 10:
_____ The Aeronautical Sectional Chart
_____ World Aeronautical Charts (WAC)
_____ VFR Terminal Area Charts
_____ Topographical Information on a Sectional Chart
_____ Relief (the sloping of terrain)
_____ Color
_____ Spot Elevation Symbols
_____ Critical Elevations
_____ Maximum Elevation Figures (MEF)
_____ Obstacles
_____ Roads
_____ Railroad Tracks
_____ Wires
_____ Shorelines, Rivers & Streams
_____ Populated Areas
_____ Airport
_____ Airways
_____ VFR Reporting Points
_____ Airborne Vehicle Symbols
_____ Park, Wildlife, Forest, Wilderness and Primitive Areas
_____ Miscellaneous

Chapter 11:
_____ Pilotage
_____ Electronic Elucidation
_____ The Big Picture
_____ VOR Stations Shown on a Sectional Chart
_____ Your VOR Equipment
_____ VORs and Airborne Freeways
_____ How to Navigate with VOR

Completion Standards:
By the following class period the student will complete the sections in *Rod Machado's Private Pilot Workbook* that pertain to the material covered in this ground lesson.

Study Assignments:
Rod Machado's Private Pilot Handbook:
Read: Chapter Eleven - Pages K10-38, *Radio Navigation*

Ground Lesson 11:

Text Reference:
Rod Machado's Private Pilot Handbook
Rod Machado's Private Pilot Workbook

Presentation Format:
Any combination of visual or oral means can be used to present the required information.

Recommended Presentation Sequence:
Chapter Eleven - Pages K10-38, *Radio Navigation (3 hours)*

Lesson Objective:
During this lesson, the student will become familiar with and develop a basic understanding of VOR course intercept, VOR tracking, DME, GPS theory, ADF operation, ADF tracking, bearing location and identification.

Lesson Content:
Chapter 11:
_____ Intercepting a VOR Course
_____ Flying from the VOR on a Selected Course
_____ Dual VORs for Position Fixing
_____ Reverse Sensing
_____ Tracking a Selected VOR Course
_____ Chasing the Needle
_____ A Nifty Technique
_____ Proper Names
_____ Distance Measuring Equipment (DME)
_____ What DME Really Tells You
_____ Position Fixing With DME
_____ Area Navigation – RNAV
_____ RNAV Based on VOR/DME
_____ Loran
_____ The Global Positioning System – GPS
_____ Automatic Direction Finding (ADF) Navigation
_____ The Radio Magnetic Indicator (RMI)
_____ ADF: Bearing Down on Homing In
_____ Tracking a Magnetic Bearing (optional)
_____ The ADF's Moveable Compass Card
_____ The ADF Fixed Compass Card
_____ Advanced ADF Navigation (optional)
_____ Another Way of Determining Your Magnetic Bearing to or from an NDB
_____ Correcting for Wind (optional)
_____ VOR Test Signal

Completion Standards:
By the following class period the student will complete the sections in *Rod Machado's Private Pilot Workbook* that pertain to the material covered in this ground lesson.

Study Assignments:
Rod Machado's Private Pilot Handbook:
Read: Chapter Twelve - Pages L1-24, *Weather Theory*

Ground Lesson 12:

Text Reference:
Rod Machado's Private Pilot Handbook
Rod Machado's Private Pilot Workbook

Presentation Format:
Any combination of visual or oral means may be used to present the required information.

Recommended Presentation Sequence:
Chapter Twelve - Pages L1-24, *Weather Theory (2:45 hours)*
Issue Stage Two Exam (:15)

Lesson Objective:
During this lesson, the student will become familiar with and develop a basic understanding of weather theory from atmospheric circulation to atmospheric stability. The Stage Two written exam will be administered during class.

Lesson Content:
Chapter 12:
_____ Introduction
_____ Atmospheric Circulation
_____ The Coriolis Force
_____ Air Pressure and Vertical Air Movement
_____ Getting Water in the Air
_____ The Water Content of Warm and Cold Air
_____ Two Ways to Cool Air
_____ Relative Humidity
_____ The Dew Point
_____ Condensation and Cloud Formation
_____ Lapse Rates and Temperature Inversions
_____ Temperature Inversions
_____ Effects of Temperature Inversions
_____ What to Expect in an Inversion
_____ Atmospheric Stability: Warm Over Cold, and Cold Over Warm
_____ The Environmental Lapse Rate
_____ Rising Parcels of Air
_____ Saturated Parcels of Rising Air, Clouds & Atmospheric Stability

Completion Standards:
By the following class period the student will complete the sections in *Rod Machado's Private Pilot Workbook* that pertain to the material covered in this ground lesson.

Study Assignments:
Rod Machado's Private Pilot Handbook:
Read: Chapter Twelve - Pages L25-58, *Weather Theory*

Ground Training Objectives

This syllabus prescribes the course of training necessary for a student to obtain the required knowledge for the Private Pilot Knowledge Exam specified in CFAR Part 61/Part 141.

Ground Training Completion Standards

Students will demonstrate by written exam that the knowledge required for the Private Pilot Knowledge Exam as prescribed in CFAR Part 61/Part 141 has been met.

Stage Three

Stage Three Objectives

In this stage, the student is introduced to the weather reporting and briefing system, weather reports, weather charts and their interpretation and flight planning, including time, distance and fuel computation, as these subjects pertain to typical light airplane operations by a private pilot. Additionally, the student learns how to determine airplane performance, compute a weight and balance, learns the physiological and psychological hazards associated with flight and acquires an understanding of flight plans and the NOTAM system. The final exam is issued at the end of Stage Three.

Stage Three Completion Standards

Completion of this stage will occur when the student takes the Stage Three final exam, completing it with a minimum passing score of 80%. The instructor will orally review each incorrect response thus ensuring an adequate understanding of the material before proceeding to the next stage.

Ground Lesson 13:

Text Reference:
Rod Machado's Private Pilot Handbook
Rod Machado's Private Pilot Workbook

Presentation Format:
Any combination of visual or oral means may be used to present the required information.

Recommended Presentation Sequence:
Chapter 12 - Pages L25-58, *Weather Theory (3 hours)*

Lesson Objective:
During this lesson, the student will become familiar with and develop a basic understanding of weather theory including pressure patterns, frontal formation and movement, fog, thunderstorms, mountains waves and wave cyclones.

Lesson Content:
Chapter 12:
_____ High and Low Pressure Areas
_____ Sea and Land Breeze Circulation
_____ Highs and Lows on Weather Maps
_____ Circulation in Highs and Lows: Going With the Flow
_____ The Answer is Flowin' in the Wind
_____ Weather Associated With Highs and Lows
_____ Ridges and Troughs
_____ Frontal Systems
_____ The Polar Front
_____ Different Types of Fronts
_____ Discontinuities Across a Front
_____ Cold Front Characteristics
_____ Two Types of Cold Fronts
_____ Warm Fronts
_____ Stationary Fronts
_____ The Jet Stream
_____ Thunderstorms
_____ The (Not So Secret) Life of a Thunderstorm
_____ Thunderstorm Types
_____ Squall Lines
_____ Thunderstorm Turbulence
_____ Virga
_____ Thunderstorm Avoidance
_____ Lightning
_____ Turbulence and Wind Shear
_____ Mountain Waves
_____ Temperature Inversions and Wind Shear
_____ Fog
_____ Radiation Fog
_____ Advection Fog
_____ Upslope Fog
_____ Precipitation-Induced Fog
_____ Ice Fog
_____ Steam Fog
_____ Weathering the Weather
_____ Advanced Weather Concepts
_____ Wave Cyclones (Frontal Waves)
_____ Wave Cyclone Weather Patterns
_____ Cold Occlusion
_____ Warm Occlusions
_____ How the Jet Stream Forms (optional)

Completion Standards:
By the following class period the student will complete the sections in *Rod Machado's Private Pilot Workbook* that pertain to the material covered in this ground lesson.

Study Assignments:
Rod Machado's Private Pilot Handbook:
Read: Chapter Thirteen - Pages M1-38, *Wx Charts/Briefings*

Ground Lesson 14:

Text Reference:
Rod Machado's Private Pilot Handbook
Rod Machado's Private Pilot Workbook

Presentation Format:
Any combination of visual or oral means may be used to present the required information.

Recommended Presentation Sequence:
Chapter Thirteen - Pages M1-38, *Wx Charts/Briefings (3 hours)*

Lesson Objective:
During this lesson, the student will become familiar with and develop a basic understanding of weather reporting services including telephone weather briefings as well as textual and graphic weather reports.

Lesson Content:
Chapter 13:
_____ Aviation Weather Services
_____ The Telephone Briefing
_____ Other Sources of Weather Information
_____ Newspapers
_____ PATWAS
_____ Telephone Information Briefing Service (TIBS)
_____ Transcribed Weather Broadcast (TWEB)
_____ Hazardous In-flight Weather Advisory Service (HIWAS)
_____ Enroute Flight Advisory Service (EFAS)
_____ Pilot Reports [An Introduction]
_____ METAR Weather Reports
_____ Automatic Weather Observing Programs
_____ ASOS
_____ AWOS
_____ Whither the Weather?
_____ Terminal Aerodrome Weather Forecasts (TAF)
_____ Area Forecasts (FA)
_____ Winds Aloft Forecasts (FD)
_____ Weather Charts: Getting the Big Picture
_____ Weather Depiction Chart
_____ Radar Summary Chart
_____ Low Level Significant Weather
_____ Prognostic Chart
_____ Surface Analysis
_____ In-flight Aviation Weather Advisories
_____ SIGMET (WS)
_____ AIRMET (WA)
_____ Convective SIGMETs (WST)
_____ Pilot Reports (PIREPS)
_____ Putting It All Together
_____ Surface Map and Weather Depiction Chart
_____ Weather Depiction Chart and Radar Summary Chart
_____ Area Forecast and Prog Charts
_____ Weather Depiction and Prog Charts
_____ METARs/Surface Map

Completion Standards:
By the following class period the student will complete the sections in *Rod Machado's Private Pilot Workbook* that pertain to the material covered in this ground lesson.

Study Assignments:
Rod Machado's Private Pilot Handbook:
Read: Chapter Fourteen - Pages N1-46 , *Flight Planning*

Ground Lesson 15:

Text Reference:
Rod Machado's Private Pilot Handbook
Rod Machado's Private Pilot Workbook

Presentation Format:
Any combination of visual or oral means may be used to present the required information.

Recommended Presentation Sequence:
Chapter Fourteen - Pages N1-46 , *Flight Planning (3 hours)*

Lesson Objective:
During this lesson, the student will become familiar with and develop a basic understanding of dead reckoning navigation, wind effects on a course, flight planning, time, speed, distance and fuel usage computation as well as density altitude, true altitude and true airspeed computation.

Lesson Content:
Chapter 14:
_____ What is Flight Planning?
_____ Measuring Direction
_____ Time Measurement
_____ A Matter of Degree: Longitude & Latitude on Sectional Charts
_____ Cross Country Navigation
_____ Flight Planning Step 1: Draw a line between airports
_____ Flight Planning Step 2: Determine the true course
_____ Flight Planning Step 3: Determine the wind correction angle
_____ The Effect of Water on a Swimmer
_____ The Effect of Wind on an Airplane
_____ Using the Wind Side of the Slide Computer (six steps)
_____ Flight Planning Step 4: Determine the true heading
_____ Flight Planning Step 5: Determining the magnetic heading
_____ Flight Planning Step 6: Determine your compass heading
_____ Return Trip From AVA/Memorial to Table Rock
_____ Planning an Actual Flight
_____ A More Accurate Flight Plan
_____ Final Words on Electronic Flight Computers
_____ The Mechanical Flight Computer
_____ Dance of the Decimals: The Number Scale
_____ Miles on the Menu: Converting Nautical and Statute Miles
_____ Time, Distance and Speed Computations
_____ Fuel Consumption Problems
_____ Finding Density Altitude
_____ Finding True Airspeed
_____ Finding Your True Altitude

Completion Standards:
By the following class period the student will complete the sections in *Rod Machado's Private Pilot Workbook* that pertain to the material covered in this ground lesson.

Study Assignments:
Rod Machado's Private Pilot Handbook:
Read: Chapter Fifteen - Pages O1-26, *Performance Charts*
Read: Chapter Sixteen - Pages P1-9, *Weight and Balance*

Ground Lesson 16:

Text Reference:
Rod Machado's Private Pilot Handbook
Rod Machado's Private Pilot Workbook

Presentation Format:
Any combination of visual or oral means may be used to present the required information.

Recommended Presentation Sequence:
Chapter Fifteen - Pages O1-26, *Performance Charts (2 hours)*
Chapter Sixteen - Pages P1-9, *Weight and Balance (1 hour)*

Lesson Objective:
During this lesson, the student will become familiar with and develop a basic understanding of airplane performance computation. Additionally, the student will become familiar with and develop an understanding of the basic terms needed to compute an airplane's weight and balance.

Lesson Content:
Chapter 15:
_____ Air Density
_____ Height
_____ Heat
_____ Humidity
_____ Density Altitude
_____ Service Ceiling
_____ Performance Charts
_____ Takeoff Concepts
_____ Best Rate and Best Angle of Climb Speeds
_____ Vx and Vy Change With Altitude
_____ Cruise Climb Speed
_____ Takeoff Distance Chart
_____ Landing Distance Performance Charts
_____ A Different Landing Distance Chart
_____ Time, Fuel and Distance to Climb Chart
_____ Cruise Performance Chart
_____ Another Variety of Cruise Performance Charts
_____ Endurance and Range Profile Charts
_____ Crosswind Component Chart

Chapter 16:
_____ Definitions
_____ Excessive Weight and Structural Damage
_____ Center of Gravity
_____ Other CG Considerations
_____ Just a Moment

Completion Standards:
By the following class period the student will complete the sections in *Rod Machado's Private Pilot Workbook* that pertain to the material covered in this ground lesson.

Study Assignments:
Rod Machado's Private Pilot Handbook:
Read: Chapter Sixteen - Pages P9-20, *Weight and Balance*
Read: Chapter Seventeen - Pages Q1-21, *Pilot Potpourri*

Ground Lesson 17:

Text Reference:
Rod Machado's Private Pilot Handbook & Workbook

Presentation Format:
Any combination of visual or oral means may be used to present the required information.

Recommended Presentation Sequence:
Chapter Sixteen - Pages P9-20, *Weight and Balance (1 hour)*
Chapter Seventeen - Pages Q1-21, *Pilot Potpourri (2 hours)*

Lesson Objective:
During this lesson, the student will become familiar with and develop a basic understanding of the knowledge necessary to compute an airplane's weight and balance. Additionally, the student will become familiar with and develop a basic understanding of the physiological and mental hazards associated with flight. The student will become familiar with and develop a basic understanding of flight plans, aeronautical publications and the NOTAM system,

Lesson Content:
Chapter 16:
_____ Don't Wait to Balance
_____ Weight Change
_____ Fuel Burn Weight and Balance
_____ Weight Shift
_____ A Different Type of Weight and Balance Chart

Chapter 17:
_____ Taking AIM: The Aeronautical Information Manual
_____ Fitness for Flight
_____ Illness
_____ Medication
_____ Alcohol-Don't Fly High
_____ Hypoxia: Low O, Two
_____ Hyperventilation
_____ CO Oh Oh
_____ Spatial Disorientation
_____ Visual Illusions
_____ Night Vision
_____ Haze and Collision Avoidance
_____ Scanning for Traffic During the Day
_____ Night Scanning For Traffic
_____ Airplane Blind Spots
_____ Filing a VFR Flight Plan
_____ FAA and Industry Publications
_____ Airport/Facility Directory
_____ Notices To Airmen (NOTAMS)
_____ NOTAM D/L & FDC NOTAMS
_____ Advisory Circulars
_____ Aviation Judgment

Completion Standards:
By the following class period the student will complete the sections in *Rod Machado's Private Pilot Workbook* that pertain to the material covered in this ground lesson.

Study Assignments:
Review: *Rod Machado's Private Pilot Handbook* as directed by instructor
Review: *Rod Machado's Private Pilot Workbook* as directed by instructor

Ground Lesson 18:

Text Reference:
Rod Machado's Private Pilot Handbook
Rod Machado's Private Pilot Workbook

Presentation Format:
Any combination of visual or oral means may be used to present the required information.

Recommended Presentation Sequence:
Review class material, discuss exam taking strategies *(:30)*
Present final (Stage Three) exam *(2:00 hours)*
Review final exam *(:30)*

Lesson Objective:
During this lesson, the instructor will review the material presented in class and the student will have the opportunity to ask questions on any topic. A final exam will be presented after the review.

Lesson Content:
Class review and exam.

Completion Standards:
The student will complete the written exam with a minimum score of 80% and the instructor will review each incorrect answer to ensure that the student understands the item missed.

Study Assignments:
The instructor will assign a specific area of review for each student based on that student's exam performance.

Stage One Exam

Rod Machado's Private Pilot Syllabus
Part 61/141

Exam Covers Chapters 1-6

1. [A6/1/2]
What do we call the part of the airplane that houses the cockpit and has the wings and engine attached to it?
A. Empennage.
B. Fuselage.
C. Undercarriage.

2. [B1/3/2]
The four forces acting on an airplane in flight are
A. lift, weight, thrust, and drag.
B. lift, weight, gravity, and thrust.
C. lift, gravity, power, and friction.

3. [B14/1/3]
When the critical angle of attack is exceeded the airplane will
A. stall.
B. ascend.
C. descend.

4. [B20/3/4]
The two basic forms of drag are:
A. parasite and induced drag.
B. planform and interference.
C. good and bad drag.

5. [B23/1/4]
Ground effect allows an airplane flying close to the runway to become or remain airborne at a slightly _____ speed.
A. lower-than-normal
B. higher-than-normal
C. higher and lower

6. [B27/1/2 & 3]
What is one purpose of wing flaps?
A. To enable the pilot to make steeper approaches to a landing without increasing the airspeed.
B. To relieve the pilot of maintaining continuous pressure on the controls.
C. To decrease wing area to vary the lift.

7. [B35/1/3]
P-factor is more likely to cause the airplane to yaw to the left
A. at low angles of attack.
B. at high angles of attack.
C. at high airspeeds.

8. [C11/1/4]
Temperature drops of as much as _____ within the carburetor's throat are not uncommon.
A. 10°F
B. 550°F
C. 70°F

9. [C13/1/2]
If an aircraft is equipped with a fixed-pitch propeller and a float-type carburetor, the first indication of carburetor ice would most likely be
A. a drop in oil temperature and cylinder head temperature.
B. engine roughness.
C. loss of RPM.

10. [C15/3/2]
With an increase in altitude the air becomes thinner and doesn't _____ as much for a given volume.
A. weigh
B. count
C. vary

11. [C18/3/3]
High cylinder temperatures also lead to something known as _____.
A. pre-ignition
B. detonation
C. combustion

12. [C21/1/1]
If present, water rests on the _____ of fuel tanks, where it is the first thing to go to the engine.
A. top
B. bottom
C. outside

13. [C25/1/2]
For internal cooling, reciprocating aircraft engines are especially dependent on
A. a properly functioning thermostat.
B. air flowing over the exhaust manifold.
C. the circulation of lubricating oil.

14. [C34/1/3]
Detonation occurs in a reciprocating aircraft engine when
A. the spark plugs are fouled or shorted out or the wiring is defective.
B. hot spots in the combustion chamber ignite the fuel/air mixture in advance of normal ignition.
C. the unburned charge in the cylinders explodes instead of burning normally.

15. [D4/3/1]
Amps are a measure of _____ flow.
A. voltage
B. current
C. water pressure

16. [D6/1/1]
Between the positive terminal of the battery and the primary bus is an ammeter, called a _____ ammeter.
A. charge-discharge
B. battery
C. load

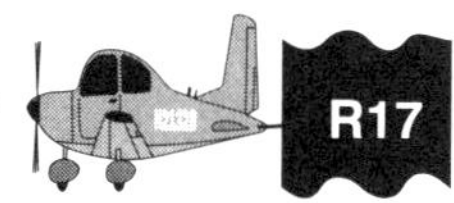

17. [D8/1/1]
A full left deflection of a load meter needle is similar to a charge-discharge ammeter reading pointing to the _____ of its scale.
A. negative side (-)
B. neutral point
C. positive side (+)

18. [D9/3/1]
Voltage regulators help alternators maintain a _____ voltage output under varying RPM conditions.
A. constant
B. low
C. high

19. [D12/1/3]
If the battery is dead, the _____ isn't going to work.
A. magneto
B. engine
C. alternator

20. [E2/3/2 & E3/3/3]
If the pitot tube and outside static vents become clogged, which instruments would be affected?
A. The altimeter, airspeed indicator, and turn-and-slip indicator.
B. The altimeter, airspeed indicator, and vertical speed indicator.
C. The altimeter, attitude indicator, and turn-and-slip indicator.

21. [E4/3/1&2]
V_{SO} is defined as the
A. stalling speed or minimum steady flight speed in the landing configuration.
B. stalling speed or minimum steady flight speed in a specified configuration.
C. stalling speed or minimum takeoff safety speed.

22. [E5/3/3]
Which V-speed represents the maneuvering speed?
A. V_a
B. V_{lo}
C. V_{ne}

23. [E9/See *TAS & IAS High Altitude Airports*]
When making an approach at a high altitude airport, you should:
A. approach at a lower than normal indicated airspeed.
B. approach at a higher than normal indicated speed.
C. approach at a normal indicated speed.

24. [E10/1/4]
What is true altitude?
A. The vertical distance of the aircraft above sea level.
B. The vertical distance of the aircraft above the surface.
C. The height above the standard datum plane.

25. [E15/2/1 & Figure 25]
If a flight is made from an area of high pressure into an area of lower pressure without the altimeter setting being adjusted, the altimeter will indicate
A. lower than the actual altitude above sea level.
B. higher than the actual altitude above sea level.
C. the actual altitude above sea level.

26. [E34/1/2]
In the northern hemisphere, if an aircraft is accelerated or decelerated, the magnetic compass will normally indicate
A. a momentary turn.
B. correctly when on a north or south heading.
C. a turn toward the south.

27. [F2/2/2]
With respect to the certification of aircraft, a class is a subdivision of _____.
A. a category
B. the number of engines
C. the category of airplane only

28. [F4/2/2]
The definition of nighttime is
A. sunset to sunrise.
B. 1 hour after sunset to 1 hour before sunrise.
C. the time between the end of evening civil twilight and the beginning of morning civil twilight.

29. [F7/2/2]
A third class medical certificate is issued to a 19 year-old pilot on August 10. To exercise the privileges of a recreational or private pilot certificate, the medical certificate expires at midnight on
A. August 10, 2 years later.
B. August 31, 2 years later.
C. August 31, 3 years later.

30. [F11/2/3]
The three takeoffs and landings that are required to act as pilot in command at night must be done during the time period from
A. sunset to sunrise.
B. 1 hour after sunset to 1 hour before sunrise.
C. the end of evening civil twilight to the beginning of morning civil twilight.

31. [F18/2/4]
According to regulations pertaining to general privileges and limitations, a private pilot may
A. be paid for the operating expenses of a flight if at least three takeoffs and three landings were made by the pilot within the preceding 90 days.
B. share the operating expenses of a flight with the passengers.
C. not be paid in any manner for the operating expenses of a flight.

32. [F23/2/3]
With respect to passengers, what obligation, if any, does a pilot in command have concerning the use of safety belts?
A. The pilot in command must instruct the passengers to keep their safety belts fastened for the entire flight.
B. The pilot in command must brief the passengers on the use of safety belts and notify them to fasten their safety belts during taxi, takeoff, and landing.
C. The pilot in command has no obligation in regard to passengers' use of safety belts.

33. [F25/1/1]
What action should the pilots of a glider and an airplane take if on a head-on collision course?
A. The airplane pilot should give way to the left.
B. The glider pilot should give way to the right.
C. Both pilots should give way to the right.

34. [F26/2/3]
Unless otherwise authorized, the maximum indicated airspeed at which aircraft may be flown when at or below 2,500 feet AGL and within 4 nautical miles of the primary airport of Class C airspace is
A. 200 knots.
B. 230 knots.
C. 250 knots.

35. [F39/1/2]
Which VFR cruising altitude is acceptable for a flight on a Victor Airway with a magnetic course of 175 degrees?
A. 4,500 feet.
B. 5,000 feet.
C. 5,500 feet.

36. [F41/1/2]
When are non-rechargeable batteries of an emergency locator transmitter (ELT) required to be replaced?
A. Every 24 months.
B. When 50 percent of their useful life expires.
C. At the time of each 100-hour or annual inspection.

37. [F41/2/3]
Except in Alaska, during what time period should lighted position lights be displayed on an aircraft?
A. End of evening civil twilight to the beginning of morning civil twilight.
B. 1 hour after sunset to 1 hour before sunrise.
C. Sunset to sunrise.

38. [F46/2/2]
Preventive maintenance has been performed on an aircraft. What paperwork is required?
A. A full, detailed description of the work done must be entered in the airframe logbook.
B. The date the work was completed, and the name of the person who did the work must be entered in the airframe and engine logbook.
C. The signature, certificate number, and kind of certificate held by the person approving the work and a description of the work must be entered in the aircraft maintenance records.

39. [F48/2/2]
If an aircraft is involved in an accident that results in substantial damage to the aircraft, the nearest NTSB field office should be notified
A. immediately.
B. within 48 hours.
C. within 7 days.

40. [F48/2/3]
Which incident requires an immediate notification to the nearest NTSB field office?
A. A forced landing due to engine failure.
B. Landing gear damage due to a hard landing.
C. Flight control system malfunction or failure.

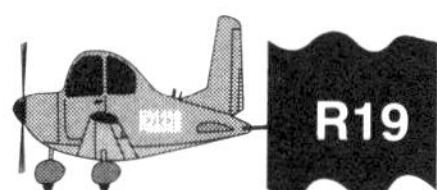

Stage Two Exam

Rod Machado's Private Pilot Syllabus
Part 61/141

Exam Covers Chapters 7-12

RUNWAY SURFACE MARKINGS

1. [G6/1/1&2] (Refer to the figure above)
According to the airport diagram, which statement is true?
A. Runway 24 is equipped at position A with emergency arresting gear to provide a means of stopping military aircraft.
B. Takeoffs may be started at position B on Runway 31, and the landing portion of this runway begins at the displaced threshold.
C. The takeoff and landing portion of Runway 6 begins at position A.

2. [G13/Figure 22] (Referring to the figure below)
The segmented circle indicates that a landing on Runway 20 will be with a
A. right-quartering headwind.
B. left-quartering headwind.
C. left-quartering tailwind.

TRAFFIC PATTERN INDICATOR

17
20
2
35

3. [G18/3/2]
Automatic Terminal Information Service (ATIS) is the continuous broadcast of recorded information concerning
A. pilots of radar-identified aircraft whose aircraft is in dangerous proximity to terrain or to an obstruction.
B. non-essential information to reduce frequency congestion.
C. non-control information in selected high-activity terminal areas.

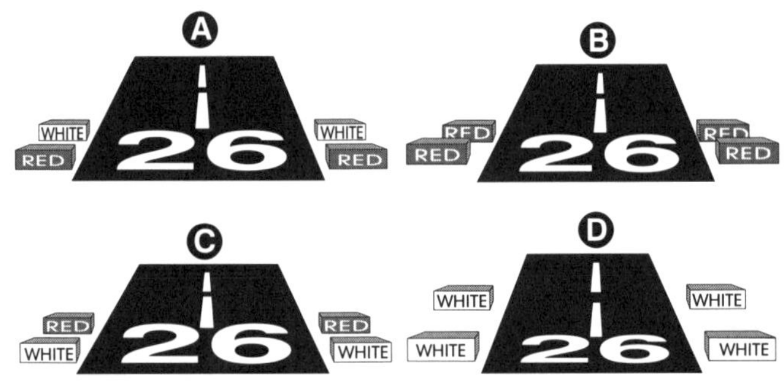

4. [G22/1/2]
VASI lights as shown by illustration D (above) indicate that the airplane is
A. off course to the left.
B. above the glideslope.
C. below the glideslope.

5. [G24/2/4 & G25/1/2]
Wingtip vortices are created only when an aircraft is
A. operating at high airspeeds.
B. heavily loaded.
C. developing lift.

6. [G28/1/2]
How should the flight controls be held while taxiing a tricycle-gear equipped airplane with a left quartering tailwind?
A. Left aileron up, elevator neutral.
B. Left aileron down, elevator down.
C. Left aileron up, elevator down.

7. [G28/See *Land and Hold Short Operations*]
Where is the "Available Landing Distance" (ALD) data published for an airport that utilizes Land and Hold Short Operations (LAHSO) published?
A. Aeronautical Information Manual (AIM).
B. 14 CFR Part 91, General Operating and Flight Rules.
C. Airport/Facility Directory (A/FD).

8. [H3/2/1]
The Federal Communications Commission (FCC) assigns frequencies ranging from _____ megahertz (MHz) to _____ MHz for aviation use.
A. 200, 850
B. 118.0, 135.975
C. 119.7, 149.325

9. [H9/1/1]
Referring to the figure below, on what frequency could you contact Riverside FSS if you're in the vicinity of Hector VOR?
A. Transmit on 122.1 MHz, listen on 112.7 MHz.
B. Transmit on 110.2 MHz, listen on 122.1 MHz.
C. Transmit on Channel 39 MHz, listen on 110.2 MHz.

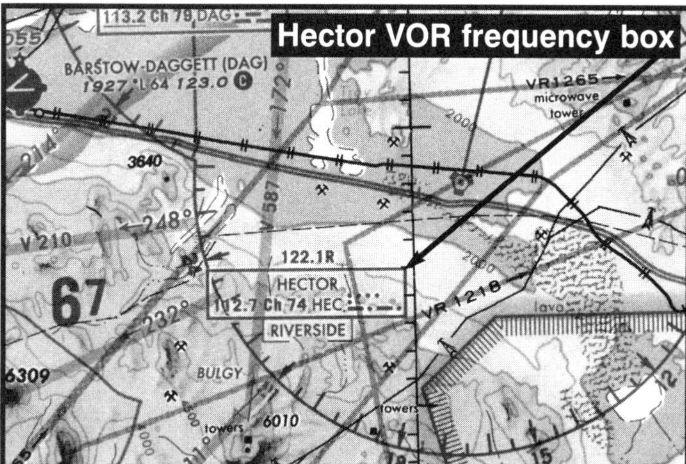

10. [H16/3/3]
An ATC radar facility issues the following advisory to a pilot flying on a heading of 090 degrees:

"TRAFFIC 3 O'CLOCK, 2 MILES, WESTBOUND..."

Where should the pilot look for this traffic?
A. East.
B. South.
C. West.

11. [H18/3/2]
TRSA Service provides
A. IFR separation (1,000 feet vertical and 3 miles lateral) between all aircraft.
B. a warning to pilots when their aircraft is in unsafe proximity to terrain, obstructions, or other aircraft.
C. sequencing and separation for participating VFR aircraft.

12. [H20/1/1]
What publication could you use to determine the stage of radar service available at an airport?
A. Airport/Facility Directory.
B. Aeronautical Information Manual.
C. Tony's Handbook of Radar Range Cooking.

13. [I7/2/1]
What minimum flight visibility is required for VFR flight operations on an airway below 10,000 feet MSL?
A. 1 mile.
B. 3 miles.
C. 4 miles.

14. [I9/1/5]
When operating at an airport having any type of surface-based controlled airspace established for it the reported ground visibility at the airport must be at least _____ statute mile(s).
A. five
B. one
C. three

15. [I11/1/2]
A SVFR clearance allows you to operate below _____ feet MSL down to the surface, within the _____ boundaries of surface-based controlled airspace
A. 10,000, lateral
B. 1,200, lateral
C. 14,500, 10 mile

16. [I20/3/1]
Normal VFR operations in Class D airspace with an operating control tower require the ceiling and visibility to be at least
A. 1,000 feet and 1 mile.
B. 1,000 feet and 3 miles.
C. 2,500 feet and 3 miles.

17. [I22/1/2]
Class C airspace is geometrically shaped like two cylinders. The surface-based inner cylinder extends upward to approximately _____ AGL and has a five nautical mile radius from the center of the _____ airport.
A. 4,000 feet, primary
B. 1,200 feet, primary
C. 1,200 feet, satellite

18. [I26/1/3]
What minimum pilot certification is required for operation within Class B airspace?
A. Private pilot certificate or student pilot certificate with appropriate logbook endorsements.
B. Commercial pilot certificate.
C. Private pilot certificate with an instrument rating.

19. [I30/2/3]
Prohibited areas are defined by _____ lines.
A. red dashed
B. red hatched
C. blue hatched

20. [J2/1/1]
Sectional charts are valid for
A. 12 months.
B. 6 months.
C. a lot of things.

21. [J7/1/2]
Maximum elevation figures (MEFs) represent the highest elevation of terrain and other obstacles (towers, trees, etc.) within _____.
A. any area on the chart
B. a quadrangle
C. a magenta bordered area

22. [J8/1/1 & J8/2/1&2]
Referring to the figure on the right, the top of the lighted obstacle approximately 2 miles north of the city of Hamburg is
A. 323 feet MSL.
B. 483 feet MSL.
C. 483 feet AGL.

23. [J11/1/1]
Airports are coded by colors on the map. Those airports colored in _____ don't have an air traffic control tower. Those shown in _____ have a tower (although it may not be in operation 24 hours a day—most aren't).
A. magenta, black
B. magenta, blue
C. blue, magenta

24. [J13/1/1]
Referring to the figure above and the airport data listed under Clinton-Sherman airport, what is the airport elevation?
A. 35 feet.
B. 1,922 feet.
C. 119.6 feet.

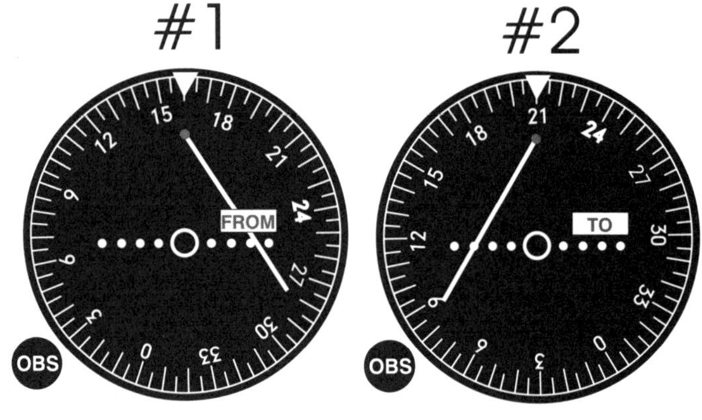

25. [K10/All & Figure 15]
Referring to VOR #1 shown above, what heading should you fly to intercept and track outbound on the 160 degree radial at a 30 degree angle?
A. 160 degrees.
B. 130 degrees.
C. 190 degrees.

26. [K16/All, K16/Figure 22 & K9/Figure 13]
Referring to VOR #2 above, what is the aircraft's position relative to the station?
A. North.
B. West.
C. South.

27. [K24/Figure 34]
How many satellites are in the GPS constellation?
A. 25
B. 24
C. 22

THE RADIO MAGNETIC INDICATOR

28. [K28/2/1]
Based on the RMI above, what radial is the airplane on from the VOR station?
A. 270 degree radial.
B. 090 degree radial.
C. 030 degree radial.

29. [K34/2/5 & Formula on K34]
Referring to the ADF and DG above, determine the magnetic bearing TO the station.
A. 330 degrees MBTS.
B. 220 degrees MBTS.
C. 190 degrees MBTS.

30. [L2/1/2]
Every physical process of weather is accompanied by, or is the result of, a
A. movement of air.
B. pressure differential.
C. heat exchange.

31. [L8/1/4]
Relative humidity is a number that tells you how much _____ the air is holding in relationship to how much it could theoretically hold at its current _____.
A. water vapor, temperature
B. water vapor, humidity
C. pressure, volume

32. [L9/3/2]
The dewpoint is a great indicator of the atmosphere's _____ content. _____ dewpoint temperatures indicate that there's a lot of water in the air. _____ dewpoint temperatures indicate that there's little water in the air.
A. pressure, Low, Low
B. water, Low, High
C. water, High, Low

33. [L10/1/1]
If the temperature/dewpoint spread is small and decreasing, and the temperature is 62 degrees F, what type weather is most likely to develop?
A. Freezing precipitation.
B. Thunderstorms.
C. Fog or low clouds.

34. [L17/3/1]
Warm air resting on top of a cold layer of air would be considered
A. a stable condition.
B. an unstable condition.
C. a neutrally stable condition.

35. [L23/Figure 38]
What is a characteristic of stable air?
A. Stratiform clouds.
B. Unlimited visibility.
C. Cumulus clouds.

36. [L32/2/1]
The boundary between two different air masses is referred to as a
A. frontolysis.
B. frontogenesis.
C. front.

37. [L40/2/4]
What conditions are necessary for the formation of thunderstorms?
A. High humidity, lifting force, and unstable conditions.
B. High humidity, high temperature, and cumulus clouds.
C. Lifting force, moist air, and extensive cloud cover.

38. [L48/Figure 82]
Possible mountain wave turbulence could be anticipated when winds of 40 knots or greater blow
A. across a mountain ridge, and the air is stable.
B. down a mountain valley, and the air is unstable.
C. parallel to a mountain peak, and the air is stable.

39. [L50/1/2&3]
A pilot can expect a wind shear zone in a temperature inversion whenever the wind speed at 2,000 to 4,000 feet above the surface is at least
A. 10 knots.
B. 15 knots.
C. 25 knots.

40. [L51/2/2]
What situation is most conducive to the formation of radiation fog?
A. Warm, moist air over low, flatland areas on clear, calm nights.
B. Moist, tropical air moving over cold, offshore water.
C. The movement of cold air over much warmer water.

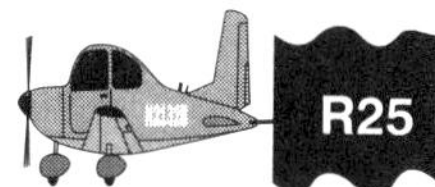

Stage Three Exam

Rod Machado's Private Pilot Syllabus
Part 61/141

Exam Covers Chapters 1-17

1. [B4/3/3]
The minimum forward speed of the airplane is called the _____ speed.
A. certified
B. stall
C. best rate of climb

2. [B28/1/6]
What force makes an airplane turn?
A. The horizontal component of lift.
B. The vertical component of lift.
C. Centrifugal force.

3. [C15/3/2]
With an increase in altitude the air becomes thinner and doesn't _____ as much for a given volume.
A. weigh
B. count
C. vary

4. [C34/3/4]
If the grade of fuel used in an aircraft engine is lower than specified for the engine, it will most likely cause
A. a mixture of fuel and air that is not uniform in all cylinders.
B. lower cylinder head temperatures.
C. detonation.

5. [D5/3/4]
While airplane batteries are rated at 12 or 24 volts, airplane electrical systems (their alternators) are rated for ____ or ____ volts.
A. 12, 24
B. 14, 28
C. 7, 14

6. [E16/2/4 & Figure 28]
Under what condition will true altitude be lower than indicated altitude?
A. In colder than standard air temperature.
B. In warmer than standard air temperature.
C. When density altitude is higher than indicated altitude.

7. [E34/3/2]
In the northern hemisphere, a magnetic compass will normally indicate initially a turn toward the east if
A. an aircraft is decelerated while on a south heading.
B. an aircraft is accelerated while on a north heading.
C. a left turn is entered from a north heading.

8. [F5/3/2]
Regulations require that you report all drug and alcohol motor vehicle actions to the FAA within _____ days.
A. 60
B. 30
C. 120

9. [F46/2/4]
An aircraft had a 100 hour inspection when the tachometer read 1259.6. When is the next 100 hour inspection due?
A. 1349.6 hours.
B. 1359.6 hours.
C. 1369.6 hours.

10. [F48/3/2]
The operator of an aircraft that has been involved in an incident is required to submit a report to the nearest field office of the NTSB
A. within 7 days.
B. within 10 days.
C. when requested.

RUNWAY SURFACE MARKINGS

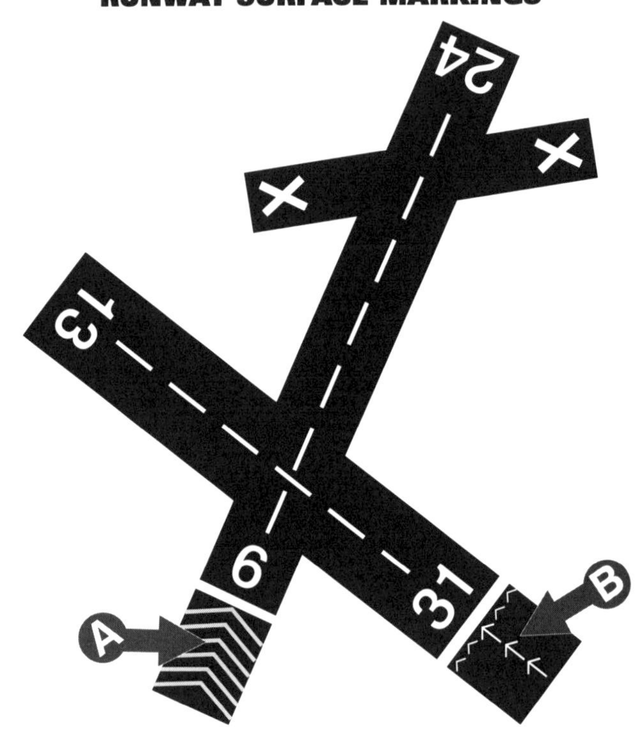

11. [G6/1/1] (Refer to the figure above)
What is the difference between area A and area B on the airport depicted?
A. "A" may be used for taxi and takeoff; "B" may be used only as an overrun.
B. "A" may be used for all operations except heavy aircraft landings; "B" may be used only as an overrun.
C. "A" may not be used at all; "B" may be used for all operations except landings.

12. [G22/1/2]
VASI lights as shown by illustration B (to the right) indicate that the airplane is
A. below the glideslope.
B. on the glideslope.
C. above the glideslope.

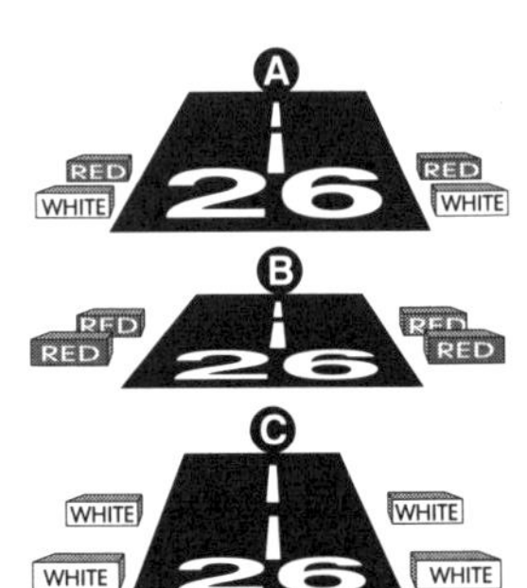

13. [G28/See *Land and Hold Short Operations*]
What is the minimum visibility for a pilot to receive a land and hold short (LAHSO) clearance?
A. 3 nautical miles.
B. 1 statute mile.
C. 3 statute miles.

14. [H16/3/3]
An ATC radar facility issues the following advisory to a pilot flying on a heading of 360 degrees:

"TRAFFIC 10 O'CLOCK, 2 MILES, SOUTHBOUND..."

Where should the pilot look for this traffic?
A. Northwest.
B. Northeast.
C. Southwest.

15. [H19/3/1]
If Air Traffic Control advises that radar service is terminated when the pilot is departing Class C airspace, the transponder should be set to code
A. 0000
B. 1200
C. 4096

16. [I7/2/1]
During operations within controlled airspace at altitudes of less than 1,200 feet AGL, the minimum horizontal distance from clouds requirement for VFR flight is
A. 1,000 feet.
B. 1,500 feet.
C. 2,000 feet.

17. [I9/1/5]
If the ground visibility isn't reported in surface-based controlled airspace, then the flight visibility during takeoff, landing or when operating in the traffic pattern must be at least _____ statute miles.
A. three
B. five
C. one

18. [I30/3/2]
Restricted areas restrict flights due to the unusual activities conducted within them. These areas often contain invisible hazards to aircraft such as the firing of _____, _____, _____.
A. artillery, aerial gunnery, guided missiles
B. artillery, lasers, rocks
C. bullets, rockets, gum wads

19. [J2/3/2]
Changes on the sectional chart occurring prior to the next publication cycle can be found in the
A. FARs.
B. POH (Pilots Operating Handbook).
C. Airport/Facility Directory

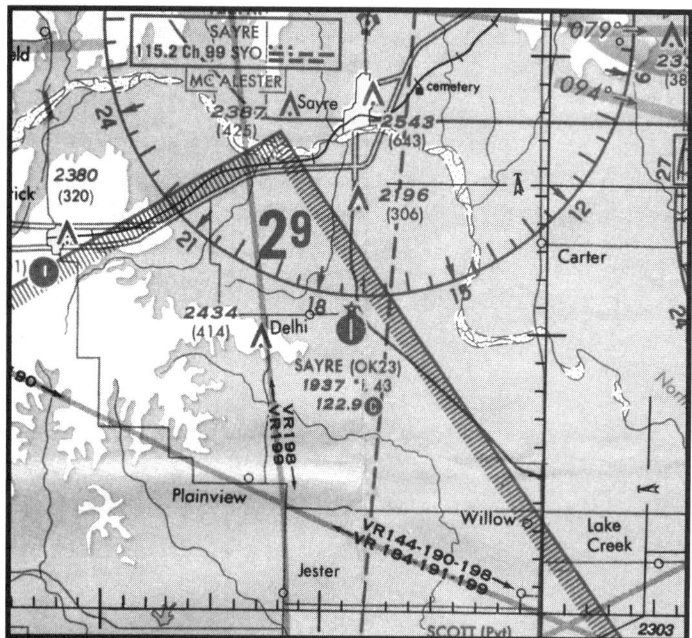

20. [J8/1/1 & J8/2/1&2]
Referring to the figure above, what minimum altitude is required to fly over the obstacle located approximately three miles west of Sayre airport? (Assume that the entire area is a congested area.)
A. 3,434 feet AGL.
B. 1,414 feet MSL.
C. 3,434 feet MSL.

21. [K10/All & Figure 15]
Referring to VOR receiver #1, shown to the right, what heading should you fly to intercept and track inbound on the 060 degree radial at a 40 degree intercept angle?
A. 020
B. 100
C. 060

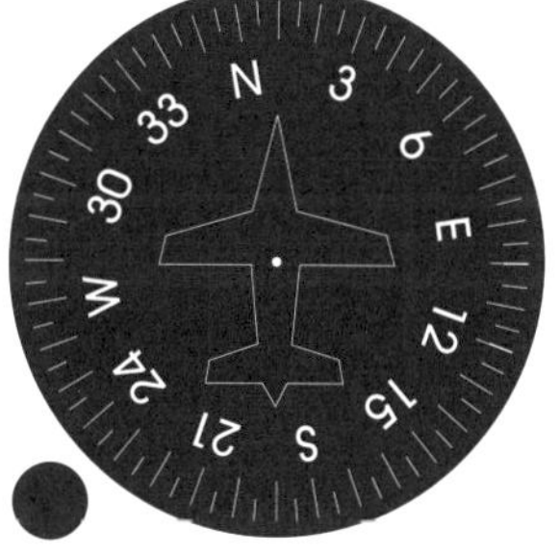

22. [K34/2/5 & Formula on K34]
Referring to the ADF and DG above, determine the magnetic bearing FROM the station (MBFS).
A. 030 degrees MBFS.
B. 010 degrees MBFS.
C. 040 degrees MBFS.

23. [L8/Figure 15]
Cooling the air _____ its relative humidity.
A. increases
B. decreases
C. doesn't affect

24. [L10/1/1]
If the temperature/dewpoint spread is small and decreasing, and the temperature is 62 degrees F, what type of weather is most likely to develop?
A. Freezing precipitation.
B. Thunderstorms.
C. Fog or low clouds.

25. [M5/1/3]
Which type weather briefing should a pilot request when departing within the hour, if no preliminary weather information has been received?
A. Outlook briefing.
B. Abbreviated briefing.
C. Standard briefing.

26. [M11/2/2]
Referring to Figure 1 below, the letters RAB34 found in the METAR for KINK indicate
A. that light rain blew at 1834 Zulu.
B. that rain began at 1934 Zulu.
C. that rain began at 1834 Zulu.

27. [M17/1/3]
Referring to the KLAX TAF below in Figure 2 below, the "FM (FROM) Group"
A. forecasts for the hours from 2200Z to 0200Z, winds of 330 degrees at 15 knots with gusts to 20 knots and a probability of a 600 foot ceiling, a 1,500 foot broken ceiling and and overcast ceiling at 2,500 feet.
B. forecasts for the hours from 0200Z to 0600Z, a ceiling of 800 feet and a 40-49% probability of 2 miles visibility between the hours of 0200Z and 0500Z.
C. forecasts for the hours from 1600Z to 1800Z, variable winds at 40 knots and visibilities less than 6 miles.

28. [M20/Figure 21]
Referring to the Area Forecast in Figure 3 below, what weather is expected in the southwestern quarter of Ohio?
A. Broken clouds at 5,000 feet MSL and tops between 6,000 and 10,000 feet AGL.
B. Broken clouds between 5,000 and 6,000 feet AGL with tops at 10,000 foot MSL and mostly VFR conditions.
C. Broken clouds between 5,000 and 6,000 feet MSL with tops at 10,000 foot MSL and an outlook for marginal VFR conditions with mist.

29. [M21/3/1]
When the term "light and variable" is used in reference to a winds aloft forecast, the coded group and wind speed are
A. 0000 and less than 7 knots.
B. 9900 and less than 5 knots.
C. 9999 and less than 10 knots.

Figure 3 **THE AREA FORECAST**

```
BOSC FA 241845
SYNOPSIS AND VFR CLDS/WX
SYNOPSIS VALID UNTIL 251300
CLDS/WX VALID UNTIL 250700...OTLK VALID 250700-251300
ME NH VT MA RI CT NY LO NJ PA OH LE WV MD DC DE VA AND
CSTL WTRS
.
SEE AIRMET SIERRA FOR IFR CONDS AND MTN OBSCN.
TS IMPLY SEV OR GTR TURB SEV ICE LLWS AND IFR CONDS.
NON MSL HGTS DENOTED BY AGL OR CIG.
.
SYNOPSIS...19Z CDFNT ALG A 160NE ACK-ENE LN...CONTG
AS A QSTNRY FNT ALG AN END-50SW MSS LN. BY 13Z...CDFNT
ALG A 140ESE ACK-HTO LN...CONTG AS A QSTNRY FNT ALG
A HTO-SYR-YYZ LN.  TROF ACRS CNTRL PA INTO NRN VA.
...REYNOLDS...
.
OH LE
NRN HLF OH LE...SCT-BKN025 OVC045. CLDS LYRD 150. SCT SHRA.
   WDLY SCT TSRA. CB TOPS FL350. 23-01Z OVC020-030. VIS 3SM BR.
   OCNL -RA. OTLK...IFR CIG BR FG.
SWRN QTR OH...BKN050-060 TOPS 100. OTLK...MVFR BR.
SERN QTR OH...SCT-BKN040 BKN070 TOPS 120. WDLY SCT -TSRA.
   00Z SCT-BKN030 OVC050. WDLY SCT -TSRA. CB TOPS FL350.
   OTLK...VFR SHRA.
```

Figure 2 **TERMINAL AERODROME WEATHER FORECAST (TAF)**

```
TAF
KLAX 121720Z 121818 20012KT 5SM HZ BKN030 PROB40 2022 1SM TRSA OVC008CB
     FM2200 33015G20KT P6SM BKN015 OVC025 PROB40 2202 3SM SHRA
     FM0200 35012KT OVC008 PROB40 0205 2SM -RASN BECMG 0608 02008KT NSW BKN012
     BECMG 1012 00000KT 3SM BR SKC TEMPO 1214 1/2SM FG
     FM1600 VRB04KT P6SM NSW SKC
```

Figure 1 **METAR WEATHER REPORTING FORMAT**

```
METAR KINK 081955Z 34016G22KT 1/2SM R30R/2400FT +SHRA OVC012 13/12 A2990 RAB34
SPECI KMKC 081936Z 20014G24KT 1/2SM R34/2600FT -SHRA OVC008 04/03 A2898 THN FG SE
```

THE LOW LEVEL SIGNIFICANT WEATHER PROGNOSTIC CHART

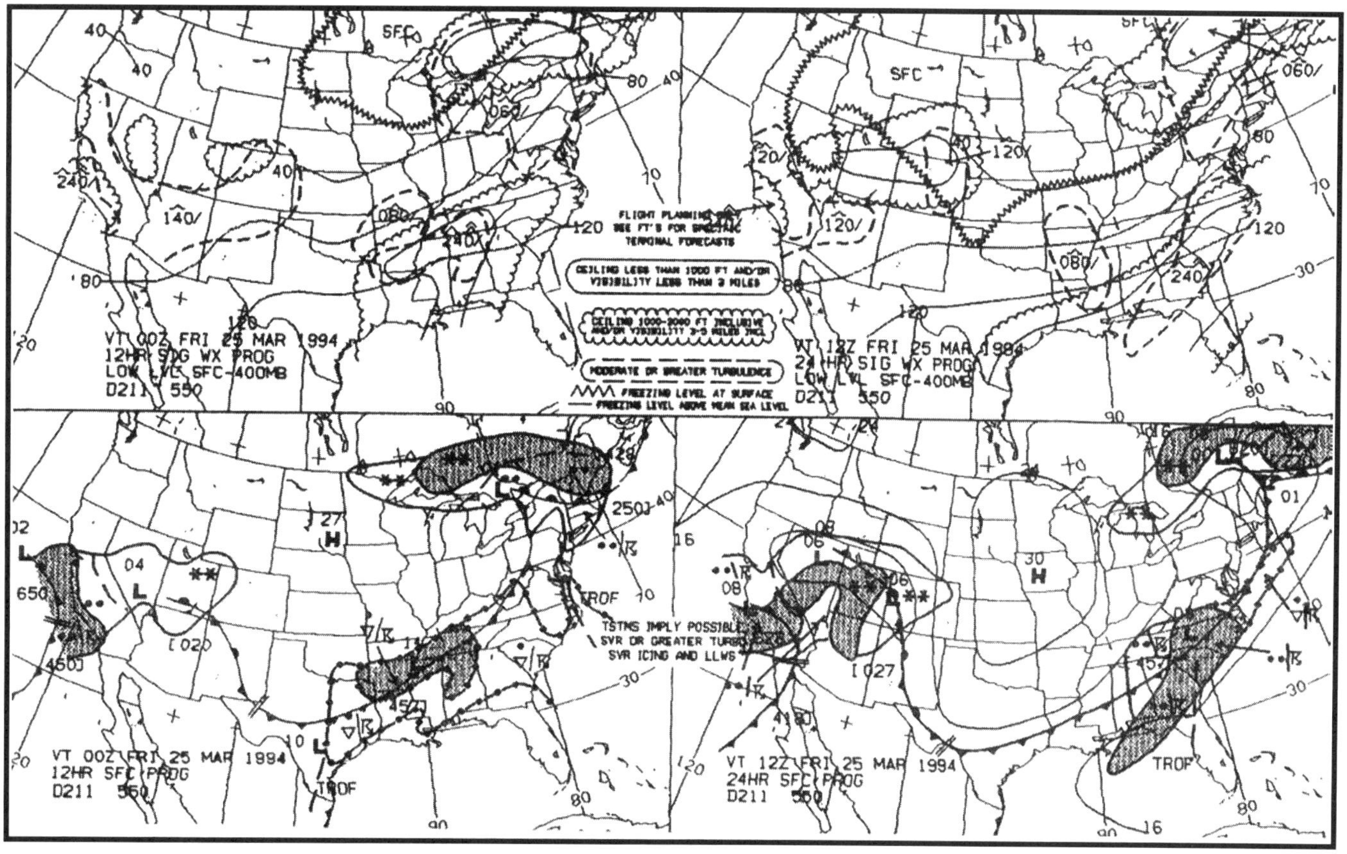

30. [M27/1/1]

Referring to the figure above, what weather is forecast for the state of Nevada during the first 12 hours?

A. Ceiling 1,000 to 3,000 feet and/or visibility 3 to 5 miles.

B. IFR conditions with ceiling less than 1,000 feet and/or visibilities less than 3 miles.

C. Moderate or greater turbulence at the surface.

31. [M29/2/5]

AIRMETs are advisories of significant weather phenomena but of lower intensities than SIGMETs and are intended for dissemination to

A. only IFR pilots.

B. only VFR pilots.

C. all pilots.

32. [N4 & N5 /All]

(Refer to the figure to the right) An aircraft departs an airport in the Mountain Standard Time zone at 1415 MST for a 2 hour 30 minute flight to an airport located in the Pacific Standard Time zone. What is the estimated time of arrival at the destination airport?

A. 1545 PST.

B. 1645 PST.

C. 1745 PST.

TO CONVERT FROM:	TO COORDINATED UNIVERSAL TIME
Eastern Standard Time	Add 5 hours
Eastern Daylight Time	Add 4 hours
Central Standard Time	Add 6 hours
Central Daylight Time	Add 5 hours
Mountain Standard Time	Add 7 hours
Mountain Daylight Time	Add 6 hours
Pacific Standard Time	Add 8 hours
Pacific Daylight Time	Add 7 hours
Yukon Standard Time	Add 9 hours
Alaska, Hawaii Standard Time	Add 10 hours
Bering Standard Time	Add 11 hours

TIME ZONES

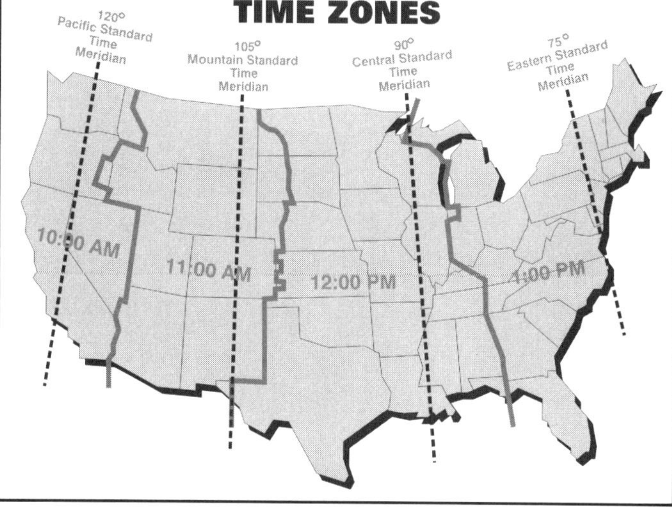

33. [Pages N6 through N9]
Referring to the chart on this page, what is the approximate latitude and longitude of Marshal Memorial (area A)?
A. 39 degrees 06'N - 94 degrees 12'W.
B. 39 degrees 06'N - 93 degrees 12'W.
C. 39 degrees 54'N - 93 degrees 48'W.

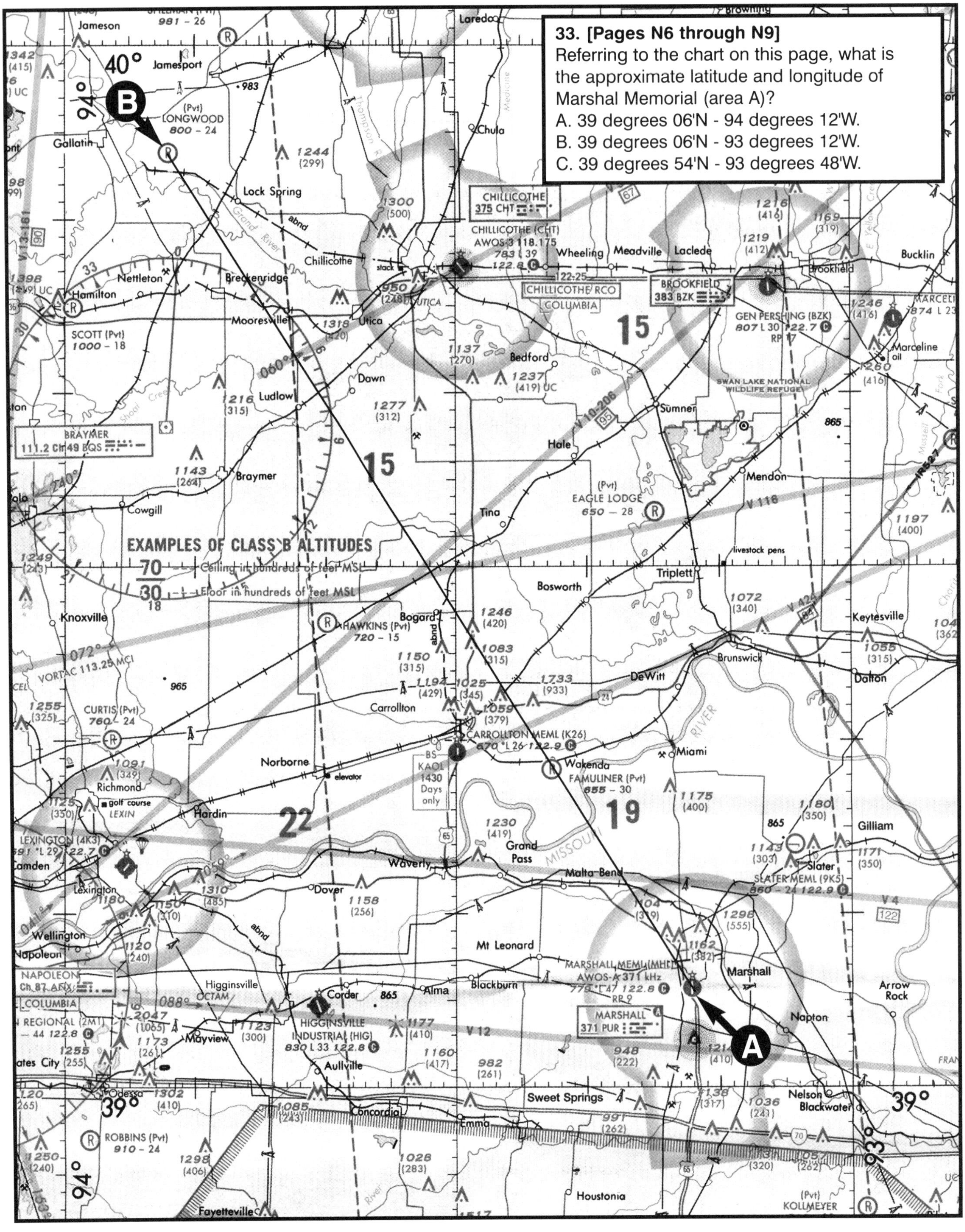

34. [N19-N22]
Referring to the chart to the left, determine the compass heading for a flight from Marshall Memorial airport (area A) to Longwood airport (area B). The wind at 4,500 feet is from 260 degrees at 17 knots, the true airspeed is 110 knots and the magnetic variation is 3 degrees east. Use the compass deviation card shown below.
A. 321 degrees.
B. 318 degrees.
C. 151 degrees.

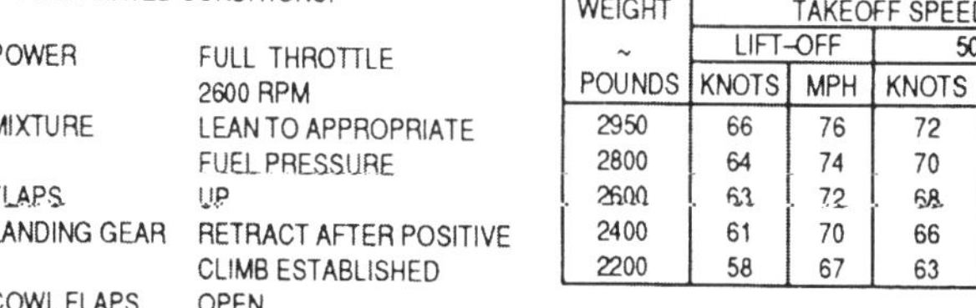

TYPICAL COMPASS DEVIATION CARD

FOR (MAGNETIC)	N	30	60	E	120	150
STEER (COMPASS)	O	28	57	86	117	148
FOR (MAGNETIC)	S	210	240	W	300	330
STEER (COMPASS)	180	212	243	274	303	332

35. [N19-N22]
Determine the fuel used on the flight described above, if the fuel consumption is 7.8 gallons per hour (add .5 gallons for taxi, takeoff and climb).
A. 4.0 gallons.
B. 4.5 gallons.
C. 4.7 gallons.

36. [N42/2/1]
What is your true airspeed if the pressure altitude is 9,000 feet, the temperature is -8 degrees Celsius and the indicated airspeed is 125 knots?
A. 124 knots.
B. 142 knots.
C. 112 knots.

37. [N42/3/2]
What is your true altitude if the pressure altitude is 7,000 feet, the indicated altitude is 8,500 feet and the outside air temperature is -10 degrees Celsius?
A. 8,155 feet.
B. 8,775 feet.
C. 7,800 feet.

38. [O10/1/1/Entire section]
Referring to the performance chart below, determine the approximate distance required to clear a 50 foot obstacle.

OAT: 80 degrees F
Pressure altitude: 2,500 ft
Takeoff weight: 2,250 lb
Headwind component: 20 kts
A. 900 feet.
B. 500 feet.
C. 700 feet.

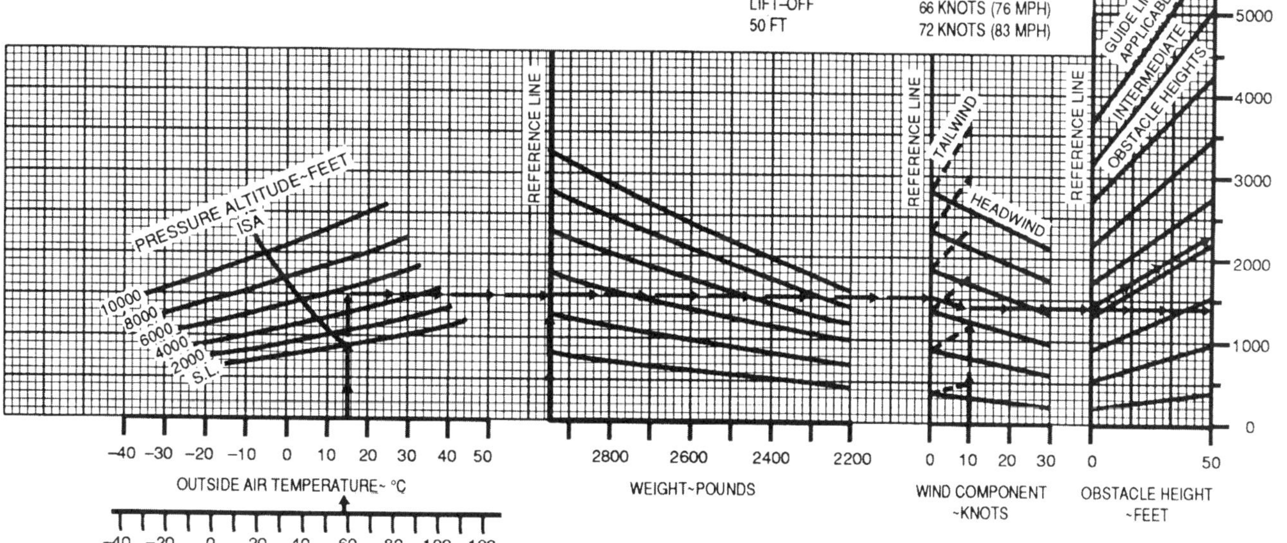

WEIGHT ~ POUNDS	TAKEOFF SPEED LIFT-OFF KNOTS	LIFT-OFF MPH	50 FT KNOTS	50 FT MPH
2950	66	76	72	83
2800	64	74	70	81
2600	63	72	68	78
2400	61	70	66	76
2200	58	67	63	73

TAKEOFF DISTANCE

SHORT FIELD

CONDITIONS:
Flaps 10°
Full Throttle Prior to Brake Release
Paved, Level, Dry Runway
Zero Wind

NOTES:
1. Short field technique as specified in Section 4.
2. Prior to takeoff from fields above 3000 feet elevation, the mixture should be leaned to give maximum RPM in a full throttle, static runup.
3. Decrease distances 10% for each 9 knots headwind. For operation with tailwinds up to 10 knots, increase distances by 10% for each 2 knots.
4. For operation on a dry, grass runway, increase distances by 15% of the "ground roll" figure.

WEIGHT LBS	TAKEOFF SPEED KIAS		PRESS ALT FT	0°C		10°C		20°C		30°C		40°C	
	LIFT OFF	AT 50 FT		GRND ROLL	TOTAL TO CLEAR 50 FT OBS	GRND ROLL	TOTAL TO CLEAR 50 FT OBS	GRND ROLL	TOTAL TO CLEAR 50 FT OBS	GRND ROLL	TOTAL TO CLEAR 50 FT OBS	GRND ROLL	TOTAL TO CLEAR 50 FT OBS
1670	50	54	S.L.	640	1190	695	1290	755	1390	810	1495	875	1605
			1000	705	1310	765	1420	825	1530	890	1645	960	1770
			2000	775	1445	840	1565	910	1690	980	1820	1055	1960
			3000	855	1600	925	1730	1000	1870	1080	2020	1165	2185
			4000	940	1775	1020	1920	1100	2080	1190	2250	1285	2440
			5000	1040	1970	1125	2140	1215	2320	1315	2525	1420	2750
			6000	1145	2200	1245	2395	1345	2610	1455	2855	1570	3125
			7000	1270	2470	1375	2705	1490	2960	1615	3255	1745	3590
			8000	1405	2800	1525	3080	1655	3395	1795	3765	1940	4195

39. [O12/2/2/Entire section]
Referring to the takeoff performance chart above, determine the total distance required for takeoff to clear a 50 foot obstacle.
OAT: 10 degrees C
Pressure altitude: 4,000 ft
Takeoff weight: 1,670 lb
Headwind component: 0 kts
Runway: dry, grass

A. 1,020 feet.
B. 1,920 feet.
C. 2,073 feet.

40. [O18/1/1/Entire section]
Referring to the cruise performance chart below, what is the expected fuel consumption for a 500 nautical mile flight under the following conditions?

Pressure altitude: 6,000 ft
Temperature: -15 degrees C
Manifold pressure: 19.8" Hg
Wind: calm

A. 31.4 gallons.
B. 37.5 gallons.
C. 44.1 gallons.

CRUISE POWER SETTINGS

65% MAXIMUM CONTINUOUS POWER (OR FULL THROTTLE)
2800 POUNDS

PRESS ALT.	ISA -20 °C (-36 °F)								STANDARD DAY (ISA)								ISA +20 °C (+36 °F)							
	IOAT		ENGINE SPEED	MAN. PRESS	FUEL FLOW		TAS		IOAT		ENGINE SPEED	MAN PRESS	FUEL FLOW		TAS		IOAT		ENGINE SPEED	MAN. PRESS	FUEL FLOW		TAS	
FEET	°F	°C	RPM	IN HG	PSI	GPH	KTS	MPH	°F	°C	RPM	IN HG	PSI	GPH	KTS	MPH	°F	°C	RPM	IN HG	PSI	GPH	KTS	MPH
SL	27	-3	2450	20.7	6.6	11.5	147	169	63	17	2450	21.2	6.6	11.5	150	173	99	37	2450	21.8	6.6	11.5	153	176
2000	19	-7	2450	20.4	6.6	11.5	149	171	55	13	2450	21.0	6.6	11.5	153	176	91	33	2450	21.5	6.6	11.5	156	180
4000	12	-11	2450	20.1	6.6	11.5	152	175	48	9	2450	20.7	6.6	11.5	156	180	84	29	2450	21.3	6.6	11.5	159	183
6000	5	-15	2450	19.8	6.6	11.5	155	178	41	5	2450	20.4	6.6	11.5	158	182	79	26	2450	21.0	6.6	11.5	161	185
8000	-2	-19	2450	19.5	6.6	11.5	157	181	36	2	2450	20.2	6.6	11.5	161	185	72	22	2450	20.8	6.6	11.5	164	189
10000	-8	-22	2450	19.2	6.6	11.5	160	184	28	-2	2450	19.9	6.6	11.5	163	188	64	18	2450	20.3	6.5	11.4	166	191
12000	-15	-26	2450	18.8	6.4	11.3	162	186	21	-6	2450	18.8	6.1	10.9	163	188	57	14	2450	18.8	5.9	10.6	163	188
14000	-22	-30	2450	17.4	5.8	10.5	159	183	14	-10	2450	17.4	5.6	10.1	160	184	50	10	2450	17.4	5.4	9.8	160	184
16000	-29	-34	2450	16.1	5.3	9.7	156	180	7	-14	2450	16.1	5.1	9.4	156	180	43	6	2450	16.1	4.9	9.1	155	178

NOTES: 1. Full throttle manifold pressure settings are approximate.
2. Shaded area represents operation with full throttle.

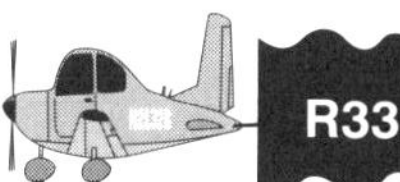

SECTION 5
PERFORMANCE

Figure 1

CESSNA
MODEL 152

CRUISE PERFORMANCE

CONDITIONS:
1670 Pounds
Recommended Lean Mixture (See Section 4, Cruise)

NOTE:
Cruise speeds are shown for an airplane equipped with speed fairings which increase the speeds by approximately two knots.

PRESSURE ALTITUDE FT	RPM	20°C BELOW STANDARD TEMP			STANDARD TEMPERATURE			20°C ABOVE STANDARD TEMP		
		% BHP	KTAS	GPH	% BHP	KTAS	GPH	% BHP	KTAS	GPH
2000	2400	- - -	- - -	- - -	75	101	6.1	70	101	5.7
	2300	71	97	5.7	66	96	5.4	63	95	5.1
	2200	62	92	5.1	59	91	4.8	56	90	4.6
	2100	55	87	4.5	53	86	4.3	51	85	4.2
	2000	49	81	4.1	47	80	3.9	46	79	3.8
4000	2450	- - -	- - -	- - -	75	103	6.1	70	102	5.7
	2400	76	102	6.1	71	101	5.7	67	100	5.4
	2300	67	96	5.4	63	95	5.1	60	95	4.9
	2200	60	91	4.8	56	90	4.6	54	89	4.4
	2100	53	86	4.4	51	85	4.2	49	84	4.0
	2000	48	81	3.9	46	80	3.8	45	78	3.7
6000	2500	- - -	- - -	- - -	75	105	6.1	71	104	5.7
	2400	72	101	5.8	67	100	5.4	64	99	5.2
	2300	64	96	5.2	60	95	4.9	57	94	4.7
	2200	57	90	4.6	54	89	4.4	52	88	4.3
	2100	51	85	4.2	49	84	4.0	48	83	3.9
	2000	46	80	3.8	45	79	3.7	44	77	3.6
8000	2550	- - -	- - -	- - -	75	107	6.1	71	106	5.7
	2500	76	105	6.2	71	104	5.8	67	103	5.4
	2400	68	100	5.5	64	99	5.2	61	98	4.9
	2300	61	95	5.0	58	94	4.7	55	93	4.5
	2200	55	90	4.5	52	89	4.3	51	87	4.2
	2100	49	84	4.1	48	83	3.9	46	82	3.8
10,000	2500	72	105	5.8	68	103	5.5	64	103	5.2
	2400	65	99	5.3	61	98	5.0	58	97	4.8
	2300	58	94	4.7	56	93	4.5	53	92	4.4
	2200	53	89	4.3	51	88	4.2	49	86	4.0
	2100	48	83	4.0	46	82	3.9	45	81	3.8
12,000	2450	65	101	5.3	62	100	5.0	59	99	4.8
	2400	62	99	5.0	59	97	4.8	56	96	4.6
	2300	56	93	4.6	54	92	4.4	52	91	4.3
	2200	51	88	4.2	49	87	4.1	48	85	4.0
	2100	47	82	3.9	45	81	3.8	44	79	3.7

Figure 2

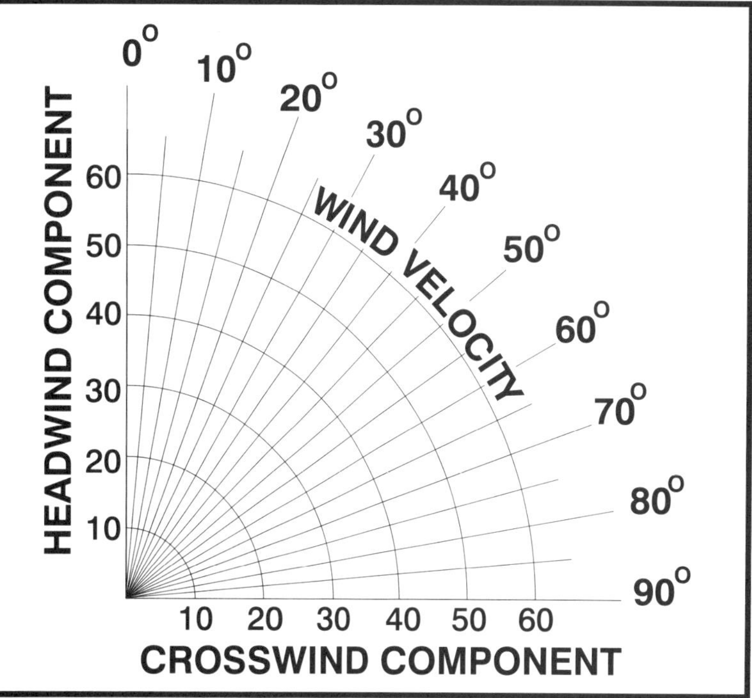

41. [O19/1/4]
Referring to Figure 1, determine the expected fuel consumption and true airspeed for a flight at a pressure altitude of 7,000 feet at 2,400 RPM under standard conditions.
A. 5.4 GPH, 101 knots.
B. 5.3 GPH, 99 knots.
C. 5.2 GPH, 105 knots.

42. [O21/Entire section]
Referring to the crosswind component chart (Figure 2), determine the maximum wind velocity for a 45 degree crosswind if the maximum crosswind component for the airplane is 20 knots.
A. 25 knots.
B. 28 knots.
C. 35 knots.

43. [O24/Postflight Briefing #15-1]
Referring to the density altitude chart (Figure 3), determine the density altitude at an airport that is 1,795 feet MSL with an altimeter setting of 29.70 and a temperature of 80 degrees F.
A. 3,800 feet MSL.
B. 2,000 feet MSL.
C. 3,500 feet MSL.

Figure 3

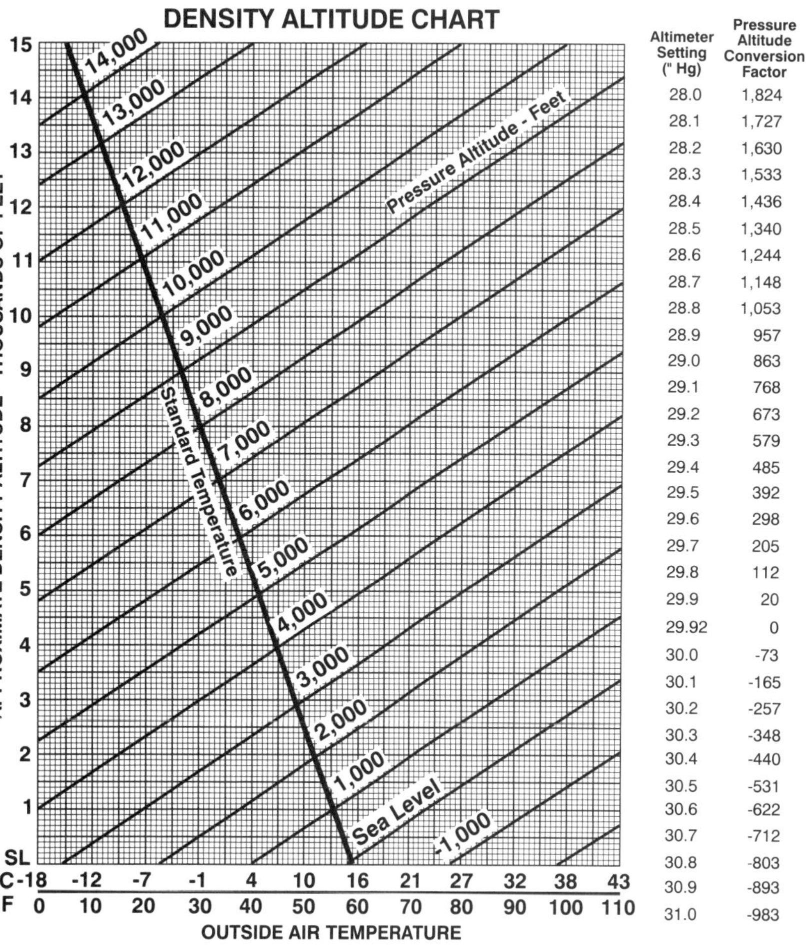

Altimeter Setting (" Hg)	Pressure Altitude Conversion Factor
28.0	1,824
28.1	1,727
28.2	1,630
28.3	1,533
28.4	1,436
28.5	1,340
28.6	1,244
28.7	1,148
28.8	1,053
28.9	957
29.0	863
29.1	768
29.2	673
29.3	579
29.4	485
29.5	392
29.6	298
29.7	205
29.8	112
29.9	20
29.92	0
30.0	-73
30.1	-165
30.2	-257
30.3	-348
30.4	-440
30.5	-531
30.6	-622
30.7	-712
30.8	-803
30.9	-893
31.0	-983

USEFUL LOAD WEIGHTS AND MOMENTS

OCCUPANTS

FRONT SEATS ARM 85 Weight	Moment/100	REAR SEATS ARM 121 Weight	Moment/100
120	102	120	145
130	110	130	157
140	119	140	169
150	128	150	182
160	136	160	194
170	144	170	206
180	153	180	218
190	162	190	230
200	170	200	242

BAGGAGE OR 5TH SEAT OCCUPANT ARM 140

Weight	Moment/100
10	14
20	28
30	42
40	56
50	70
60	84
70	98
80	112
90	126
100	140
110	154
120	168
130	182
140	196
150	210
160	224
170	238
180	252
190	266
200	280
210	294
220	308
230	322
240	336
250	350
260	364
270	378

USABLE FUEL

MAIN WING TANKS ARM 75

Gallons	Weight	Moment/100
5	30	22
10	60	45
15	90	68
20	120	90
25	150	112
30	180	135
35	210	158
40	240	180
44	264	198

AUXILIARY WING TANKS ARM 94

Gallons	Weight	Moment/100
5	30	28
10	60	56
15	90	85
19	114	107

***OIL**

Quarts	Weight	Moment/100
10	19	5

*Included in basic Empty Weight

Basic **Empty Weight** ~ 2015

MOM / 100 ~ 1554

MOMENT LIMITS vs WEIGHT

Moment limits are based on the following weight and center of gravity limit data (landing gear down).

WEIGHT CONDITION	FORWARD CG LIMIT	AFT CG LIMIT
2950 lb (takeoff or landing)	82.1	84.7
2525 lb	77.5	85.7
2475 lb or less	77.0	85.7

MOMENT LIMITS vs WEIGHT (Continued)

Weight	Minimum Moment/100	Maximum Moment/100
2100	1617	1800
2110	1625	1808
2120	1632	1817
2130	1640	1825
2140	1648	1834
2150	1656	1843
2160	1663	1851
2170	1671	1860
2180	1679	1868
2190	1686	1877
2200	1694	1885
2210	1702	1894
2220	1709	1903
2230	1717	1911
2240	1725	1920
2250	1733	1928
2260	1740	1937
2270	1748	1945
2280	1756	1954
2290	1763	1963
2300	1771	1971
2310	1779	1980
2320	1786	1988
2330	1794	1997
2340	1802	2005
2350	1810	2014
2360	1817	2023
2370	1825	2031
2380	1833	2040
2390	1840	2048
2400	1848	2057
2410	1856	2065
2420	1863	2074
2430	1871	2083
2440	1879	2091
2450	1887	2100
2460	1894	2108
2470	1902	2117
2480	1911	2125
2490	1921	2134
2500	1932	2143
2510	1942	2151
2520	1953	2160
2530	1963	2168
2540	1974	2176
2550	1984	2184
2560	1995	2192
2570	2005	2200
2580	2016	2208
2590	2026	2216

Weight	Minimum Moment/100	Maximum Moment/100
2600	2037	2224
2610	2048	2232
2620	2058	2239
2630	2069	2247
2640	2080	2255
2650	2090	2263
2660	2101	2271
2670	2112	2279
2680	2123	2287
2690	2133	2295
2700	2144	2303
2710	2155	2311
2720	2166	2319
2730	2177	2326
2740	2188	2334
2750	2199	2342
2760	2210	2350
2770	2221	2358
2780	2232	2366
2790	2243	2374
2800	2254	2381
2810	2265	2389
2820	2276	2397
2830	2287	2405
2840	2298	2413
2850	2309	2421
2860	2320	2428
2870	2332	2436
2880	2343	2444
2890	2354	2452
2900	2365	2460
2910	2377	2468
2920	2388	2475
2930	2399	2483
2940	2411	2491
2950	2422	2499

44. [P3/1/1]
If an airplane will return, unassisted, to level flight after its controls are disturbed, it is said to have _____ dynamic stability.
A. negative
B. positive
C. neutral

45. [P3/Figure 3]
The term used to describe the airplane's pitching motion is known as _____ stability.
A. vertical
B. longitudinal
C. lateral

46. [P8/2/6]
An aircraft is loaded 110 pounds over maximum certificated gross weight. If fuel (gasoline) is drained to bring the aircraft weight within limits, how much fuel should be drained?
A. 15.7 gallons.
B. 16.2 gallons.
C. 18.4 gallons.

Weight & Balance Problem

Determine if the airplane's weight and balance are within safe limits.

Pilot & front seat occupants.......	340 lb
Rear seat occupants.................	295 lb
Fuel (main & aux tanks both full)	44 gal
Baggage..................................	36 lb

47. [P14/1/2]
Based on the conditions listed above and the weight and balance charts shown above, what is the airplane's center of gravity location and is the airplane within proper CG limits?
A. The airplane is over gross weight, within the CG aft of aft limits.
B. The airplane is at gross weight and the CG is within the limits.
C. The airplane is under gross weight with the CG forward of the forward limit.

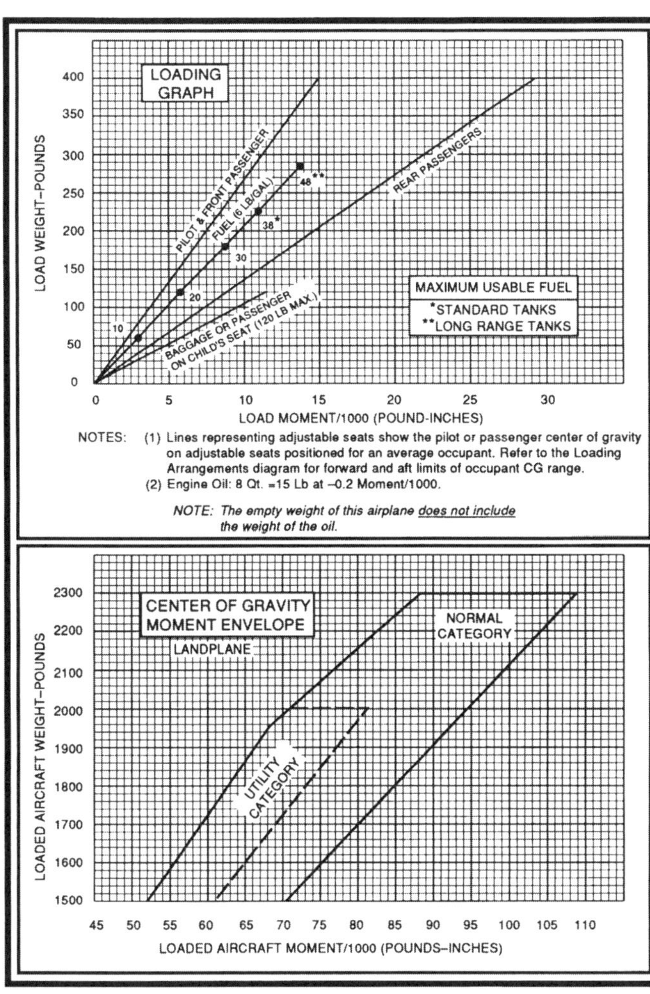

	Weight (lb)	Moment (lb-in)
Empty weight	1,350.0	51.5
Pilot & front passenger	400.0	?
Rear passengers	400.0	?
Fuel (std tanks)	?	?
Oil (8 qts.)	15.0	-.2
Baggage	17.0	1.7
Total	?	?

48. [P20/1/1]
Using the airplane loading information and the weight and balance charts shown above, determine the maximum amount of fuel that can be carried aboard the aircraft.
A. 13.3 gallons.
B. 19.6 gallons.
C. 15.5 gallons.

49. [Q3/2/2]
What happens to the percentage of oxygen available in the atmosphere as altitude increases?
A. It decreases dramatically.
B. It actually increases slightly.
C. It remains the same.

50. [Q4/1/3]
Which would most likely result in hyperventilation?
A. Emotional tension, anxiety, or fear.
B. The excessive consumption of alcohol.
C. An extremely slow rate of breathing and insufficient oxygen.

51. [Q5/1/4]
Ear problems common to pilots usually involve a little flaccid tube that connects the middle ear to the back of the throat. This tube is known as the _____ tube.
A. throat
B. eustachian
C. middle ear tube

52. [Q8/1/2]
At night, a blending of the earth and sky is often responsible for creating an indiscernible _____, resulting in near-instrument flight conditions. This is most prevalent on moonless nights when stars take on the appearance of _____ and city lights appear to be stars.
A. star map, planets
B. horizon, city lights
C. horizon, the sky

53. [Q10/1/6]
What effect does haze have on the ability to see traffic or terrain features during flight?
A. Haze causes the eyes to focus at infinity.
B. The eyes tend to overwork in haze and do not detect relative movement easily.
C. All traffic or terrain features appear to be farther away than their actual distance.

54. [Q13/Figure 15]
During a night flight, you observe steady red and green lights ahead and at the same altitude. What is the general direction of movement of the other aircraft?
A. The other aircraft is crossing to the left.
B. The other aircraft is flying away from you.
C. The other aircraft is approaching head-on.

55. [Q17/3/2] (Refer to the A/FD excerpt at the top of page R34)
Which type of radar service is provided to VFR aircraft at Lincoln Municipal?
A. Sequencing to the primary Class C airport and standard separation.
B. Sequencing to the primary Class C airport and conflict resolution so that radar targets do not touch, or 1,000 feet vertical separation.
C. Sequencing to the primary Class C airport, traffic advisories, conflict resolution, and safety alerts.

NEBRASKA

LINCOLN MUNI (LNK) 4 NW UTC–6(–5DT) N40°51.05′ W96°45.55′ **OMAHA**
1218 B S4 **FUEL** 100LL, JET A TPA—2218(1000) ARFF Index B **H–1E, 3F, 4F, L–11B**
RWY 17R–35L: H12901X200 (ASPH–CONC–GRVD) S–100, D–200, DT–400 HIRL **IAP**
RWY 17R: MALSR. VASI(V4L)—GA 3.0° TCH 55′. Rgt tfc. 0.4% down.
RWY 35L: MALSR. VASI(V4L)—GA 3.0° TCH 55′.
RWY 14–32: H8620X150 (ASPH–CONC–GRVD) S–80, D–170, DT–280 MIRL
RWY 14: REIL. VASI(V4L)—GA 3.0° TCH 48′.
RWY 32: VASI(V4L)—GA 3.0° TCH 53′. Thld dsplcd 431′. Pole. 0.3% up.
RWY 17L–35R: H5400X100 (ASPH–CONC–AFSC) S–49, D–60 HIRL 0.8% up N
RWY 17L: PAPI(P4L)—GA 3.0° TCH 33′. **RWY 35R:** PAPI(P4L)—GA 3.0° TCH 40′. Pole. Rgt tfc.
AIRPORT REMARKS: Attended continuously. Birds in vicinity of arpt. Twy D clsd between taxiways S and H indef. For MALSR Rwy 17R and Rwy 35L ctc twr. When twr clsd MALSR Rwy 17R and Rwy 35L preset on med ints, and REIL Rwy 14 left on when wind favor. NOTE: See Land and Hold Short Operations Section.
WEATHER DATA SOURCES: ASOS (402) 474–9214. LLWAS
COMMUNICATIONS: CTAF 118.5 **ATIS** 118.05 **UNICOM** 122.95
COLUMBUS FSS (OLU) TF 1–800–WX–BRIEF. NOTAM FILE LNK.
RCO 122.65 (COLUMBUS FSS)
Ⓡ **APP/DEP CON** 124.0 (170°–349°) 124.8 (350°–169°) (1130–0630Z‡)
Ⓡ **MINNEAPOLIS CENTER APP/DEP CON** 128.75 (0630–1130Z‡)
TOWER 118.5 125.7 (1130–0630Z‡) **GND CON** 121.9 **CLNC DEL** 120.7
AIRSPACE: CLASS C svc 1130–0630Z‡ ctc **APP CON** other times CLASS E.
RADIO AIDS TO NAVIGATION: NOTAM FILE LNK. VHF/DF ctc FSS.
(H) VORTACW 116.1 LNK Chan 108 N40°55.43′ W96°44.52′ 181° 4.5 NM to fld. 1370/9E
POTTS NDB (MHW/LOM) 385 LN N40°44.83′ W96°45.75′ 355° 6.2 NM to fld. Unmonitored when twr clsd.
ILS 111.1 I–OCZ Rwy 17R. MM and OM unmonitored.
ILS 109.9 I–LNK Rwy 35L LOM POTTS NDB. MM unmonitored. LOM unmonitored when twr clsd.
COMM/NAVAID REMARKS: Emerg frequency 121.5 not available at tower.

56. [Q20/3/2]
How might you identify the FSS having jurisdiction over your destination airport?
A. Look in the FDC NOTAMS.
B. Look in the A/FD.
C. Look in the Advisory Circulars.

57. [Q24/1/4]
What often leads to spatial disorientation or collision with ground/obstacles when flying under Visual Flight Rules (VFR)?
A. Getting behind the aircraft.
B. Duck-under syndrome.
C. Continued flight into instrument conditions.

58. [Postflight Briefing #17-2/Q25]
What is the antidote when a pilot has a hazardous attitude, such as "anti-authority"?
A. Follow the rules.
B. Rules do not apply in this situation.
C. I know what I am doing.

59. [Postflight Briefing #17-2/Q25]
What is the antidote when a pilot has a hazardous attitude, such as "invulnerability"?
A. It could happen to me.
B. It cannot be that bad.
C. It will not happen to me.

60. [Q25/3/1]
What is the one factor common to most preventable accidents?
A. Human error.
B. Mechanical difficulties.
C. Luck.

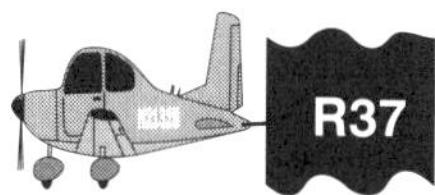

Stage One Exam Answers

1. B
2. A
3. A
4. A
5. A
6. A
7. B
8. C
9. C
10. A
11. B
12. B
13. C
14. C
15. B
16. A
17. A
18. A
19. C
20. B
21. A
22. A
23. C
24. A
25. B
26. B
27. A
28. C
29. C
30. B
31. B
32. B
33. C
34. A
35. C
36. B
37. C
38. C
39. A
40. C

Stage Two Exam Answers

1. B
2. C
3. C
4. B
5. C
6. B
7. C
8. B
9. A
10. B
11. C
12. A
13. B
14. C
15. A
16. B
17. A
18. A
19. C
20. B
21. B
22. B
23. B
24. B
25. C
26. A
27. B
28. B
29. C
30. C
31. A
32. C
33. C
34. A
35. A
36. C
37. A
38. A
39. C
40. A

Note: To ensure that you have the most current answers to these questions, please check the *Book & Slide Updates* section at Rod Machado's web site: www.rodmachado.com

Stage Three Exam Answers

1. B
2. A
3. A
4. C
5. B
6. A
7. C
8. A
9. B
10. C
11. C
12. A
13. C
14. A
15. B
16. C
17. A
18. A
19. C
20. C
21. B
22. C
23. A
24. C
25. C
26. B
27. B
28. C
29. B
30. A
31. C
32. A
33. B
34. B
35. B
36. B
37. A
38. A
39. C
40. B
41. B
42. B
43. A
44. B
45. B
46. C
47. B
48. B
49. C
50. A
51. B
52. B
53. C
54. C
55. C
56. B
57. C
58. A
59. A
60. A

Note: To ensure that you have the most current answers to these questions, please check the *Book & Slide Updates* section at Rod Machado's web site: www.rodmachado.com

Airport/Facility Directory Legend

GENERAL INFORMATION

TABLE OF CONTENTS

General Information .... Inside Front Cover
Abbreviations 1
Legend, Airport/Facility Directory 2
Airport/Facility Directory 13
Seaplane Landing Areas 227
Notices 229
Land and Hold Short Operations (LAHSO) 243
FAA and National Weather Service Telephone Numbers 245
Air Route Traffic Control Centers/Flight Service Station Communication Frequencies 252
FSDO Addresses/Telephone Numbers 257
Preferred IFR Routes/VFR Waypoints 258
VOR Receiver Check 267
Parachute Jumping Areas 271
Aeronautical Chart Bulletin 274
Tower Enroute Control (TEC) 281
National Weather Service (NWS) Upper Air Observing Stations 288
Enroute Flight Advisory Service (EFAS) .... Inside Back Cover

ABBREVIATIONS

The following abbreviations are those commonly used within this Directory. Other abbreviations may be found in the Legend and are not duplicated below. The abbreviations presented are intended to represent grammatical variations of the basic form. (Example–"req" may mean "request", "requesting", "requested", or "requests").

abv	above	lgts	lights
acft	aircraft	med	medium
AER	approach end rwy	MSL	mean sea level
AFSS	Automated Flight Service Station	MSAW	minimum safe altitude warning
AGL	above ground level	NFCT	non-federal control tower
apch	approach	ngt	night
arpt	airport	npi	non precision instrument
avbl	available	NSTD	nonstandard
bcn	beacon	ntc	notice
blo	below	opr	operate, operator, operational
byd	beyond	ops	operations
clsd	closed	OTS	out of service
ctc	contact	ovrn	overrun
dalgt	daylight	PAEW	personnel and equipment working
dsplcd	displaced	p-line	power line
durn	duration	PPR	prior permission required
eff	effective	PRM	Precision Runway Monitoring
emerg	emergency	req	request
extd	extend	rgt tfc	right traffic
FBO	fixed-base operator	rqr	require
FCT	FAA Contract Tower	RSRS	reduced same runway separation
fld	field	rwy	runway
FSS	Flight Service Station	SPB	Seaplane Base
hr	hour	SR	sunrise
indef	indefinite	SS	sunset
ints	intensity	svc	service
invof	in the vicinity of	tfc	traffic
LAA	Local Airport Advisory	thld	threshold
ldg	landing	tkf	take-off
lgtd	lighted	tmpry	temporary
		twr	tower
		twy	taxiway

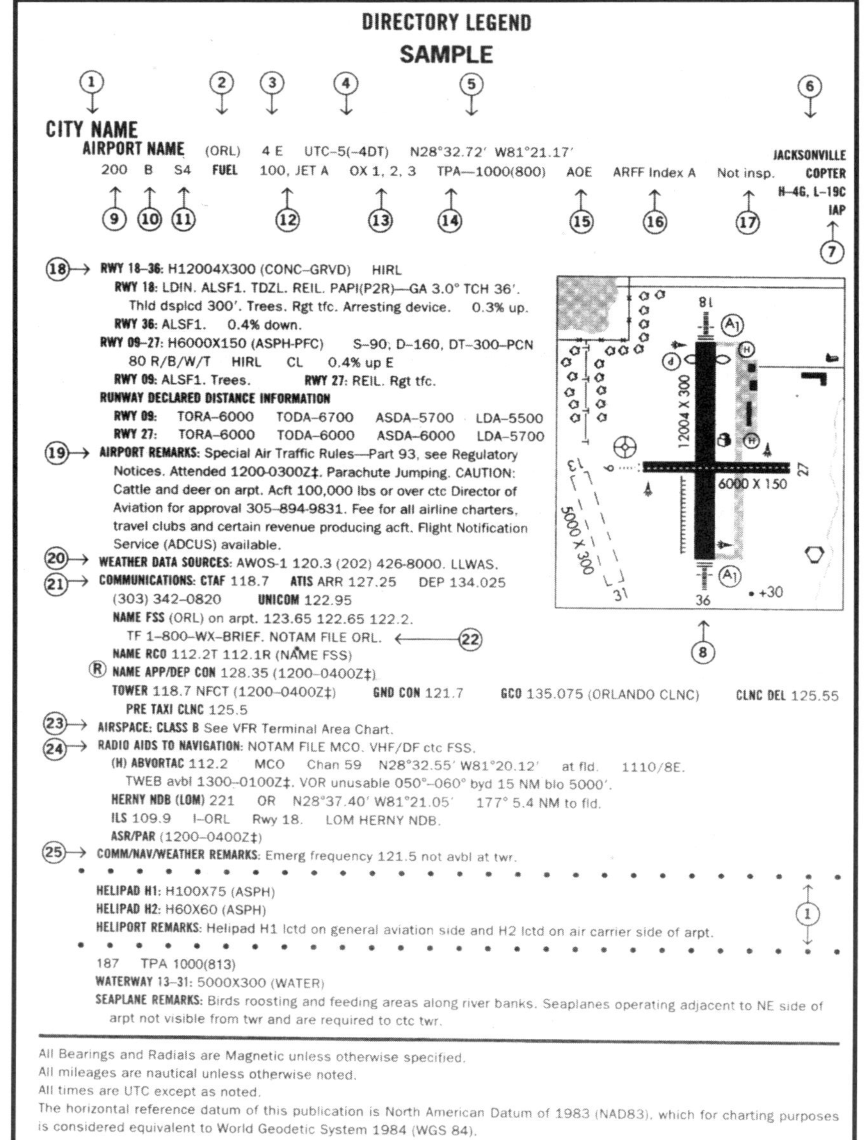

DIRECTORY LEGEND

SAMPLE

(1) CITY NAME
AIRPORT NAME (2) (ORL) (3) 4 E (4) UTC–5(–4DT) (5) N28°32.72′ W81°21.17′ (6) JACKSONVILLE
(9) 200 (10) B (11) S4 FUEL (12) 100, JET A (13) OX 1, 2, 3 (14) TPA—1000(800) (15) AOE (16) ARFF Index A (17) Not insp. COPTER
H–4G, L–19C
IAP (7)

(18)→ **RWY 18–36:** H12004X300 (CONC–GRVD) HIRL
RWY 18: LDIN. ALSF1. TDZL. REIL. PAPI(P2R)—GA 3.0° TCH 36′. Thld dsplcd 300′. Trees. Rgt tfc. Arresting device. 0.3% up.
RWY 36: ALSF1. 0.4% down.
RWY 09–27: H6000X150 (ASPH-PFC) S–90, D–160, DT–300–PCN 80 R/B/W/T HIRL CL 0.4% up E
RWY 09: ALSF1. Trees. **RWY 27:** REIL. Rgt tfc.
RUNWAY DECLARED DISTANCE INFORMATION
RWY 09: TORA–6000 TODA–6700 ASDA–5700 LDA–5500
RWY 27: TORA–6000 TODA–6000 ASDA–6000 LDA–5700

(19)→ **AIRPORT REMARKS:** Special Air Traffic Rules—Part 93, see Regulatory Notices. Attended 1200-0300Z‡. Parachute Jumping. CAUTION: Cattle and deer on arpt. Acft 100,000 lbs or over ctc Director of Aviation for approval 305–894-9831. Fee for all airline charters, travel clubs and certain revenue producing acft. Flight Notification Service (ADCUS) available.

(20)→ **WEATHER DATA SOURCES:** AWOS-1 120.3 (202) 426-8000. LLWAS.
(21)→ **COMMUNICATIONS: CTAF** 118.7 **ATIS** ARR 127.25 DEP 134.025 (303) 342–0820 **UNICOM** 122.95
NAME FSS (ORL) on arpt. 123.65 122.65 122.2.
TF 1–800–WX–BRIEF. NOTAM FILE ORL. ←(22)
NAME RCO 112.2T 112.1R (NAME FSS)
(R) **NAME APP/DEP CON** 128.35 (1200–0400Z‡)
TOWER 118.7 NFCT (1200–0400Z‡) **GND CON** 121.7 **GCO** 135.075 (ORLANDO CLNC) **CLNC DEL** 125.55
PRE TAXI CLNC 125.5
(23)→ **AIRSPACE: CLASS B** See VFR Terminal Area Chart.
(24)→ **RADIO AIDS TO NAVIGATION:** NOTAM FILE MCO. VHF/DF ctc FSS.
(H) ABVORTAC 112.2 MCO Chan 59 N28°32.55′ W81°20.12′ at fld. 1110/8E.
TWEB avbl 1300–0100Z‡. VOR unusable 050°–060° byd 15 NM blo 5000′.
HERNY NDB (LOM) 221 OR N28°37.40′ W81°21.05′ 177° 5.4 NM to fld.
ILS 109.9 I–ORL Rwy 18. LOM HERNY NDB.
ASR/PAR (1200–0400Z‡)
(25)→ **COMM/NAV/WEATHER REMARKS:** Emerg frequency 121.5 not avbl at twr.

HELIPAD H1: H100X75 (ASPH)
HELIPAD H2: H60X60 (ASPH)
HELIPORT REMARKS: Helipad H1 lctd on general aviation side and H2 lctd on air carrier side of arpt. (1)

187 TPA 1000(813)
WATERWAY 13–31: 5000X300 (WATER)
SEAPLANE REMARKS: Birds roosting and feeding areas along river banks. Seaplanes operating adjacent to NE side of arpt not visible from twr and are required to ctc twr.

All Bearings and Radials are Magnetic unless otherwise specified.
All mileages are nautical unless otherwise noted.
All times are UTC except as noted.
The horizontal reference datum of this publication is North American Datum of 1983 (NAD83), which for charting purposes is considered equivalent to World Geodetic System 1984 (WGS 84).

DIRECTORY LEGEND

LEGEND

This Directory is an alphabetical listing of data on record with the FAA on all airports that are open to the public, associated terminal control facilities, air route traffic control centers and radio aids to navigation within the conterminous United States, Puerto Rico and the Virgin Islands. Airports are listed alphabetically by associated city name and cross referenced by airport name. Facilities associated with an airport, but with a different name, are listed individually under their own name, as well as under the airport with which they are associated.

The listing of an airport in this directory merely indicates the airport operator's willingness to accommodate transient aircraft, and does not represent that the facility conforms with any Federal or local standards, or that it has been approved for use on the part of the general public.

The information on obstructions is taken from reports submitted to the FAA. It has not been verified in all cases. Pilots are cautioned that objects not indicated in this tabulation (or on charts) may exist which can create a hazard to flight operation.

Detailed specifics concerning services and facilities tabulated within this directory are contained in Aeronautical Information Manual, Basic Flight Information and ATC Procedures.

The legend items that follow explain in detail the contents of this Directory and are keyed to the circled numbers on the sample on the preceding pages.

① CITY/AIRPORT NAME

Airports and facilities in this directory are listed alphabetically by associated city and state. Where the city name is different from the airport name the city name will appear on the line above the airport name. Airports with the same associated city name will be listed alphabetically by airport name and will be separated by a dashed rule line. All others will be separated by a solid rule line. (Designated Helipads and Seaplane Landing Areas (Water) associated with a land airport will be separated by a dotted line.)

② LOCATION IDENTIFIER

A three or four character code assigned to airports. These identifiers are used by ATC in lieu of the airport name in flight plans, flight strips and other written records and computer operations.

③ AIRPORT LOCATION

Airport location is expressed as distance and direction from the center of the associated city in nautical miles and cardinal points, i.e., 4 NE.

④ TIME CONVERSION

Hours of operation of all facilities are expressed in Coordinated Universal Time (UTC) and shown as "Z" time. The directory indicates the number of hours to be subtracted from UTC to obtain local standard time and local daylight saving time UTC–5(–4DT). The symbol ‡ indicates that during periods of Daylight Saving Time effective hours will be one hour earlier than shown. In those areas where daylight saving time is not observed that (–4DT) and ‡ will not be shown. Daylight savings time is in effect from 0200 local time the first Sunday in April to 0200 local time the last Sunday in October. All states observe daylight savings time except Arizona, Hawaii and that portion of Indiana in the Eastern Time Zone and Puerto Rico and the Virgin Islands.

⑤ GEOGRAPHIC POSITION OF AIRPORT

Positions are shown in degrees, minutes and hundredths of a minute and represent the approximate center of mass of all usable runways.

⑥ CHARTS

The Sectional Chart and Low and High Altitude Enroute Chart and panel on which the airport or facility is located. Helicopter Chart locations will be indicated as, i.e., COPTER.

⑦ INSTRUMENT APPROACH PROCEDURES

IAP indicates an airport for which a prescribed (Public Use) FAA Instrument Approach Procedure has been published.

⑧ AIRPORT SKETCH

The airport sketch, when provided, depicts the airport and related topographical information as seen from the air and should be used in conjunction with the text. It is intended as a guide for pilots in VFR conditions. Symbology that is not self-explanatory will be reflected in the sketch legend. The airport sketch will be oriented with True North at the top. Airport sketches will be added incrementally.

⑨ ELEVATION

The highest point of an airport's usable runways measured in feet from mean sea level. When elevation is sea level it will be indicated as (00). When elevation is below sea level a minus (–) sign will precede the figure.

⑩ ROTATING LIGHT BEACON

B indicates rotating beacon is available. Rotating beacons operate dusk to dawn unless otherwise indicated in AIRPORT REMARKS.

⑪ SERVICING

S1: Minor airframe repairs.
S2: Minor airframe and minor powerplant repairs.
S3: Major airframe and minor powerplant repairs.
S4: Major airframe and major powerplant repairs.

DIRECTORY LEGEND

⑫ FUEL

CODE	FUEL
80	Grade 80 gasoline (Red)
100	Grade 100 gasoline (Green)
100LL	100LL gasoline (low lead) (Blue)
115	Grade 115 gasoline
A	Jet A—Kerosene freeze point–40° C.
A1	Jet A-1—Kerosene freeze point–47°C.
A1+	Jet A-1—Kerosene with icing inhibitor, freeze point–47° C.
B	Jet B—Wide-cut turbine fuel, freeze point–50° C.
B+	Jet B—Wide-cut turbine fuel with icing inhibitor, freeze point–50° C.
J8	(JP-8 Military specification) Jet A-1, kerosene with icing inhibitor, freeze point –47°C.
J8+100	(JP-8 Mil spec) Jet A-1, Kerosene with FS-II (Fuel System Icing Inhibitor), FP (Freeze Point) minus 47°C, with fuel additive package that improves thermo stability characteristics of JP-8.
MOGAS	Automobile gasoline which is to be used as aircraft fuel.

NOTE: Automobile Gasoline. Certain automobile gasoline may be used in specific aircraft engines if a FAA supplemental type certificate has been obtained. Automobile gasoline which is to be used in aircraft engines will be identified as "MOGAS", however, the grade/type and other octane rating will not be published.

Data shown on fuel availability represents the most recent information the publisher has been able to acquire. Because of a variety of factors, the fuel listed may not always be obtainable by transient civil pilots. Confirmation of availability of fuel should be made directly with fuel dispensers at locations where refueling is planned.

⑬ OXYGEN

OX 1 High Pressure
OX 2 Low Pressure
OX 3 High Pressure—Replacement Bottles
OX 4 Low Pressure—Replacement Bottles

⑭ TRAFFIC PATTERN ALTITUDE

Traffic Pattern Altitude (TPA)—The first figure shown is TPA above mean sea level. The second figure in parentheses is TPA above airport elevation.

⑮ AIRPORT OF ENTRY, LANDING RIGHTS, AND CUSTOMS USER FEE AIRPORTS

U.S. CUSTOMS USER FEE AIRPORT—Private Aircraft operators are frequently required to pay the costs associated with customs processing.

AOE—Airport of Entry—A customs Airport of Entry where permission from U.S. Customs is not required, however, at least one hour advance notice of arrival must be furnished.

LRA—Landing Rights Airport—Application for permission to land must be submitted in advance to U.S. Customs. At least one hour advance notice of arrival must be furnished.

NOTE: Advance notice of arrival at both an AOE and LRA airport may be included in the flight plan when filed in Canada or Mexico, where Flight Notification Service (ADCUS) is available the airport remark will indicate this service. This notice will also be treated as an application for permission to land in the case of an LRA. Although advance notice of arrival may be relayed to Customs through Mexico, Canadian, and U.S. Communications facilities by flight plan, the aircraft operator is solely responsible for insuring that Customs receives the notification. (See Customs, Immigration and Naturalization, Public Health and Agriculture Department requirements in the International Flight Information Manual for further details.)

⑯ CERTIFICATED AIRPORT (FAR 139)

Airports serving Department of Transportation certified carriers and certified under FAR, Part 139, are indicated by the ARFF index; i.e., ARFF Index A, which relates to the availability of crash, fire, rescue equipment.

FAR–PART 139 CERTIFICATED AIRPORTS

INDICES AND AIRCRAFT RESCUE AND FIRE FIGHTING EQUIPMENT REQUIREMENTS

Airport Index	*Required No. Vehicles*	*Aircraft Length*	*Scheduled Departures*	*Agent + Water for Foam*
A	1	<90′	≥1	500#DC or HALON 1211 or 450#DC + 100 gal H_2O
B	1 or 2	≥90′, <126′	≥5	Index A + 1500 gal H_2O
		≥126′, <159′	<5	
C	2 or 3	≥126′, <159′	≥5	Index A + 3000 gal H_2O
		≥159′, <200′	<5	
D	3	≥159′, <200′	≥5	Index A + 4000 gal H_2O
		>200′	<5	
E	3	≥200′	≥5	Index A + 6000 gal H_2O

> Greater Than; < Less Than; ≥ Equal or Greater Than; ≤ Equal or Less Than; H_2O–Water; DC–Dry Chemical.

NOTE: The listing of ARFF index does not necessarily assure coverage for non-air carrier operations or at other than prescribed times for air carrier. ARFF Index Ltd.—indicates ARFF coverage may or may not be available, for information contact airport manager prior to flight.

DIRECTORY LEGEND

⑰ FAA INSPECTION

All airports not inspected by FAA will be identified by the note: Not insp. This indicates that the airport information has been provided by the owner or operator of the field.

⑱ RUNWAY DATA

Runway information is shown on two lines. That information common to the entire runway is shown on the first line while information concerning the runway ends are shown on the second or following line. Lengthy information will be placed in the Airport Remarks.

Runway direction, surface, length, width, weight bearing capacity, lighting, slope and appropriate remarks are shown for each runway. Direction, length, width, lighting and remarks are shown for sealanes. The full dimensions of helipads are shown, i.e., 50X150.

RUNWAY SURFACE AND LENGTH

Runway lengths prefixed by the letter "H" indicate that the runways are hard surfaced (concrete, asphalt). If the runway length is not prefixed, the surface is sod, clay, etc. The runway surface composition is indicated in parentheses after runway length as follows:

(AFSC)—Aggregate friction seal coat	(GRVD)—Grooved	(RFSC)—Rubberized friction seal coat
(ASPH)—Asphalt	(GRVL)—Gravel, or cinders	(TURF)—Turf
(CONC)—Concrete	(PFC)—Porous friction courses	(TRTD)—Treated
(DIRT)—Dirt	(PSP)—Pierced steel plank	(WC)—Wire combed

RUNWAY WEIGHT BEARING CAPACITY

Runway strength data shown in this publication is derived from available information and is a realistic estimate of capability at an average level of activity. It is not intended as a maximum allowable weight or as an operating limitation. Many airport pavements are capable of supporting limited operations with gross weights of 25-50% in excess of the published figures. Permissible operating weights, insofar as runway strengths are concerned, are a matter of agreement between the owner and user. When desiring to operate into any airport at weights in excess of those published in the publication, users should contact the airport management for permission. Add 000 to figure following S, D, DT, DDT, AUW, etc., for gross weight capacity:

- S—Single-wheel type landing gear. (DC-3), (C-47), (F-15), etc.
- D—Dual-wheel type landing gear. (DC-6), etc.
- T—Twin-wheel type landing gear. (DC-6), (C-9A), etc.
- ST—Single-tandem type landing gear. (C-130).
- SBTT—Single-belly twin tandem landing gear (KC-10).
- DT—Dual-tandem type landing gear, (707), etc.
- TT—Twin-tandem type (includes quadricycle) landing gear (707), (B-52), (C-135), etc.
- TRT—Triple-tandem landing gear, (C-17)
- DDT—Double dual-tandem landing gear. (E4A/747).
- TDT—Twin delta-tandem landing gear. (C-5, Concorde).
- AUW—All up weight. Maximum weight bearing capacity for any aircraft irrespective of landing gear configuration.
- SWL—Single Wheel Loading. (This includes information submitted in terms of Equivalent Single Wheel Loading (ESWL) and Single Isolated Wheel Loading). SWL figures are shown in thousands of pounds with the last three figures being omitted.
- PSI—Pounds per square inch. PSI is the actual figure expressing maximum pounds per square inch runway will support, e.g., (SWL 000/PSI 535).

Quadricycle and dual-tandem are considered virtually equal for runway weight bearing consideration, as are single-tandem and dual-wheel. Omission of weight bearing capacity indicates information unknown.

The ACN/PCN System is the ICAO method of reporting pavement strength for pavements with bearing strengths greater than 12,500 pounds. The Pavement Classification Number (PCN) is established by an engineering assessment of the runway. The PCN is for use in conjunction with an Aircraft Classification Number (ACN). Consult the Aircraft Flight Manual or other appropriate source for ACN tables or charts. Currently, ACN data may not be available for all aircraft. If an ACN table or chart is available, the ACN can be calculated by taking into account the aircraft weight, the pavement type, and the subgrade category. For runways that have been evaluated under the ACN/PCN system, the PCN will be shown as a five part code (e.g. PCN 80 R/B/W/T). Details of the coded format are as follows:

(1) The PCN NUMBER—The reported PCN indicates that an aircraft with an ACN equal or less than the reported PCN can operate on the pavement subject to any limitation on the tire pressure.

(2) The type of pavement:
- R — Rigid
- F — Flexible

(3) The pavement subgrade category:
- A — High
- B — Medium
- C — Low
- D — Ultra-low

(4) The maximum tire pressure authorized for the pavement:
- W — High, no limit
- X — Medium, limited to 217 psi
- Y — Low, limited to 145 psi
- Z — Very low, limited to 73 psi

(5) Pavement evaluation method:
- T — Technical evaluation
- U — By experience of aircraft using the pavement

NOTE: Prior permission from the airport controlling authority is required when the ACN of the aircraft exceeds the published PCN or aircraft tire pressure exceeds the published limits.

DIRECTORY LEGEND

RUNWAY LIGHTING

Lights are in operation sunset to sunrise. Lighting available by prior arrangement only or operating part of the night only and/or pilot controlled and with specific operating hours are indicated under airport remarks. Since obstructions are usually lighted, obstruction lighting is not included in this code. Unlighted obstructions on or surrounding an airport will be noted in airport remarks. Runway lights nonstandard (NSTD) are systems for which the light fixtures are not FAA approved L-800 series: color, intensity, or spacing does not meet FAA standards. Nonstandard runway lights, VASI, or any other system not listed below will be shown in airport remarks.

Temporary, emergency or limited runway edge lighting such as flares, smudge pots, lanterns or portable runway lights will also be shown in airport remarks. Types of lighting are shown with the runway or runway end they serve.

- NSTD—Light system fails to meet FAA standards.
- LIRL—Low Intensity Runway Lights.
- MIRL—Medium Intensity Runway Lights.
- HIRL—High Intensity Runway Lights.
- RAIL—Runway Alignment Indicator Lights.
- REIL—Runway End Identifier Lights.
- CL—Centerline Lights.
- TDZL—Touchdown Zone Lights.
- ODALS—Omni Directional Approach Lighting System.
- AF OVRN—Air Force Overrun 1000' Standard Approach Lighting System.
- LDIN—Lead-In Lighting System.
- MALS—Medium Intensity Approach Lighting System.
- MALSF—Medium Intensity Approach Lighting System with Sequenced Flashing Lights.
- MALSR—Medium Intensity Approach Lighting System with Runway Alignment Indicator Lights.
- SALS—Short Approach Lighting System.
- SALSF—Short Approach Lighting System with Sequenced Flashing Lights.
- SSALS—Simplified Short Approach Lighting System.
- SSALF—Simplified Short Approach Lighting System with Sequenced Flashing Lights.
- SSALR—Simplified Short Approach Lighting System with Runway Alignment Indicator Lights.
- ALSAF—High Intensity Approach Lighting System with Sequenced Flashing Lights.
- ALSF1—High Intensity Approach Lighting System with Sequenced Flashing Lights, Category I, Configuration.
- ALSF2—High Intensity Approach Lighting System with Sequenced Flashing Lights, Category II, Configuration.
- VASI—Visual Approach Slope Indicator System.

NOTE: Civil ALSF-2 may be operated as SSALR during favorable weather conditions.

VISUAL GLIDESLOPE INDICATORS

APAP—A system of panels, which may or may not be lighted, used for alignment of approach path.

PNIL	APAP on left side of runway	PNIR	APAP on right side of runway

PAPI—Precision Approach Path Indicator

P2L	2-identical light units placed on left side of runway	P4L	4-identical light units placed on left side of runway
P2R	2-identical light units placed on right side of runway	P4R	4-identical light units placed on right side of runway

PVASI—Pulsating/steady burning visual approach slope indicator, normally a single light unit projecting two colors.

PSIL	PVASI on left side of runway	PSIR	PVASI on right side of runway

SAVASI—Simplified Abbreviated Visual Approach Slope Indicator

S2L	2-box SAVASI on left side of runway	S2R	2-box SAVASI on right side of runway

TRCV—Tri-color visual approach slope indicator, normally a single light unit projecting three colors.

TRIL	TRCV on left side of runway	TRIR	TRCV on right side of runway

VASI—Visual Approach Slope Indicator

V2L	2-box VASI on left side of runway	V6L	6-box VASI on left side of runway
V2R	2-box VASI on right side of runway	V6R	6-box VASI on right side of runway
V4L	4-box VASI on left side of runway	V12	12-box VASI on both sides of runway
V4R	4-box VASI on right side of runway	V16	16-box VASI on both sides of runway

NOTE: Approach slope angle and threshold crossing height will be shown when available; i.e., -GA 3.5° TCH 37'.

PILOT CONTROL OF AIRPORT LIGHTING

Key Mike	Function
7 times within 5 seconds	Highest intensity available
5 times within 5 seconds	Medium or lower intensity (Lower REIL or REIL-Off)
3 times within 5 seconds	Lowest intensity available (Lower REIL or REIL-Off)

Available systems will be indicated in the Airport Remarks, as follows:

ACTIVATE MALSR Rwy 07, HIRL Rwy 07–25–122.8 (or CTAF).
or
ACTIVATE MIRL Rwy 18–36–122.8 (or CTAF).
or
ACTIVATE VASI and REIL, Rwy 07–122.8 (or CTAF).

Where the airport is not served by an instrument approach procedure and/or has an independent type system of different specification installed by the airport sponsor, descriptions of the type lights, method of control, and operating frequency will be explained in clear text. See AIM, "Basic Flight Information and ATC Procedures," for detailed description of pilot control of airport lighting.

RUNWAY SLOPE

Runway slope will be shown only when it is 0.3 percent or more. On runways less than 8000 feet: When available the direction of the slope upward will be indicated, ie., 0.3% up NW. On runways 8000 feet or greater: When available the slope will be shown on the runway end line, ie., RWY 13: 0.3% up., RWY 21: Pole. Rgt tfc. 0.4% down.

RUNWAY END DATA

Lighting systems such as VASI, MALSR, REIL; obstructions; displaced thresholds will be shown on the specific runway end. "Rgt tfc"—Right traffic indicates right turns should be made on landing and takeoff for specified runway end.

DIRECTORY LEGEND

RUNWAY DECLARED DISTANCE INFORMATION

TORA—Take-off Run Available
TODA—Take-off Distance Available
ASDA—Accelerate-Stop Distance Available
LDA—Landing Distance Available

(19) AIRPORT REMARKS

Landing Fee indicates landing charges for private or non-revenue producing aircraft, in addition, fees may be charged for planes that remain over a couple of hours and buy no services, or at major airline terminals for all aircraft.
Remarks—Data is confined to operational items affecting the status and usability of the airport.
Parachute Jumping.—See "PARACHUTE" tabulation for details.
Unless otherwise stated, remarks including runway ends refer to the runway's approach end.

(20) WEATHER DATA SOURCES

ASOS—Automated Surface Observing System. Reports the same as an AWOS-3 plus precipitation identification and intensity, and freezing rain occurrence (future enhancement).
AWOS—Automated Weather Observing System
AWOS-A—reports altimeter setting.
AWOS-1—reports altimeter setting, wind data and usually temperature, dewpoint and density altitude.
AWOS-2—reports the same as AWOS-1 plus visibility.
AWOS-3—reports the same as AWOS-1 plus visibility and cloud/ceiling data.
See AIM, Basic Flight Information and ATC Procedures for detailed description of AWOS.
HIWAS—See RADIO AIDS TO NAVIGATION
LAWRS—Limited Aviation Weather Reporting Station where observers report cloud height, weather, obstructions to vision, temperature and dewpoint (in most cases), surface wind, altimeter and pertinent remarks.
LLWAS—indicates a Low Level Wind Shear Alert System consisting of a center field and several field perimeter anemometers.
SAWRS—identifies airports that have a Supplemental Aviation Weather Reporting Station available to pilots for current weather information.
SWSL—Supplemental Weather Service Location providing current local weather information via radio and telephone.
TDWR—indicates airports that have Terminal Doppler Weather Radar.
TWEB—See RADIO AIDS TO NAVIGATION
When the automated weather source is broadcast over an associated airport NAVAID frequency (see NAVAID line), it shall be indicated by a bold ASOS, AWOS, HIWAS or TWEB, followed by the frequency, identifier and phone number, if available.

(21) COMMUNICATIONS

Communications will be listed in sequence in the order shown below:
Common Traffic Advisory Frequency (CTAF), Automatic Terminal Information Service (ATIS) and Aeronautical Advisory Stations (UNICOM) along with their frequency is shown, where available, on the line following the heading "COMMUNICATIONS." When the CTAF and UNICOM is the same frequency, the frequency will be shown as CTAF/UNICOM freq.
Flight Service Station (FSS) information. The associated FSS will be shown followed by the identifier and information concerning availability of telephone service, e.g., Direct Line (DL), Local Call (LC-384-2341), Toll free call, dial (TF 800–852–7036 or TF 1–800–227–7160), Long Distance (LD 202-426-8800 or LD 1-202-555-1212) etc. The airport NOTAM file identifier will be shown as "NOTAM FILE IAD." Where the FSS is located on the field it will be indicated as "on arpt" following the identifier. Frequencies available will follow. The FSS telephone number will follow along with any significant operational information. FSS's whose name is not the same as the airport on which located will also be listed in the normal alphabetical name listing for the state in which located. Remote Communications Outlet (RCO) providing service to the airport followed by the frequency and name of the Controlling FSS.
FSS's provide information on airport conditions, radio aids and other facilities, and process flight plans. Local Airport Advisory Service is provided on the CTAF by FSS's located at non-tower airports or airports where the tower is not in operation.
(See AIM, Para 4–1–9 Traffic Advisory Practices at airports where a tower is not in operation or AC 90 - 42C.)
Aviation weather briefing service is provided by FSS specialists. Flight and weather briefing services are also available by calling the telephone numbers listed.
Remote Communications Outlet (RCO)—An unmanned air/ground communications facility, remotely controlled and providing UHF or VHF communications capability to extend the service range of an FSS.
Civil Communications Frequencies—Civil communications frequencies used in the FSS air/ground system are now operated simplex on 122.0, 122.2, 122.3, 122.4, 122.6, 123.6; emergency 121.5; plus receive-only on 122.05, 122.1, 122.15, and 123.6.

a. 122.0 is assigned as the Enroute Flight Advisory Service channel at selected FSS's.
b. 122.2 is assigned to most FSS's as a common enroute simplex service.
c. 123.6 is assigned as the airport advisory channel at non-tower FSS locations, however, it is still in commission at some FSS's collocated with towers to provide part time Local Airport Advisory Service.
d. 122.1 is the primary receive-only frequency at VOR's. 122.05, 122.15 and 123.6 are assigned at selected VOR's meeting certain criteria.
e. Some FSS's are assigned 50 kHz channels for simplex operation in the 122-123 MHz band (e.g. 122.35). Pilots using the FSS A/G system should refer to this directory or appropriate charts to determine frequencies available at the FSS or remoted facility through which they wish to communicate.

Part time FSS hours of operation are shown in remarks under facility name.
Emergency frequency 121.5 is available at all Flight Service Stations, Towers, Approach Control and RADAR facilities, unless indicated as not available.
Frequencies published followed by the letter "T" or "R", indicate that the facility will only transmit or receive respectively on that frequency. All radio aids to navigation frequencies are transmit only.

DIRECTORY LEGEND

TERMINAL SERVICES

CTAF—A program designed to get all vehicles and aircraft at uncontrolled airports on a common frequency.
ATIS—A continuous broadcast of recorded non-control information in selected areas of high activity.
UNICOM—A non-government air/ground radio communications facility utilized to provide general airport advisory service.
APP CON —Approach Control. The symbol Ⓡ indicates radar approach control.
TOWER—Control tower.
GND CON—Ground Control.
GCO—GROUND COMMUNICATION OUTLET—An unstaffed, remotely controlled, ground/ground communications facility. Pilots at uncontrolled airports may contact ATC and FSS via VHF to a telephone connection to obtain an instrument clearance or close a VFR or IFR flight plan. They may also get an updated weather briefing prior to takeoff. Pilots will use four "key clicks" on the VHF radio to contact the appropriate ATC facility or six "key clicks" to contact the FSS. The GCO system is intended to be used only on the ground.
DEP CON—Departure Control. The symbol Ⓡ indicates radar departure control.
CLNC DEL—Clearance Delivery.
PRE TAXI CLNC—Pre taxi clearance.
VFR ADVSY SVC—VFR Advisory Service. Service provided by Non-Radar Approach Control.
Advisory Service for VFR aircraft (upon a workload basis) ctc APP CON.
TOWER, APP CON and DEP CON RADIO CALL will be the same as the airport name unless indicated otherwise.

(22) NOTAM SERVICE

All public use landing areas are provided NOTAM "D" (distant dissemination) and NOTAM "L" (local dissemination) service. Airport NOTAM file identifier is shown following the associated FSS data for individual airports, e.g. "NOTAM FILE IAD". See AIM, Basic Flight Information and ATC Procedures for detailed description of NOTAM's.

(23) AIRSPACE

CLASS B—Radar Sequencing and Separation Service for all aircraft in CLASS B airspace
TRSA—Radar Sequencing and Separation Service for participating VFR Aircraft within a Terminal Radar Service Area
Class C, D, and E airspace described in this publication is that airspace usually consisting of a 5 NM radius core surface area that begins at the surface and extends upward to an altitude above the airport elevation (charted in MSL for Class C and Class D).
When CLASS C airspace defaults to CLASS E, the core surface area becomes CLASS E. This will be formatted as: **AIRSPACE: CLASS C** svc "times" ctc **APP CON** other times CLASS E.
When Class C airspace defaults to Class G, the core surface area becomes Class G up to but not including the overlying controlled airspace. There are Class E airspace areas beginning at either 700′ or 1200′ AGL used to transition to/from the terminal or enroute environment. This will be formatted as: **AIRSPACE: CLASS C** svc "times" ctc **APP CON** other times CLASS G, CLASS E 700′ (or 1200′) AGL & abv.
NOTE: AIRSPACE SVC "TIMES" INCLUDE ALL ASSOCIATED EXTENSIONS. Arrival extensions for instrument approach procedures become part of the primary core surface area. These extensions may be either Class D or Class E airspace and are effective concurrent with the times of the primary core surface area.
(See CLASS AIRSPACE in the Aeronautical Information Manual for further details)

(24) RADIO AIDS TO NAVIGATION

The Airport Facility Directory lists by facility name all Radio Aids to Navigation, except Military TACANS, that appear on National Ocean Service Visual or IFR Aeronautical Charts and those upon which the FAA has approved an Instrument Approach Procedure. All VOR, VORTAC ILS and MLS equipment in the National Airspace System has an automatic monitoring and shutdown feature in the event of malfunction. Unmonitored, as used in this publication for any navigational aid, means that FSS or tower personnel cannot observe the malfunction or shutdown signal. The NAVAID NOTAM file identifier will be shown as "NOTAM FILE IAD" and will be listed on the Radio Aids to Navigation line. When two or more NAVAIDS are listed and the NOTAM file identifier is different than shown on the Radio Aids to Navigation line, then it will be shown with the NAVAID listing. NOTAM file identifiers for ILS's and their components (e.g., NDB (LOM) are the same as the identifiers for the associated airports and are not repeated. Automated Surface Observing System (ASOS), Automated Weather Observing System (AWOS), Hazardous Inflight Weather Advisory Service (HIWAS) and Transcribed Weather Broadcast (TWEB) will be shown where this service is broadcast over selected NAVAIDs.
NAVAID information is tabulated as indicated in the following sample:

TACAN/DME Channel — Geographical Position — Site Elevation — Transcribed Weather Broadcast

NAME (L) ABVORTAC 117.55 ABE Chan 122(Y) N40°43.60′ W75°27.30′ 180°4.1 NM to fld. 1110/8E. **AWOS. HIWAS. TWEB.**

Class — Frequency — Identifier — Bearing and distance facility to center of airport — Magnetic Variation — Automated Weather Observing System — Hazardous Inflight Weather Advisory Service

VOR unusable 020°–060° byd 26 NM blo 3,500′

Restriction within the normal altitude/range of the navigational aid (See primary alphabetical listing for restrictions on VORTAC and VOR/DME).

Note: Those DME channel numbers with a (Y) suffix require TACAN to be placed in the "Y" mode to receive distance information.

The Senior Editor: Brian Weiss

Brian Weiss is the owner of WORD'SWORTH, a marketing communications and design company. WORD'SWORTH provides services for all forms and formats of communications materials. Capabilities include the creation of brochures, books, newsletters, direct response letters, advertising, catalog sheets, slide shows and videos.

WORD'SWORTH clients have included the FAA, Bank of America, Xerox Corporation, the National Childhood Cancer Foundation, Childrens Hospital (Los Angeles), the University of California (Irvine and Los Angeles campuses), Health Valley Foods, McGraw-Hill/CRM Films, Saint Joseph Hospital, *American Health* magazine, *Psychology Today* magazine, *Equity Quarterly*, Saint John's Hospital, *Aviation Safety* magazine, Long Beach and Santa Monica airports, and many others in a wide variety of fields.

In addition to general business expertise, WORD'SWORTH provides specialized background and knowledge in the areas of medicine and health care, science and technology, aviation and fundraising.

WORD'SWORTH also offers consulting on marketing strategies, direct mail campaigns, and fundraising proposals.

A little about Brian Weiss:

Founder (1977) and owner, WORD'SWORTH
Former editor, *Baja Explorer* magazine
Former associate editor, *Psychology Today* magazine
Served on the faculties of UCLA and the University of Michigan teaching introductory courses and advanced seminars in departments of human behavior, geography, and journalism.
Author for six years of a nationally syndicated consumer newspaper column (*FREEBIES*), and creator of a national magazine of the same name
Former medical/science editor, *Aviation Safety* magazine
Member (and former Board of Directors member), of Angel Flight, a not-for-profit community service organization.
Brian has been a pilot since 1980. He packs a private pilot certificate with an instrument rating and is the proud owner of a Cessna 172. He's one of the organizers of Flight Log, a group which provides information for pilots flying in Baja and throughout Mexico.

WORD'SWORTH
3675 Greenhill Rd.
Pasadena, CA 91107
(626) 510-9180
FAX (213) 477-2189

"Brian is one of the most talented, energetic and intelligent people with whom I've had the pleasure of working. His advice and dedication to this project were simply invaluable!"

Rod Machado

The Aviation Speakers Bureau

Providing Quality Service Since 1986

The Aviation Speakers Bureau features speakers for your banquet, educational seminar, safety standdown, convention, conference, forum, trade show, keynote, corporate training, airshow, safety program or association meeting.

We guarantee a perfect match for your needs and objectives,
and recommend only the very best in speakers.
Our professionals shine and make YOU look good every time!

For speaker information call
(949) 498-2498
(800) 247-1215

P.O. Box 6030
San Clemente, CA 92674-6030
***
Read speaker biographies and view video clips:
www.aviationspeakers.com
***

IASB
INTERNATIONAL ASSOCIATION OF SPEAKERS BUREAUS

"We will help you find the perfect speaker for your budget and there is never a charge for our service."

The Ongoing Editor: Diane Titterington

Diane Titterington

Learning to fly in 1973, Diane holds a commercial certificate with an instrument rating. Her logbook is a Heinz 57 mixture of different makes and models. Flights include ferrying aircraft from the factory, flying fire patrol and a number of air races. Until 1981, she worked as a radar qualified air traffic controller on the high/low sectors at Houston Center. Diane has been a passenger on a carrier landing and takeoff, flown a T-38, rode dozens of airline jumpseats and logged a few blimp flights.

Her father, who worked at WPAFB, told tales of test pilots Bob Hoover, Scott Crossfield and Chuck Yeager when she was young. Diane never dreamed that she would later work with such aviation greats. As the President of The Aviation Speakers Bureau, Diane supplies speakers for hundreds of safety seminars, banquets and conventions. She places aviation speakers, celebrities and specialists at events across the United States, Canada and other countries.

"It is a pure pleasure to work with dozens of brilliant and gifted individuals such as Sean Tucker, Col. Joe Kittinger, Brian Udell, Alec Cody, Dave Gwinn, Dr. Jerry Cockrell and Ralph Hood. With their unusual experiences, unique delivery styles, and vast knowledge of aviation our speakers are the most sought after in the business. By providing inspiring speakers for aviation events, we help motivate and educate. And we help to keep the skies safe too." Visit web site: www.aviationspeakers.com

Diane has been the ongoing editor of *Rod Machado's Private Pilot Handbook* and *Rod Machado's Instrument Pilot's Survival Manual*. She is the designer, compiler, managing editor and producer of *Speaking of Flying*, a book of stories from 44 aviation speakers.

Professional Images (slides) on CD ROM for Flight/Ground Instructors

for Use With *Rod Machado's Private Pilot Handbook*

Flight & ground instructors, here are 811 beautiful, full color detailed images to use for your Private Pilot Ground School.

This CD contains 16 Microsoft Power Point 97 files each representing Chapters 2 through 17 of *Rod Machado's Private Pilot Handbook*. All pictures used in these Power Point presentations are also provided in 16 separate folders so you can build your own slideshow. Additionally, 16 separate chapter files are also provided in HTML format which allows ease of picture previewing with a web browser. If you do not have Microsoft PowerPoint 97, this CD contains a Microsoft Power Point Viewer program that you can easily install on your computer for viewing and presenting Power Point files.

Technical Information: Pictures are GIF files in a 256 color format. Picture resolution is 740 X 480 (7.6 X 5 inches @ 96 DPI). These images are made from the pictures, text and illustrations found in Rod Machado's Private Pilot Handbook and are designed to follow its curriculum.

$149.95 without purchase of private pilot books
$89.95 with purchase of 10 private pilot books at wholesale price

How to order: CD ordering ONLY call (800) 247-1215
All other products call (800) 437-7080

The Aviation Speakers Bureau Presents

Speaking of Flying

Air Tales Told by Top Aviation Speakers and Storytellers

One of aviation's greatest traditions is talking about flying, a tradition that continues in *Speaking of Flying*. Represented in this collection of recollections are some of aviation's most notable participants: Scott Crossfield, Julie Clark, Gen. Robin Olds, Al Haynes, Cliff Robertson, Barry Schiff, Bill Cox, Bob Gilliland, Burt Rutan, John Nance, Col. Bob Morgan, Shanda Lear, Rich Stowell, Mark Grady, Major Dee Brasseur, Rich Graham, Captain Jerry Coffee, Jim Slade, Danny Mortensen, Kent Jackson, Bill Lishman and Ken Dravis.

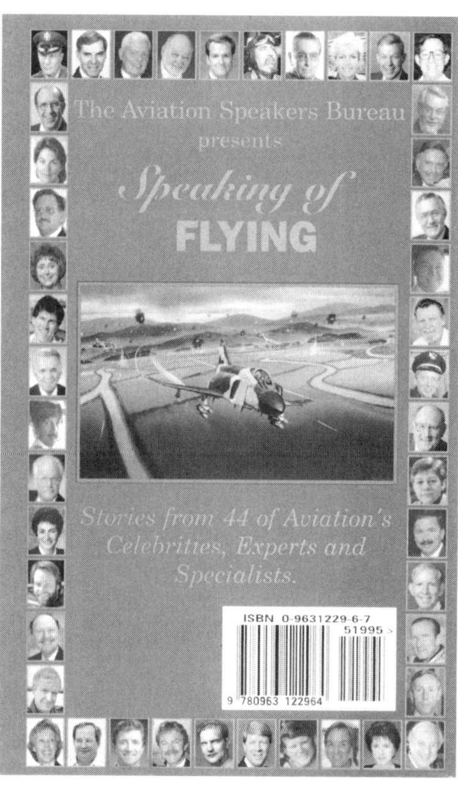

Imagine inviting 44 of aviation's celebrities and experts to the ultimate hangar flying session. Then record (in print) the unique stories, lace them with more than 200 pictures providing windows into aviation's past, and voilà: *Speaking of Flying*. This book has true stories, tales of wartime heroism, selfless sacrifice and recollections of some of flying's funnier incidents. Each of the 44 authors is also a professional speaker with **The Aviation Speakers Bureau**.

438 page, gold-embossed hardcover book with colorful dust jacket. $19.95.

Defensive Flying Video

Watch with over 300 pilots in this live, entertaining and educational video presentation as Rod Machado discusses how pilots can learn to think defensively. As Bill Wagstaff from *Aviation International News* says, "... Machado's humor serves as the glue that keeps his message together. Source for most of his guiles are real-world incidents, the essence of hangar-flying tales told with style by Machado. ...he makes us smile while we learn, a rare ability. *Defensive Flying* is an excellent tape for pilots to share." Learn about acknowledging your own limitations and a pilot's psychological predators as well as never underestimating aerial enemies. Listen to an actual, hair-raising, in-flight emergency as two professional pilots exercise one of the most important skills in Defensive Flying. This presentation contains many stories and humor not previously heard on Rod's audio tapes.

Defensive Flying - $ 29.95 Approximate Length 1:45

Aviation Humor Video

Laugh along with over 2,000 pilots as Rod delights and entertains his audience with some of the best of his aviation stories. As a professional humorist, Machado has always been known for his ability to move people off the edge of their seats and onto the floor with his fast paced, humorous presentations. As Scott Spangler, Editor of *Flight Training* magazine says, ".. Get a copy of Aviation Humor. It's Rod Machado at his best. His humor is effective and funny because it strikes at the truth pilots seldom admit, such as the pride a new pilot feels when he uses his certificate as identification when cashing a check and the clerk asks, 'What is that?' If you're in need of a good laugh, get this video. You won't regret it. "After so many requests for a video version of his very popular audio tapes, this video of Rod's is sure to be a popular addition to your library. Funny, Funny stuff!

Aviation Humor - $29.95 Approximate Length 1:00

Rod Machado's Instrument Pilot's Survival Manual

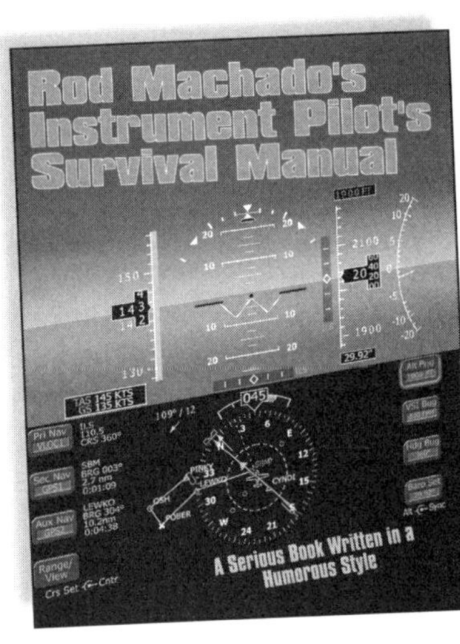

This unique book contains a wealth of interesting and exciting information on instrument flying. Rod's 26 years of flying experience makes this an important resource. Chapters include: Scanning, Ice & Thunderstorm Avoidance, Jep & NACO Chart Usage, Pro Thinking Skills, IFR Departure Skills, Cockpit Management, GPS and much more.

This book is unlike any other book on instrument flying you have encountered. Illustrated with humorous drawings and containing some of the most spectacular reports of pilots gripped by the problems of instrument flight, it's sure to educate and entertain you. Written in a humorous manner, this 426 page book will prepare you to be a more confident and proficient pilot. Excellent for any IFR student, experienced professional pilot or as an IFR refresher. $34.95

IFR Flying *Tips & Techniques*

If you've enjoyed Rod's presentation in the *ABC's Wide World Of Flying* and *Wonderful World Of Flying* video magazine, you're sure to enjoy this tape. Rod has incorporated six of his most popular IFR segments into this educational video and interspersed live, humorous clips from one of his popular seminars between each story segment.

Rod shows you:

1. *A unique, multi-step instrument scan*
2. *Single pilot IFR cockpit management*
3. *How to really use the approach lighting system*
4. *Techniques to make safe IFR departures*
5. *A two step method for flying NDB approaches*
6. *What maneuvering speed means and how to use it*

IFR Flying Tips & Techniques - $29.95 Appx. Length 1:30

Rod Machado's Plane Talk
The Mental Art of Flying an Airplane
You'll Learn, You'll Laugh, You'll Remember!

Welcome to a collection of Rod Machado's most popular aviation articles and stories from the last 15 years. *Rod Machado's Plane Talk* contains nearly 100 flights of fun and knowledge that will stimulate your aviation brain and tickle your funny bone. In addition to the educational topics listed below, you'll read about higher learning, the value of aviation history, aviation literature, aviation art and how an artist's perspective can help you better understand weather. You'll also find more than a few articles written just to make you laugh.

In this book you'll discover...

How to Assess and Manage Aviation Risks
Learn how safe pilots think, how to apply the safety strategy used by General Jimmy Doolittle (known as the master of the calculated risk), how famed gunfighter Wyatt Earp can help you cope with aviation's risks, how misleading aviation statistics can be and why flying isn't as dangerous as some folks say it is.

Several Techniques for Making Better Cockpit Decisions
Discover how to use your inner copilot in the cockpit and the value of one good question asked upside down.

New Ways to Help You Cope With Temptation
Fly safer by developing an aviation code of ethics, understand how human nature can trick you into flying beyond your limits, why good pilots are prejudiced and how a concept like honor will protect you while aloft.

How to Use Your Brain for a Change
You can learn faster by understanding how the learning curve—the brain's performance chart—is affected by the little lies we tell ourselves, the mistakes we need to make, our need to please our instructors, and simulator and memory training.

The Truth About Flying, Anxiety and Fear
Learn why it's often the safest of pilots that make excuses instead of flights, why anxiety should be treated as a normal part of flying, and a three-step process to avoiding panic in the cockpit.

How to Handle First Time Flyers and Anxious Passengers
Discover how to behave around new passengers, how to avoid most common mistakes that scare passengers in airplanes and how to reduce the cockpit stress between pilot and spouse.

Favorite Skills Used By Good Pilots
Learn why good pilots scan behind an airplane as well as ahead of it, are sometimes rough and bully-like on the flight controls, occasionally fly without using any of the airplane's electronic navigation equipment, don't worry about turbulence breaking their airplanes, master airspeed control as a means of making better landings and much more.

$29.95

What They Didn't Teach You InFlight School *Audio Tapes*

These unique audio tape albums each contain 7.5 hours of aviation material recorded *live* before pilot audiences with Machado at his best. Using his unique "Laugh & Learn" style, Rod entertains his audience as he teaches. As Wayman Dunlap, Editor of *Pacific Flyer Aviation News* says, "...There is valuable safety information which, because of the way it is presented, makes more of an impact than a dry, strictly technical lecture could ever do. I carried my tapes with me in the car and listened to them constantly on the way to work, on sales trips and during the boring one hour drive to the printers. Besides being entertained, I have the satisfying feeling of actually learning a few things... But it was the little aviation stories (whether true or not) that pepper the tapes that kept me listening and laughing. In fact, this may have been the first set of educational tapes I ever listened to that I hated to see end." $49.95 Per Album.

Volume One:

Tape 1 - Thinking Like The Pros (50/50)*
Tape 2 - Secrets Of The Test Pilots (40/60)*
Tape 3 - Understanding Weather (30/70)*
Tape 4 - Handling In-flight Emergencies 1 (10/90)*
Tape 5 - Handling In-flight Emergencies 2 (10/90)*
Tape 6 - Handling In-flight Emergencies 3 (10/90)*

Volume Two:

Tape 1 - Laugh Your Empennage Off (100/0)* Samurai Airmanship (30/70)*
Tape 2 - Reducing Cockpit Stress Between The Sexes (60/40)*
Tape 3 - Cockpit Management (40/60)* An Aviation Sense Of Humor (95/5)*
Tape 4 - Pilots & Their Tribes (99/1)* The Adventure Of Flight (99/1)*
Tape 5 - Decision Making Psychology (10/90)* Advanced Aviation Topics *
Tape 6 - Creative Solutions To The Aviator's Common Problems (40/60)*

*(% Humor content / % Information content)

"Rod has a wonderful sense of humor that will keep you in stitches and when you're done laughing, you'll be amazed at how much you've learned."
John & Martha King
King Video

"Mary Poppins once sang, 'A spoon full of sugar helps the medicine go down.' Rod Machado goes one step further by sugar coating invaluable aeronautical wisdom with entertaining wit and humor."
Barry Schiff
TWA Captain/Noted Author

COPY & MAIL COMPLETED FORM TO: THE AVIATION SPEAKERS BUREAU, P.O. BOX 6030 SAN CLEMENTE, CA 92674-6030

Order Form

Credit card ordering 24 hours - 7 days a week
Call: **(800) 437-7080** for ordering only

Title	Quantity	X =	Total
Defensive Flying Video		$29.95	
Aviation Humor Video		$29.95	
What They Didn't Teach You	Vol 1	$49.95	
In Flight School (Audio Tapes)	Vol 2	$49.95	
IFR Flying Tips & Techniques Video		$29.95	
Instrument Pilot's Survival Manual		$34.95	
Rod Machado's Private Pilot Handbook		$34.95	
Rod Machado's Private Pilot Workbook		$24.95	
Rod Machado's Plane Talk		$29.95	
Speaking of Flying		$19.95	
	Product Subtotal		
Sales Tax - CA Residents only	Product Subtotal x .0775		
Product shipping & handling (US)	———	———	$3.00
International shipping: $10 surface $27 air ($5 each add. item) Canada: $6 surface $10 air ($5 each add. item) Please indicate if ordering PAL videos			
		Total	

For more information and secure internet ordering visit *www.rodmachado.com*
Questions: products@rodmachado.com
FAX: (888) MACHADO (622-4236)

Name ____________________

Address ____________________

City ____________________ State ______

Zip ____________________

Phone (____)____________________

Check # __________ or:

We accept checks, money orders, MasterCard, VISA, American Express and Discover.

Credit Card # ____________________

Expiration Date: ____________________

Authorized Signature ____________________

About Our Book Cover

The Photographer: Erik Hildebrandt

Erik Hildebrandt is one of the nation's foremost aviation photographers. An airplane owner and licensed pilot himself, Hildebrandt's photography and articles have been featured in aviation journals such as *Aviation Week*, *Air & Space Smithsonian* and *FLIGHT Journal*. His images have been used to illustrate national advertising campaigns by such major corporations, as the Boeing Company and the Discovery Wings channel.

Erik Hildebrandt

A native of Long Island, Hildebrandt was educated at the State University of New York where he earned a bachelor's degree in American history. As a freelancer, he also studied editorial photography under renowned National Geographic photo essayist David Alan Harvey at the Maine Photographic Workshop. Hildebrandt now lives in Minneapolis with his wife, Christine Hurley.

Erik crisscrossed the United States for a year photographing 15 of aviation's best events. Showcasing his spectacular work, his book *Front Row Center, Inside the Great American Air Show*, captures the premier air show performers and venues. This master project is a colorful, large-format coffee table book that brings us along on an entire season of airshow glory. Our cover shot of Mike Goulian is a sample of the heart stopping aerobatic photos including Sean D. Tucker, Patty Wagstaff, and the U.S. Navy Blue Angels. www.vulturesrow.com

The subject: Aerobatic Pilot, Mike Goulian

Mike Goulian has been performing in air shows since age 18. Mike specializes in performing and teaching aerobatics with the characteristic precision demanded of top level competition flying.

Mike grew up in a flying family in Arlington, Massachusetts. He now owns and operates Executive Flyers Aviation, the FAA Part 141 flight school his father started in 1964. Mike learned to fly before he could drive a car, then focused on aerobatics with a passion. By age 17 he won his first aerobatic title performing in his first air show.

After a shower of regional titles, in 1990, at age 22, he became U.S. National Champion in the Advanced Category. He won the prestigious Fond du Lac Cup invitational competition and by 1992 was the top ranked U.S. male aerobatic pilot and silver medalist in the unlimited category. He repeated this achievement in 1993. He earned a spot on the 1994 U.S. National Aerobatic Team, representing the U.S. at the World Aerobatic Championship held in Hungary. In 1995, Mike reached the pinnacle of American aerobatics by becoming the U.S. National Champion in the unlimited category. He has been a member of the 1994, 1996, and 1998 U.S. Aerobatic Teams. Mike is co-author of *Basic Aerobatics* and *Advanced Aerobatics* published by McGraw Hill. www.executive flyers.com